Advances in
PARASITOLOGY

VOLUME 38

Advances in
PARASITOLOGY

Edited by

J.R. BAKER

Royal Society of Tropical Medicine and Hygiene,
London, England

R. MULLER

International Institute of Parasitology,
St Albans, England

and

D. ROLLINSON

The Natural History Museum,
London, England

VOLUME 38

ACADEMIC PRESS

Harcourt Brace & Company, Publishers
San Diego   London   New York   Boston
Sydney   Tokyo   Toronto

ACADEMIC PRESS INC.
525 B Street, Suite 1900, San Diego,
California 92101-4495, USA

ACADEMIC PRESS LIMITED
24/28 Oval Road
LONDON NW1 7DX

A catalogue record for this book is available from the British Library

ISBN 0-12-031738-9

Typeset by J&L Composition Ltd, Filey, North Yorkshire
Printed in Great Britain by Hartnolls Ltd, Bodmin, Cornwall

CONTRIBUTORS TO VOLUME 38

J.C. ALLAN, *Department of Biological Sciences, University of Salford, Salford, M5 4WT, UK*

O. BILLKER, *Molecular and Cellular Parasitology Research Group, Infection and Immunity Section, Department of Biology, Imperial College of Science, Technology and Medicine, Prince Consort Road, South Kensington, London, SW7 2BB, UK*

G.A. BUTCHER, *Molecular and Cellular Parasitology Research Group, Infection and Immunity Section, Department of Biology, Imperial College of Science, Technology and Medicine, Prince Consort Road, South Kensington, London, SW7 2BB, UK*

P.S. CRAIG, *Department of Biological Sciences, University of Salford, Salford, M5 4WT, UK*

S.L. FLECK, *Molecular and Cellular Parasitology Research Group, Infection and Immunity Section, Department of Biology, Imperial College of Science, Technology and Medicine, Prince Consort Road, South Kensington, London, SW7 2BB, UK*

B. FRIED, *Department of Biology, Lafayette College, Easton, Pennsylvania 18402, USA*

D.I. GROVE, *Department of Clinical Microbiology and Infectious Diseases, The Queen Elizabeth Hospital, Adelaide, South Australia*

J.E. HUFFMAN, *Department of Biological Sciences, East Stroudsberg University, East Stroudsberg, PA 18301, USA*

N. LE BRUN, *Laboratoire de Parasitologie Comparée, URA CNRS 698, Université Montpellier II, Place Eugène Bataillon, 34095 Montpellier, Cédex 05, France*

J. MAUËL, *Institute of Biochemistry, Ch. des Boveresses, CH-1066 Epalinges, Switzerland*

C. MOULIA, *Laboratoire de Parasitologie Comparée, URA CNRS 698, Université Montpellier II, Place Eugène Bataillon, 34095 Montpellier, Cédex 05, France*

F. RENAUD, *Laboratoire de Parasitologie Comparée, URA CNRS 698, Université Montpellier II, Place Eugène Bataillon, 34095 Montpellier, Cédex 05, France*

M.T. Rogan, *Department of Biological Sciences, University of Salford, Salford, M5 4WT, UK*

R.E. Sinden, *Molecular and Cellular Parasitology Research Group, Infection and Immunity Section, Department of Biology, Imperial College of Science, Technology and Medicine, Prince Consort Road, South Kensington, London, SW7 2BB, UK*

PREFACE

The volume starts with two reviews dealing with parasitic protozoa. First, Jacques Mauël, from the Institute of Biochemistry in Lausanne, describes the means by which the intracellular parasites *Leishmania, Toxoplasma* and *Trypanosoma cruzi* enter and survive in their host's phagocytic cells, which would normally be expected to destroy the invader. Each of these three organisms has evolved different means of entry into, and survival within, its host cells, and these are reviewed in detail, together with a consideration of how the established intracellular parasites obtain their nutrients.

Second, Bob Sinden and his colleagues at Imperial College, London, deal with the factors regulating the infectivity to mosquitoes of the gametocytes of malaria parasites. After an initial description of the process of gametocytogenesis, the authors discuss both intrinsic and extrinsic (host-mediated) factors which may influence the maturation and subsequent infectivity of the gametocytes. The process of gametogenesis is then described, including the impressively rapid process of microgametogenesis or "exflagellation", and the mechanisms which induce it. The review concludes with a consideration of fertilization and the subsequent development of the parasite within its mosquito vector.

The broad and complicated subject of the interactions between parasites and their hosts has been tackled succinctly by Catherine Moulia, Nathalie Le Brun and François Renaud of Montpellier University. They explore the pitfalls involved in the use of experimental, unnatural murine hosts, which ignores the genetic polymorphism of the natural host, and describe ingenious studies investigating the relationships of various parasites to populations of two mouse subspecies (and hybrids) in the wild in relation to their genetic make-up. These studies provide fascinating insights into the presence of so-called "wormy" individuals. The authors discuss the genetic basis of resistance or susceptibility to parasites and suggest a new approach, which they have termed "evolutionary immunology", to provide an essential bridge between laboratory and field-based studies.

Drawing on a wealth of experience gained from working in many endemic areas, Philip Craig and his colleagues from the University of Salford present a comprehensive and modern account of the detection and epidemiology of the tapeworms within the family Taeniidae which are of public health significance. These include the zoonotic species that cause larval (metacestode) infection in human tissues, *Taenia solium, Echinococcus granulosus* and *E. multilocularis*. Although global figures of infection are difficult to estimate, transmission of at least one of the species occurs in almost every region of the globe. The review covers many aspects of the biology of the organisms relating to their life cycle and transmission dynamics. Pathology of metacestode infection is variable and usually relates to pressure effects exerted by the developing cyst, *E. multilocularis* being one of the most pathogenic of human parasitic infections. Advances in detection of taeniid larval cystic infection, including various scans and immunodiagnostic tests, are described and details of community studies in selected countries are presented. Control is difficult to achieve but the significant progress in understanding the epidemiology of cestode infection detailed in this review should help to improve future control strategies.

David Grove, who works at Queen Elizabeth Hospital, Adelaide, reviews recent advances in the knowledge of human strongyloidiasis since the publication of a book on the subject he edited in 1989. This parasite differs in many ways from other worms inhabiting the human body, particularly in its ability to replicate in the human host. Most aspects of the nematode and its relationship to the host are reviewed, including an interesting hypothesis on the genesis of disseminated strongyloidiasis. The author concludes that much remains to be learned about the biology, host–parasite relationships, immunology, chemotherapy and diagnosis of this still enigmatic parasite.

Finally, Bernard Fried from Lafayette College and Jane Huffman at East Stroudberg University have reviewed the intestinal echinostome trematode *Echinostoma caproni*. The same two authors reviewed the literature on various species of *Echinostoma* six years ago (in volume 29 of *Advances in Parasitology*), but the present review concentrates particularly on the use of this species as a useful laboratory model of an intestinal digenean. *E. caproni* can be cycled easily between the snail *Biomphalaria glabrata*, widely kept for schistosome studies, and laboratory mice or hamsters. Recent studies of experimental procedures for maintenance, host–parasite relationships, pathology and immunology are included.

JOHN BAKER
RALPH MULLER
DAVID ROLLINSON

CONTENTS

CONTRIBUTORS TO VOLUME 38 v
PREFACE vii

Intracellular Survival of Protozoan Parasites with Special Reference to *Leishmania* spp., *Toxoplasma gondii* and *Trypanosoma cruzi*

J. Mauël

1. Introduction 1
2. Entering the Host Cell 2
3. Avoiding Intracellular Killing Mechanisms 11
4. Meeting Nutritional Requirements 28
5. Concluding Remarks 34
   Acknowledgements 35
   References 35

Regulation of Infectivity of *Plasmodium* to the Mosquito Vector

R.E. Sinden, G.A. Butcher, O. Billker and S.L. Fleck

1. Introduction 54
2. Gametocytogenesis 55
3. Biology of the Mature Gametocyte 58
4. Gametogenesis 79
5. Fertilization 92
6. Post-fertilization Development 95
   Acknowledgements 99
   References 99

Mouse–Parasite Interactions: from Gene to Population

C. Moulia, N. Le Brun and F. Renaud

1. Introduction 119
2. One Kind of Immune Response, One Set of T Cells 122
3. One Gene, One Cell, One Phenotype 127
4. Involvement of the Major Histocompatibility Complex . . . 133
5. The Mouse Hybrid Zone: a Natural Research Laboratory . . 140
6. Conclusion and Perspectives 153
   Acknowledgements 155
   References 155

Detection, Screening and Community Epidemiology of Taeniid Cestode Zoonoses: Cystic Echinococcosis, Alveolar Echinococcosis and Neurocysticercosis

P.S. Craig, M.T. Rogan and J.C. Allan

1. Introduction 170
2. Life-cycle Biology and Transmission Dynamics 185
3. Measurement of Human Infection with Taeniid Cestodes in
   Endemic Communities 189
4. Detection of Taeniid Infection in Animal Definitive and
   Intermediate Hosts 200
5. Cystic Echinococcosis in Endemic Communities 203
6. Alveolar Echinococcosis in Endemic Communities 215
7. *T. solium* Cysticercosis and Taeniasis in Communities . . . 223
8. Conclusions 229
   Acknowledgements 230
   References 231

Human Strongyloidiasis

D.I. Grove

1. Introduction 252
2. History 252
3. Species of *Strongyloides* 255
4. Morphology of *Strongyloides stercoralis* 255
5. Other Taxonomic Parameters 260

6. Cryopreservation 260
7. Life Cycle of *Strongyloides stercoralis* 261
8. The Host–Parasite Relationship 267
9. Clinical Features 276
10. Diagnosis 280
11. Treatment 290
12. Prevention and Control 295
13. *Strongyloides fuelleborni* 296
14. Future Directions 297
    References 297

The Biology of the Intestinal Trematode *Echinostoma caproni*

B. Fried and J.E. Huffman

1. Introduction to the Biology of *Echinostoma caproni* 312
2. Maintenance of the Cycle in the Laboratory 315
3. Adults 319
4. Eggs and Miracidia 326
5. Sporocysts and Rediae 330
6. Cercariae and Metacercariae 333
7. Infectivity of *E. caproni* in Intermediate Hosts 337
8. Echinostomes, Schistosomes and Snails 340
9. Infectivity, Growth and Development of *E. caproni* in
   Definitive Hosts 341
10. Concurrent Studies Using *E. caproni* in Other Helminths . . 343
11. Effects of *E. caproni* Infection on Pregnancy in Mice . . . 344
12. Excystation, Implantation and *In Vitro* and *In Ovo* Cultivation
    of *E. caproni* 345
13. Pathobiochemical Effects of Larval and Adult *E. caproni* on
    Their Hosts 347
14. Gross, Histopathological and Clinical Effects of Adult
    E. caproni on Their Hosts 350
15. Immunobiology of *E. caproni* in Definitive Hosts 354
16. Electrophoretic and Polymerase Chain Reaction Studies on
    E. caproni 357
17. Summary and Concluding Remarks 359
    References 360

    Index 369

Intracellular Survival of Protozoan Parasites with Special Reference to *Leishmania* spp., *Toxoplasma gondii* and *Trypanosoma cruzi*

Jacques Mauël

Institute of Biochemistry, Ch. des Boveresses, CH-1066 Epalinges, Switzerland

1. Introduction 1
2. Entering the Host Cell 2
   2.1. Conventional phagocytosis: the *Leishmania* case 3
   2.2. Active invasion 7
3. Avoiding Intracellular Killing Mechanisms 11
   3.1. Phagolysosome formation and the microbicidal activities of host cells . . . 11
   3.2. Resistance to lysosomal toxicity 14
   3.3. Interference with the respiratory burst of phagocytes 21
   3.4. Inhibition of nitric oxide synthesis 26
   3.5. Other inhibitory effects 27
4. Meeting Nutritional Requirements 28
   4.1. Different strategies for different parasites 28
   4.2. Metabolic adaptation to intracellular life 29
   4.3 Acquisition of nutrients 30
5. Concluding Remarks 34
Acknowledgements 35
References 35

1. INTRODUCTION

This review is concerned with mechanisms whereby certain parasites are able to survive and proliferate within cells of their hosts. New insights have

ADVANCES IN PARASITOLOGY VOL 38
ISBN 0–12–031738–9

been gained in this domain in recent years, thanks to detailed ultrastructural and biochemical investigations of parasite invasion processes. These studies have uncovered fascinating new properties of parasites regarding the strategies developed by them to adapt to intracellular life. Much of this novel information has been obtained through investigations conducted on parasites of the genus *Leishmania*, as well as on *Toxoplasma gondii* and *Trypanosoma cruzi*. Although these are not the only intracellular parasites of interest, it is striking that each one of these microorganisms has evolved a peculiar way of infecting cells and a distinct procedure to survive intracellularly. Moreover, the amount of information on these processes is now considerable, and therefore this review will be restricted to these microorganisms.

These parasites are also pathogenic to humans, and studies on the molecular mechanisms that govern the parasite–host cell interaction therefore represent more than an academic exercise. Indeed, by providing a deeper understanding of the infectious process, they are expected to provide fresh leads for the formulation of new therapeutic approaches.

2. ENTERING THE HOST CELL

Analysis of the mechanisms of entry into host cells indicates that different parasites have developed entirely different invasion procedures. As a result, not only do parasites of different species display markedly different cellular host ranges, but also they may eventually become established in different intracellular compartments. Hence, their physiological environment within cells may be quite different, and thus they have evolved distinct survival strategies.

A broad distinction can be made between those microorganisms which lack a specialized mechanism for entering cells and therefore depend on the phagocytic potential of prospective host cells, and those parasites which have developed specific organelles or molecular devices that enable them to play an active role in the invasion process and thereby gain access into both phagocytic and non-phagocytic cells. *Leishmania* spp. belong to the first category, whereas *Tox. gondii* and *Tryp. cruzi* are members of the second group. The mechanisms of intracellular survival may critically depend on the route of entry into host cells, and the main features of the invasion process of these parasites are therefore briefly reviewed below.

2.1. Conventional Phagocytosis: the *Leishmania* Case

2.1.1. *General Considerations*

(a) *Receptors involved in phagocytosis.* The most straightforward mechanism for entering a cell, one that requires no particular device on the part of the invader, consists of relying on the phagocytic activity of host cells. As a corollary, only professional phagocytes will usually undergo infection by these parasites. However, these cells are also best equipped to combat infection; parasites that display little sophistication in their invasion mechanisms must therefore devise particularly ingenious tools to escape intracellular killing.

Phagocytosis of microbes by macrophages requires preliminary attachment of the particle to the phagocyte surface through specific ligand–receptor recognition (for a review, see Mosser, 1994). Macrophage membrane structures that may take part in these interactions include (i) receptors for fragments of the third component of complement, C3b and iC3b, which bind preferentially to complement receptors 1 and 3 (CR1, CR3 [or Mac-1]), respectively (Brown, 1992; Krych *et al.*, 1992); (ii) receptors for the Fc portion of certain immunoglobulins (mainly IgG, IgE, IgA and, depending on the state of activation of the cell, IgM) (Schreiber *et al.*, 1992; Hulett and Hogarth, 1994); (iii) receptors of the lectin type, which mediate specific interactions with native or modified carbohydrate moieties on polysaccharides, glycoproteins or proteoglycans (McWilliam *et al.*, 1992; Stahl, 1992); (iv) molecules belonging to the integrin family, that recognize structures determined by specific amino acid sequences (in particular the Arg-Gly-Asp or RGD motif found in fibronectin, vitronectin and other adhesive molecules) (Hynes, 1992).

Experimental models *in vitro* are commonly used to assess the importance of a defined ligand–receptor couple for the internalization of a given parasite by a given set of phagocytes. It will be appreciated that interaction of microorganisms with phagocytes within the host is presumably far more complex than can be deduced from such studies. Identification of critical receptors involved in parasite recognition is nevertheless very important. Indeed, engagement of a given receptor dictates many subsequent intracellular events, and may in fact spell the difference between life and death for the microorganism. For instance, when binding of *Tox. gondii* to macrophages occurs through appropriate ligand–receptor interactions (see Section 2.2.2), parasite entry is accompanied by only a weak respiratory burst and the microbe readily survives within such cells (Jones and Hirsch, 1972; Jones *et al.*, 1972; Wilson *et al.*, 1980). When the microorganism has been coated with specific antibody, however, binding to macrophages engages Fc receptors, leading to a strong metabolic burst with ensuing superoxide production and parasite killing (Wilson *et al.*, 1980).

A parasite that relies on "simple" phagocytosis for setting foot within a host is *Leishmania*, an obligate intracellular protozoan that lives exclusively within mononuclear phagocytes. Considerable effort has been made to identify receptors and ligands involved in attachment of *Leishmania* to the macrophage surface. Not surprisingly, variations in experimental conditions such as the use of different parasite species and sources of macrophages, the presence or absence of serum, etc., have led to somewhat different conclusions (for more comprehensive discussions on this topic, see Alexander and Russell, 1992; Mosser and Rosenthal, 1993; Titus *et al.*, 1994).

(b) *The major surface molecules of* Leishmania. The surface of *Leishmania* promastigotes bears two major molecular species, both anchored via a glycosyl phosphatidylinositol (GPI) moiety, the promastigote surface protease (PSP), also known as gp63 or leishmanolysin (Bordier, 1987; Schneider *et al.*, 1992a; Bouvier *et al.*, 1995), and a lipophosphoglycan (LPG; for reviews, see Turco and Descoteaux, 1992; McConville and Ferguson, 1993).

PSP is a zinc-containing metalloprotease present on all species of *Leishmania* that have been examined (Etges, 1992). In *L. major*, it accounts for 70% of the surface proteins accessible to iodination, 15% of the $[^{35}S]$methionine-labelled membrane proteins, and up to 1% of the total cellular protein of the promastigote. PSP is able to hydrolyse a wide range of denatured polypeptides, but little is known about its relevant substrates *in vivo*.

LPG is the major glycoconjugate of the *Leishmania* promastigote, covering its entire surface, including the flagellum. The structure of LPG consists of four domains: (i) a phosphatidylinositol anchor, (ii) a glycan core containing galactose, mannose and glucosamine residues, (iii) a chain of repeating phosphodisaccharide units, and (iv) a small oligosaccharide cap. Modifications in the structure of LPG are concomitant with acquisition of infective potential through the process of metacyclogenesis (Sacks *et al.*, 1985; Sacks and da Silva, 1987). A most striking feature of these modifications is the elongation of LPG in the metacyclic parasite, resulting from an approximate doubling in the number of repeating phosphorylated saccharide moieties (Sacks *et al.*, 1990). In addition, terminal galactosyl residues that branch off the disaccharide backbone in non-infectious, logarithmically growing *L. major* (see Sacks and Perkins, 1984; Sacks *et al.*, 1985) are capped by D-arabinose and to a lesser extent by β-glucose (McConville and Ferguson, 1993). These changes are thought to account for the loss of agglutinability of infective, metacyclic leishmanias by peanut agglutinin since, contrary to galactose, arabinose and glucose are not ligands for this lectin (Sacks *et al.*, 1985).

Both the promastigote surface metalloprotease and LPG are downregulated in the amastigote stage. LPG is undetectable in amastigotes of *L. donovani* and *L. mexicana*; in *L. major* amastigotes, a structurally and antigenically distinct LPG is present, although in much lesser amounts than found in promastigotes (McConville and Blackwell, 1991; Moody *et al.*, 1991, 1993; Turco and Sacks, 1991). However, *Leishmania* parasites also synthesize a family of low molecular mass glycosyl-inositol phospholipids known as GIPLs (McConville *et al.*, 1990). GIPLs are comparatively short glycolipids whose structure closely resembles that of the phosphatidylinositol–phosphosaccharide core of LPG; their fine composition may differ between *Leishmania* species (McConville and Blackwell, 1991; Schneider *et al.*, 1993). In contrast to LPG, there is evidence that GIPLs are synthesized in high copy number in the amastigote as well as the promastigote stages; the abundance and surface localization of these molecules suggest that they perform functions of their own, in addition to that of putative precursors of GPI-anchored proteins and LPG (Schneider *et al.*, 1993).

2.1.2. *Binding Through Interaction with Serum Factors*

There is little doubt that complement receptors are important in mediating parasite attachment to macrophages, as shown by the severalfold enhancement obtained by incubating promastigotes from different *Leishmania* species with mouse serum deficient in the fifth component of complement (to prevent formation of a lytic membrane attack complex) (Mosser and Rosenthal, 1993). Both gp63 (Russell, 1987) and LPG (Puentes *et al.*, 1988) have been shown to function as C3 acceptor sites on *Leishmania* promastigotes, and both C3b (Puentes *et al.*, 1988) and iC3b (Mosser and Edelson, 1985) have been identified on the parasite surface. It has been suggested that formation of iC3b from C3b might be facilitated by the parasite surface metalloprotease itself (Mosser and Rosenthal, 1993). Based on experiments using affinity-purified or recombinant CR3, stationary phase promastigotes of *L. amazonensis*, *L. donovani* and *L. major* were shown to bind in a complement-dependent manner to the iC3b-binding site of this receptor (Mosser *et al.*, 1992). Inhibition studies using antibodies reacting with different moieties of the complement receptors also identified CR1 as participating in the binding of opsonized metacyclic promastigotes of *L. major* to human macrophages (da Silva *et al.*, 1989).

The LPG covering the *Leishmania* promastigote surface, as well as the GIPLs that are highly expressed on amastigotes, bind the serum mannan-binding protein (MBP) (Green *et al.*, 1994). MBP functions as a C1q-like molecule and can activate complement through the classical cascade (Ohta *et al.*, 1990), thus providing a possible mechanism for antibody-independent complement activation and deposition of C3 cleavage fragments on the parasite surface. This might be particularly relevant in the case of amastigotes, which display little or no surface protease or LPG.

It should be noted that the complement cascade fails to lyse infective metacyclic forms of *L. major*, apparently because the C5-9 membrane attack complex is shed from the promastigote surface (Puentes *et al.*, 1990), concomitant with developmental modifications and elongation of LPG (Sacks *et al.*, 1990; McConville *et al.*, 1992; see also Section 3.2.1). Interaction with macrophages through complement receptors may constitute an important protective mechanism for this parasite: contrary to Fc receptors, engagement of complement receptors is known not to trigger the respiratory burst (Wright and Silverstein, 1983) and indeed opsonization by complement has been shown to confer a survival advantage on *L. major* infecting murine macrophages *in vitro* (Mosser and Edelson, 1987).

Still other serum factors may promote binding of *Leishmania* to macrophages, including antibodies (Mosser and Rosenthal, 1993) and fibronectin (Wyler *et al.*, 1985), presumably interacting with Fc and fibronectin receptors, respectively (see also below).

2.1.3. *Serum-independent Binding*

Binding of *Leishmania* to macrophages can also occur through direct interaction of parasite surface structures with macrophage receptors. Parasite molecules responsible for this effect include LPG (Kelleher *et al.*, 1992), the gp63 surface protease, and other surface components (see below). Experiments using inhibitory oligosaccharides and glyco- or neoglycoproteins indicated that non-opsonized *L. donovani* promastigotes can attach to macrophages by the mannose/fucose receptor (Blackwell *et al.*, 1985; Wilson and Pearson, 1986, 1988). The receptor for advanced glycosylation end-products (AGE) on murine macrophages has also been found to recognize promastigotes of *L. major*, as determined by competition experiments using AGE–bovine serum albumin (Mosser *et al.*, 1987).

Several groups have proposed that the macrophage CR3 is capable of binding *Leishmania* promastigotes in the absence of serum (Blackwell *et al.*, 1985; Mosser and Edelson, 1985; Wozencraft *et al.*, 1986; Cooper *et al.*, 1988; Russell and Wright, 1988; Wilson and Pearson, 1988; Talamas-Rohana *et al.*, 1990). This phenomenon might depend on synthesis of complement components by the macrophage itself, followed by complement activation and deposition of C3 fragments on the parasite surface (Wozencraft *et al.*, 1986). Alternatively, it has been suggested that the parasite might interact directly with macrophage CR3 receptors. The latter proposition has been disputed, however, and, using CR3 expressed in non-haematopoietic cells or adsorbed on to plastic, virtually no direct binding of three different species of *Leishmania* promastigotes could be detected

(Mosser et al., 1992). CR3 is a member of the β_2 integrin family and as such would be expected to interact with molecules bearing an Arg-Gly-Asp (RGD) sequence. However, gp63 lacks an RGD tripeptide (Miller et al., 1990). Nevertheless, anti-RGDS (Arg-Gly-Asp-Ser) antibodies were found to precipitate gp63 of L. chagasi promastigotes, and the F(ab')2 fragment of an anti-fibronectin antibody inhibited attachment of the parasite to macrophages (Rizvi et al., 1988). These experiments suggested that gp63 shares strong structural homologies with serum fibronectin. Studies using overlapping tetracosapeptides covering the whole sequence of L. major gp63 identified the peptide Ser-Arg-Tyr-Asp (SRYD) as a moiety most probably functioning as a ligand for adhesion to macrophages in the absence of serum (Soteriadou et al., 1992). This peptide bears structural resemblance to RGDS of fibronectin and is recognized by antibodies to RGDS-containing fibronectin peptides. These observations are consistent with the hypothesis that Leishmania promastigotes may attach to macrophages by interaction of gp63 with integrin-type receptors.

Interestingly, amastigotes may engage different receptors from those used by promastigotes. Thus, a heparin-binding activity has been described in amastigotes of L. amazonensis, that interacts with heparan sulphate proteoglycans of the macrophage surface (Love et al., 1993). Stage-specific glycosphingolipids (GSLs) have also been described in amastigotes of L. amazonensis by Strauss et al. (1993). As shown by using Fab fragments of specific anti-GSL antibodies, these molecules also mediate attachment to, and infection of, macrophages by amastigotes, hence they are presumably important for the maintenance of an established infection. The macrophage receptors interacting with GSLs have not yet been characterized.

2.2. Active Invasion

2.2.1. Induced Phagocytosis: Trypanosoma cruzi

(a) Attachment to host cells. Tryp. cruzi can invade a variety of vertebrate cells. As with other microorganisms, cell invasion involves a mandatory interaction between complementary moieties on the host cell and on the parasite surface. Both protein and carbohydrate structures are involved in recognition of Tryp. cruzi by macrophages (Hall, 1993). Fibronectin may act as a molecular bridge between the parasite and target cells (Ouaissi et al., 1984; Peyrol et al., 1987). Indeed, exogenous fibronectin increased attachment and uptake of Tryp. cruzi amastigotes by murine macrophages and human monocytes in vitro, an effect that was inhibited by pretreatment of the parasite with the RGDS tetrapeptide of the fibronectin cell attachment site (Noisin and Villalta, 1989). Laminin also binds to *Tryp. cruzi* trypomastigotes in a saturable manner, and appears to be involved in adhesion of the parasite to its host cell, since antilaminin antibodies inhibited host cell invasion (Giordano *et al.*, 1994). Based on the use of mutant cell lines deficient in surface sialic acid and therefore resistant to infection by *Tryp. cruzi*, Schenkman *et al.* (1993) and Ming *et al.* (1993) demonstrated that host cell sialic acid was involved in the invasion process. Of interest, invasion of the mutant cells could be restored to levels seen in control cells by re-sialylation with *Tryp. cruzi* trans-sialidase (Section 3.2.2) in the presence of a sialyl donor, pointing to the role of this enzyme as a virulence factor. The sialyl receptor on the parasite might be the trans-sialidase itself (Ming *et al.*, 1993). Recognition of other sugar moieties presumably plays a role in parasite binding to target cells and, indeed, carbohydrate-binding proteins were recently identified on *Tryp. cruzi* trypomastigotes and epimastigotes (Bonay and Fresno, 1995). These molecules recognize mannose and galactose residues and bind with high affinity to cells susceptible to infection by this parasite.

Another protein exposed on the surface of *Tryp. cruzi*, and called penetrin (Ortega-Barria and Pereira, 1991), has been shown to promote adhesion of trypomastigotes to components of the extracellular matrix. The purified molecule inhibited infection of non-phagocytic Vero cells by the parasite. Similarly, *Escherichia coli* cells expressing the recombinant protein on their surface adhered to, and penetrated, Vero cells, pointing to a probable role of this molecule in the process of cellular invasion by *Tryp. cruzi*.

Different cell types may vary with regard to the type of ligand–receptor interactions involved in parasite binding, and it is striking that myoblasts and heart muscle sarcolemma membranes were shown to interact with *Tryp. cruzi* through their cholinergic and β-adrenergic receptors (Von Kreuter *et al.*, 1995). This particular interaction also led to reduced cyclic adenosine monophosphate (cAMP) levels in the parasite, an observation whose significance remains to be clarified.

(b) *Entry of* Trypanosoma cruzi *into host cells.* Much of our current understanding of the invasion process of *Tryp. cruzi* comes from studies by N.W. Andrews and co-workers. As shown by earlier investigators (Schenkman *et al.*, 1991b), cytochalasin D, a classical inhibitor of phago-cytosis which acts by blocking actin polymerization, failed to prevent and even enhanced (Tardieux *et al.*, 1992) trypomastigote entry into several cell types. This finding raised the question of the origin of the membrane necessary to form the vacuole around *Tryp. cruzi* during internalization. It was then discovered that, upon infection, lysosomes rapidly leave the perinuclear area and fuse with the cell membrane at its contact point with the parasites, thus apparently providing the membrane required for formation of the parasitophorous vacuole (Tardieux *et al.*, 1992). This proposition was further supported by the observation that agents known to induce migration of lysosomes from the perinuclear area to the cell periphery, such as brefeldin A or cAMP analogues, promoted cell invasion by *Tryp. cruzi* trypomastigotes, whereas blocking lysosomal fusion by sucrose loading reduced invasion. That the membrane of the parasitophorous vacuole is formed by a process other than invagination of the host cell plasma membrane is shown by the finding that few plasmalemmal constituents are incorporated into the vacuolar membrane surrounding *Tryp. cruzi* (see Meirelles *et al.*, 1983, 1984, 1986; Meirelles and de Souza, 1986; Hall *et al.*, 1991). On the contrary, lysosomal membrane glycoproteins are well represented in such vacuoles.

The mechanism of clustering and fusion of lysosomes at the site of parasite invasion is not yet fully understood, but interesting clues have already been obtained. Access of organelles such as lysosomes to the plasma membrane is normally prevented by the cortical layer of polymerized actin filaments that forms a dense scaffold just beneath the plasma membrane of all cells (Bretscher, 1991). Actin filaments are continually synthesized and broken down to accommodate various dynamic cell functions such as locomotion, endocytosis and secretion. The balance between assembly and disassembly can be disturbed by certain drugs, such as cytochalasins, which prevent actin polymerization; this may explain the enhancing effect of cytochalasin D on cell invasion by trypomastigotes (see above). The process of cell penetration by *Tryp. cruzi* presents striking analogies with that of regulated secretion, since the latter phenomenon also involves rearrangement of the cortical cytoskeleton and movement of organelles to the cell periphery. Cytochalasin D, presumably by its capacity to disassemble the subplasmalemmal actin network, promotes degranulation in chromaffin secretory cells (Aunis and Bader, 1988). Under physiological conditions secretion is known to be regulated at least in part by changes in the cytosolic concentration of the ubiquitous second messenger Ca^{2+}. It has now been observed that infective trypomastigotes induce repetitive cytosolic free Ca^{2+}-transients in rat kidney fibroblasts, through a G-protein-coupled pathway (Tardieux *et al.*, 1994). A requirement for increased cytosolic Ca^{2+} in both the parasite and its prospective host cell has also been noted by other investigators (Moreno *et al.*, 1994). These effects may be due to a soluble trypomastigote factor, which also triggers phosphoinositide hydrolysis and reorganization of the host cell F-actin network, presumably in relation to movement of lysosomes to the cell periphery (Rodriguez *et al.*, 1995).

Several aspects of host cell invasion by *Tryp. cruzi* still remain unclear. In particular, a possible role of parasite proteases in the infection process has been suggested by the observation that inhibitors of the major cystein protease, cruzipain, markedly reduced infection of Vero cells and subsequent intracellular multiplication of the microorganism (Franke de Cazzulo *et al.*, 1994). The precise effect of these molecules on the invasion process has yet to be determined.

2.2.2. *Forcible Penetration:* Toxoplasma gondii

Toxoplasma gondii can infect many animal species and, *in vitro*, tachyzoites invade and multiply within most nucleated cells. This can be achieved only because host cell penetration by *Toxoplasma* is an active event. The latter involves a mechanism quite distinct from induced phagocytosis as used by *Tryp. cruzi* (Section 2.2.1); what follows is a brief account of the invasion process, many aspects of which are still conjectural. The interested reader is referred to papers by Joiner and Dubremetz (1993), Sibley (1993), and Kasper and Mineo (1994) for more details.

(a) *Attachment of* Toxoplasma gondii *to host cells.* As with phagocytosis, invasion requires preliminary attachment of the parasite through interaction of parasite ligands and cell surface receptors. Host laminin deposited on to the parasite surface (Furtado *et al.*, 1992a, b), surface lectins (Robert *et al.*, 1991), and a major parasite surface protein (P30, SAG-1) (Grimwood and Smith, 1992; Mineo and Kasper, 1994) may participate in this process. Structures on the host cell that are capable of interacting with these molecules have yet to be fully characterized. The laminin receptor on macrophages (Furtado *et al.*, 1992a) and the $\alpha6/\beta1$ integrin receptor on human fibroblasts and Chinese ovary cells (Furtado *et al.*, 1992b) have been implicated in mediating attachment of laminin-coated *Toxoplasma*, and indeed laminin has been suggested to function as a ubiquitous bridge between the parasite and host cells (Kasper and Mineo, 1994). The fact that most cells can be infected also suggests that common membrane constituents such as certain lipids or the carbohydrate portion of glycoproteins might be involved in this initial binding phase.

(b) *Penetration of* Toxoplasma gondii *into host cells.* Attachment of *Tox. gondii* permits orientation of the parasite, such that its anterior end comes in contact with the membrane of the prospective host cell. Invasion then proceeds rapidly by a "sliding penetration" process and can be complete within 10–15 seconds. This process involves extension of the conoid (a cone-shaped organelle), and secretion by the rhoptries and micronemes (both specialized vesicles) of factors that alter properties of the host cell membrane. Several rhoptry proteins have been cloned, but the precise function of these molecules remains obscure. Of particular interest in this regard is ROP 1, a 61 kDa rhoptry protein which is secreted at the time of invasion and associates with the membrane of the parasitophorous vacuole (Saffer *et al.*, 1992), perhaps a result of the charge asymmetry of the molecule (Ossorio *et al.*, 1992). The importance of ROP 1 in the invasion process is illustrated by the observation that antibodies against this molecule block host cell penetration by *Toxoplasma* (see Schwartzman, 1986; Schwartzman and Krug, 1989).

Following attachment of the parasite to its target cell, it is assumed that an annular junction forms between parasite and host cell membranes, through which the parasite forcibly penetrates the host cell while apparently pulling the cell membrane around itself. The driving force for this remarkable process might be akin to that allowing capping of antibodies bound to cell surface proteins. In the end, the parasite is enclosed within a parasitophorous vacuole whose membrane derives from that of the host cell, like that of a phagocytic vacuole formed by bona fide phagocytosis. As discussed below, however, phagocytic vacuoles (such as those containing *Leishmania* in macrophages) and parasitophorous vacuoles resulting from *Tox. gondii* infection display quite different properties, which also explains why these parasites have evolved distinct intracellular survival strategies.

3. AVOIDING INTRACELLULAR KILLING MECHANISMS

3.1. Phagolysosome Formation and the Microbicidal Activities of Host Cells

Whether entry into cells occurs via phagocytosis or by active penetration, parasites eventually find themselves enclosed in a vacuole. Vesicles resulting from true endocytic activity, and in particular phagosomes created when professional phagocytes engulf bacteria or unicellular eukaryotic parasites, normally undergo fusion with primary lysosomes originating from the Golgi apparatus of the cell, or with secondary lysosomes formed by fusion of primary lysosomes with pinocytic vesicles. The resulting organelles are rich in hydrolytic enzymes and other microbicidal proteins and peptides (Nathan and Gabay, 1992). These vesicular bodies may assume different morphologies and be ascribed different names. When formed by phagocytosis of microbes, they are called phagolysosomes.

3.1.1. *Acidification of Endocytic Vesicles*

The use of pH-sensitive fluorescent probes has shown that the pH of endocytic vesicles, formed when macrophages phagocytize particulate material such as yeast cells, normally falls to around 5 within minutes of ingestion (Geisow *et al.*, 1981). Phagosome acidification may involve the activity of the membrane NADPH (β-nicotinamide adenine dinucleotide phosphate, reduced form) oxidase (see below), but is mainly regulated by an adenosine triphosphate (ATP)-dependent proton pump described in endocytic vesicles of various cell types (Fogac *et al.*, 1983; Galloway *et al.*, 1983), and particularly active in macrophages (Lukacs *et al.*, 1990, 1991; Pitt *et al.*, 1992).

3.1.2. *Formation of Reactive Metabolites of Oxygen*

Ingestion of particles by professional phagocytes usually induces a metabolic response known as the respiratory burst. The respiratory burst is characterized by increased uptake of molecular oxygen, which is reduced to superoxide (O_2^-) by electrons derived from the reduced pyridine nucleotide NADPH in a reaction catalysed by the plasma membrane enzyme NADPH oxidase (for a review, see Segal and Abo, 1993 and Robinson and Badwey, 1994). $NADP^+$ is recycled back to NADPH through the rapid oxidation of glucose via the hexosemonophosphate shunt. Superoxide formed by the univalent reduction of oxygen is further dismutated to hydrogen peroxide (H_2O_2), and the generation of the hydroxyl radical $(HO^\cdot)$ and of singlet oxygen $(^1O_2)$ may then proceed by interaction of O_2^- and H_2O_2 (Robinson and Badwey, 1994; Rosen *et al.*, 1995). Production of these reactive oxygen intermediates (ROI) is generally triggered by phagocytosis, and microbes are therefore exposed to these agents as they are internalized. ROIs are highly toxic molecules and are thought to constitute an essential part of the defence mechanisms used by phagocytes to destroy invading microorganisms (Hughes, 1988). In this connection, it is interesting to note that the mature human macrophage does not seem to generate ROIs intracellularly (Johansson *et al.*, 1995); how these molecules are delivered to the phagosome, and how they exercise their toxicity towards ingested microorganisms, therefore remains to be determined. The generation of oxygen metabolites is further enhanced when phagocytes have been activated by interferon γ (IFN-γ), endotoxin, or other stimuli (Johnston, 1978; Murray and Cohn, 1980; Kaku *et al.*, 1983; Nathan *et al.*, 1983; Murray *et al.*, 1985).

3.1.3. *Formation of Oxidized Derivatives of Nitrogen*

Nitric oxide (NO), and possibly other oxidized derivatives of nitrogen (Rosen *et al.*, 1995), form another important group of cytotoxic molecules whose production is triggered by activation of phagocytes from certain animal species. NO is generated by different cell types upon conversion of

L-arginine into L-citrulline by the enzyme nitric oxide synthase (NOS). This molecule is thought to play a role in several important physiological processes such as neurotransmission and smooth muscle relaxation (Marletta, 1993; Moncada and Higgs, 1993). Murine macrophages can be induced to upregulate the expression of an inducible NOS (iNOS) by various agonists including IFN-γ, tumour necrosis factor α (TNF-α) and lipopolysaccharide, with concomitant release of large amounts of NO. Production of NO appears to constitute one of the main microbicidal mechanisms of murine macrophages (Nathan and Hibbs, 1991). NO has been shown to be instrumental in the destruction of *Leishmania* parasites (Green *et al.*, 1990; Mauël *et al.*, 1991), as well as *Tox. gondii* (see Adams *et al.*, 1990) and *Tryp. cruzi* (see Gazzinelli *et al.*, 1992; Munoz-Fernandez *et al.*, 1992), by activated murine macrophages *in vitro*. The importance of NO as a microbicidal agent whose function is critical to host survival is best illustrated by the observation that mutant mice lacking the inducible NO synthase gene displayed high susceptibility to *L. major* infection (Wei *et al.*, 1995); additionally, oral treatment of normal mice with N^G-monomethyl-L-arginine, an iNOS inhibitor, considerably decreased their resistance to infection by *L. major* and could even lead to death of the animals as a result of an otherwise curable infection (Evans *et al.*, 1993). Inhibition of NO production *in vivo* has also been shown by Petray *et al.* (1994) to increase the susceptibility of experimental animals to infection by *Tryp. cruzi*. However, it is important to note that human macrophages from various origins, and stimulated by a variety of agonists, have been reported to release little or no NO (Murray and Teitelbaum, 1992; Schneeman *et al.*, 1993), in spite of their ability to generate iNOS messenger ribonucleic acid and protein (Weinberg *et al.*, 1995). Concomitantly, and although some investigators have reported NO-mediated parasite killing in human macrophages (Munoz-Fernandez *et al.*, 1992; Vouldoukis *et al.*, 1995), other groups consider microbicidal mechanisms in such cells to be largely independent of the production of NO. The physiological role of iNOS expression in human macrophages, and whether deficient NO production by these cells *in vitro*, as generally observed, is due to inadequacies of the experimental approaches, remain to be fully determined.

In view of their wealth of degradative enzymes, microbicidal peptides and reactive metabolites of oxygen and nitrogen, intracellular spaces (particularly within macrophages) are thought to constitute a rather inhospitable milieu for microorganisms. Notable differences can be observed, however, depending on the parasite and host cell under consideration. As a consequence of the diverse mechanisms used by different parasites to invade cells, the properties of the vacuoles in which they become entrapped are also different, confronting the microorganism with different physicochemical and biochemical environments. Parasites have therefore developed different adaptive strategies to thrive in the normally hostile intracellular milieu.

3.2. Resistance to Lysosomal Toxicity

3.2.1. *Survival Within Lysosomes:* Leishmania

Early studies on the interaction of *Leishmania* with host macrophages identified the phagolysosome as the final intracellular site in which this parasite survives and multiplies (Alexander and Vickerman, 1975; Chang and Dwyer, 1976). The demonstration that extracellularly added electron-dense material is eventually detectable within parasite-harbouring phago-lysosomes indicates continuous accessibility to the extracellular environ-ment via the vacuolar apparatus of the host cell, a finding perhaps significant in terms of parasite nutrition (Section 4) and chemotherapy.

(a) *The* Leishmania *lipophosphoglycan and its importance in intracel-lular survival.* Several lines of evidence suggest that LPG is a molecule of major importance to promastigotes, both in their insect vector and when infecting their vertebrate hosts (for a review, see Turco and Descoteaux, 1992). As discussed in Section 2.1.2, LPG appears to be instrumental in protecting *Leishmania* from complement-mediated damage. In addition, it clearly promotes intracellular survival of promastigotes early after macro-phage infection, as shown by experiments using LPG-deficient strains. LPG (together with other phosphoglycans) is released by promastigotes in a soluble form formerly called the excreted factor (EF). That EF might function as a virulence factor, perhaps by protecting *Leishmania* promas-tigotes against early intralysosomal degradation, was first suggested by the observation that growth of *L. enriettii* in otherwise non-permissive mouse macrophages was promoted by the addition of EF from *Leishmania* species that could proliferate in such host cells (Handman and Greenblatt, 1977). Similarly, a strain of *L. major* that failed to survive in mouse or hamster macrophages *in vitro* was found to lack a full surface LPG; coating the parasite with LPG from a virulent strain restored its capacity to infect those cells (Handman *et al.*, 1986). Very similar results were reported in a study of human monocyte infection by glycosylation variants of *L. donovani* that express deficient LPG (McNeely and Turco, 1990).

In addition to its putative role as a molecule protecting promastigotes from early lysosomal degradation (see above), LPG is thought to contribute to the infective potential of *Leishmania* by scavenging oxygen radicals formed upon phagocytosis (Section 3.3.1), and by dampening the respiratory burst and interfering with signal transduction pathways in the infected cell (Section 3.3.2).

(b) *Living at acid pH.* The mechanisms by which *Leishmania* can thrive in the environment of the phagolysosome remain a major unknown factor in the biology of this parasite. The *Leishmania*-harbouring vacuole is known to acidify (Antoine *et al.*, 1990), which evidently does not harm the parasite. The cytosolic pH of the amastigote is maintained close to neutrality at an external pH as low as 4.0 (Glaser *et al.*, 1988), which is effected presumably through a Mg^{2+}-dependent, vanadate-sensitive adenosine triphosphatase (ATPase) identified in the parasite plasma membrane (Zilberstein and Dwyer, 1988). The ATPase of *L. donovani* promastigotes is located on the cytoplasmic side of the membrane, has a pH optimum of 6.5 (i.e., correlating with the cytoplasmic pH) and clearly functions as a proton pump. Although expression of this ATPase in amastigotes remains to be formally established, it is significant that two tandemly linked *L. donovani* ATPase genes have been cloned (Meade *et al.*, 1987, 1989). These genes are most closely related to the *Saccharomyces cerevisiae* and *Neurospora crassa* H^+-ATPase genes. Whereas transcripts of the upstream gene are expressed to a similar level in both extracellular promastigotes and intracellular amastigotes, the second gene is expressed predominantly in the intracellular stage of the parasite life cycle. Thus it is tempting to speculate that the protein product of the latter gene indeed plays a role in the regulation of pH homeostasis in the intralysosomal *Leishmania* amastigote.

Leishmania amastigotes appear to have become exquisitely adapted to living in an acid environment. Indeed, they are metabolically more active at acid than at neutral pH (Mukkada *et al.*, 1985; Glaser *et al.*, 1988) and are thought to take advantage of the proton gradient formed across their plasma membrane (outside acid, inside near neutral) to drive the active inward transport of glucose and amino acids (Zilberstein and Dwyer, 1985, 1988; Glaser *et al.*, 1988, 1992; see also Section 4.3), a process best achieved at a pH between 5.0 and 5.5 (Mukkada *et al.*, 1985; Glaser and Mukkada, 1992).

The parasitophorous vacuole resulting from infection by *Tox. gondii* does not acidify, a point further discussed in Section 3.2.3.

(c) *Resistance against lysosomal enzymes.* The phagosome formed after internalization of *Leishmania* by host macrophages readily fuses with lysosomes (Alexander and Vickerman, 1975; Chang and Dwyer, 1976), and the newly formed phagolysosome has been shown to contain cathepsins B, D, H and L, and dipeptidyl peptidases I and II (Prina *et al.*, 1990). How *Leishmania* resists the degradative action of these and other enzymes remains subject to discussion, and the evidence is sometimes contradictory.

In early experiments it was observed that the parasite *Leptomonas costoris* was destroyed at a slower rate in macrophages co-infected with *L. donovani* than in macrophages infected with *Leptomonas costoris* alone, suggesting that *Leishmania* inhibited macrophage degradative processes (Kutish and Janovy, 1981). Two *Leishmania* surface molecules might be implicated in these effects, the gp63 surface protease and LPG. The surface protease of *L. amazonensis* promastigotes, reported to be active at acid pH (Chaudhuri and Chang, 1988), was shown to inhibit the degradation of liposome-encapsulated proteins by macrophages (Chaudhuri *et al.*, 1989), suggesting that it might function as a virulence factor capable of inactivating microbicidal molecules of the phagolysosomes. Discrepant results regarding the pH optimum of the protease were obtained by Etges *et al.* (1986) and Bouvier *et al.* (1990), who reported maximum proteolytic activity in the neutral to basic pH range, thus in effect ruling out an activity of the enzyme at the acid pH of the phagolysosome. These differences may have been related to the types of substrates used by the different investigators (Tzinia and Soteriadou, 1991).

The role of gp63 in intracellular survival of *Leishmania* remains highly conjectural. Indeed, results from studies on the synthesis of the protease by different developmental stages of *L. major* indicated that its expression is reduced at least 300-fold in amastigotes relative to that in promastigotes (Schneider *et al.*, 1992b). These findings make it unlikely that gp63 plays a critical role in intramacrophage survival of *Leishmania*, either as infecting promastigotes or as established amastigotes. Of interest, other species of gp63 proteases have recently been described. *L. mexicana* amastigotes express a lysosomal protease that is antigenically related to surface gp63 and displays a definitely acidic pH optimum (Ilg *et al.*, 1993). Highly infectious (stationary phase) promastigotes of *L. chagasi* generate, in addition to bona fide gp63, a parent 59 kDa protease whose expression decreases as the parasite becomes attenuated following long-term cultivation *in vitro* (Roberts *et al.*, 1995). The possible function of these various molecules is unknown.

Amastigotes of *L. mexicana* are characterized by the presence of an unusually large lysosome-like organelle, called the megasome (Pupkis *et al.*, 1986), that is rich in cysteine proteinases. Homozygous null mutants for one of these proteinases (LmCPb) were produced by targetted gene disruption (J.C. Mottram, personal communication). The mutant parasite exhibited a fivefold reduction in infectivity for murine macrophages. Virulence was restored to wild-type levels by re-expression of the *lmcpb* gene, while expression of a homologous *Tryp. brucei* cysteine proteinase did not restore infectivity, clearly pointing to LmCPb as a molecule capable of promoting intracellular survival of *L. mexicana*. Further studies will be necessary to delineate more precisely the function of this interesting molecule. Amastigotes of *L. mexicana* also release in the parasitophorous vacuole a high molecular weight phosphoglycan complexed with protein (Ilg et al., 1994). The function of this compound remains to be determined.

In a study of the effect of EF on several lysosomal enzymes, the compound was found to display strong anti-β-galactosidase activity, whereas acid phosphatase, β-glucuronidase and N-acetyl-β-glucosaminidase were not affected (El-On et al., 1980). EF is negatively charged, as are other polyanionic inhibitors of lysosomal enzymes such as heparin and chondroitin sulphate, which are thought to bind to positively charged hydrolases so as to control lysosomal activity (Avila and Convit, 1976). EF also strongly complexes Ca^{2+} ions (Eilam et al., 1985). Furthermore, coating red blood cells with LPG increased their resistance against cytolysis by macrophages (Eilam et al., 1985). Although these various observations point to LPG as a factor capable of protecting Leishmania against lysosomal degradation, the mechanisms of this effect remain unclear and may not involve inhibition of lysosomal hydrolases. Indeed, Leishmania-infected macrophages were found to hydrolyse ovalbumin and serum albumin as efficiently as uninfected control cells (Prina et al., 1990), suggesting unimpaired lysosomal enzyme activity.

It will be appreciated that, as already indicated, the gp63 metalloprotease and LPG are much less abundant in amastigotes than in promastigotes. Therefore, whatever their mechanism of action, and although these two molecules may play a role in helping promastigotes initiate the infection, they are unlikely to contribute to its maintenance.

3.2.2. Leaving the Parasitophorous Vacuole: Trypanosoma cruzi

Over twenty years ago, ultrastructural studies indicated that trypomastigotes of Tryp. cruzi infecting murine macrophages and fibroblasts could occasionally be observed to lie free in the cytoplasmic matrix of the host cell, as shown by the conspicuous absence of a membrane around the parasite body (Kress et al., 1975; Tanowitz et al., 1975; Nogueira and Cohn, 1976). As Tryp. cruzi was always seen within vacuoles during the initial stages of the infection, it was concluded that the parasite was capable of escaping from the vacuolar spaces to reach the host cell cytoplasm where its multiplication could proceed unhindered. It is noteworthy that, when ingested by activated macrophages, trypomastigotes were rapidly destroyed, suggesting that they were killed in the parasitophorous vacuoles before they had a chance to reach a safer intracellular location (Tanowitz et al., 1975).

(a) Phagosome membrane disruption. Understanding this mechanism of escape has progressed considerably in recent years. Disruption of the membrane of the parasitophorous vacuole occurs within one hour of infection and can first be recognized by the appearance of discrete discontinuities in the membrane (Ley *et al.*, 1990), followed by its complete disappearance two hours after cell invasion. Importantly, destruction of the vacuolar membrane correlates with release by *Tryp. cruzi* of a pore-forming protein toxin (Andrews *et al.*, 1990). This toxin, Tc-Tox, is immunologically cross-reactive with human C9, the terminal component of the membrane attack complex formed by complement, as well as with mouse perforin, the pore-forming protein of cytolytic T lymphocytes and NK cells (Andrews *et al.*, 1990). Tc-Tox associates with the membrane of the parasitophorous vacuole. It also inserts into artificial membranes, where it induces discrete conductance fluctuations characteristic of the formation of transmembrane ion channels. Furthermore, Tc-Tox is lytic *in vitro* for both red blood cells and nucleated cells (Andrews and Whitlow, 1989), but only at acid pH. Vacuoles containing *Tryp. cruzi* do indeed acidify, and buffering agents inhibit parasite escape from the intravacuolar spaces (Ley *et al.*, 1990). All of these observations are consistent with the hypothesis that Tc-Tox is at least partly responsible for disruption of the membrane of the parasitophorous vacuole and escape of *Tryp. cruzi* into the host cell cytoplasm. Whether other *Tryp. cruzi*-derived molecules, such as penetrin (Ortega-Barria and Pereira, 1991; see Section 2.2.1), also play a role in this process remains to be determined.

(b) *The role of the* Trypanosoma cruzi *trans-sialidase.* If indeed Tc-Tox functions as a pore-forming molecule and mediates disruption of the parasitophorous vacuole membrane, why does the molecule not attack the membrane of the parasite itself? A possible answer to this question, as well as further clues to the mechanisms of Tc-Tox action, have been provided by the results of studies on *Tryp. cruzi* trans-sialidase.

Tryp. cruzi is unable to synthesize sialic acid (Schauer *et al.*, 1983), but expresses trans-sialidase/neuraminidase (TS/N). This enzyme catalyses the reversible transfer of sialyl groups from donor glycoproteins to acceptors containing terminal β-galactosyl residues (the trans-sialidase activity) or, depending on acceptor concentration, the irreversible transfer of sialyl moieties to water, thus generating free sialic acid (the neuraminidase activity) (Previato *et al.*, 1985; Zingales *et al.*, 1987; Schenkman *et al.*, 1991a; Schenkman *et al.*, 1993; for reviews, see Shenkman and Eichinger, 1993 and Schenkman *et al.*, 1994). Infective trypomastigotes express high levels of the enzyme, which enables the parasite to sialylate its surface proteins. The enzyme is linked to the parasite membrane by a glycosyl-phosphatidylinositol anchor (Rosenberg *et al.*, 1991), and is released at the acid pH of the parasitophorous vacuole, presumably through the action of a phospholipase C (Frevert *et al.*, 1992; Hall *et al.*, 1992).

The luminal side of the lysosomal membrane is lined with heavily glycosylated glycoproteins, which may play a role in protecting the membrane from attack by lysosomal hydrolases (Holtzman, 1989). Lysosomal membrane glycoproteins can serve as targets for *Tryp. cruzi* TS/N, and several types of evidence suggest that desialylation of these molecules might facilitate subsequent disruption of the membrane by Tc-Tox. Thus, when sialylation-deficient cells were infected by *Tryp. cruzi*, escape of the parasites from the parasitophorous vacuole into the cytoplasm was faster than in control cells (Hall *et al.*, 1992). Moreover, neuraminidase treatment of erythrocytes renders the cells more susceptible to lysis by Tc-Tox. These results are compatible with the hypothesis that sialyl residues decrease the activity of Tc-Tox, and that removing these residues allows Tc-Tox to insert more efficiently into the lipid bilayer. At the same time, the parasite surface would become sialylated and would concomitantly acquire a higher degree of resistance against its own pore-forming protein. Similarly, sialic acid has been implicated as a protective device against perforin, and its removal from T cell surface proteins enhances their susceptibility to perforin-mediated lysis (Jiang *et al.*, 1990).

Trans-sialidase appears to play still another and more complex role in the infectivity of *Tryp. cruzi* than can be deduced from the above observations. Indeed, a single dose of trans-sialidase injected into the connective tissue of mice greatly enhanced the parasitaemia and mortality rate following challenge of the animals with *Tryp. cruzi*, even when the two inoculation sites were entirely distinct (Chuenkova and Pereira, 1995). No such enhanced virulence was observed in severe combined immunodeficiency mice, suggesting that trans-sialidase exercised its activity through an effect on cells of the host immune system. The precise mechanisms of this interesting phenomenon remain to be established.

3.2.3. *Creating a Modified Parasitophorous Vacuole:* Toxoplasma

(a) *The* Toxoplasma gondii-*harbouring parasitophorous vacuole fails to fuse with lysosomes.* Early experiments dealing with the mechanisms of intracellular survival of *Tox. gondii* in mouse macrophages uncovered a remarkable feature of the *Toxoplasma*-harbouring vacuole: parasite survival correlated with failure of such vesicles to undergo fusion with surrounding lysosomes (Jones and Hirsch, 1972; Jones *et al.*, 1972). Morphological studies indicated that strips of endoplasmic reticulum and mitochondria came to lie along the cytoplasmic side of the vacuolar membrane, apparently hindering access of lysosomes (Jones and Hirsch, 1972; de Melo *et al.*, 1992). Other modifications of the host cell architecture included rearrangement of the vimentin network around the parasitophorous vacuole, which might help the vacuole to become attached to the host cell nuclear envelope (Halonen and Weidner, 1994). It should be stressed that the evidence that failure to form phagolysosomes constitutes a protective device remains circumstantial. This notion is nevertheless supported by the observation that defective fusion depends on integrity of the microorganism: parasites damaged by heat or coated with specific antibody are ingested by classical phagocytosis, lose their capacity to inhibit lysosome–phagosome fusion, and become susceptible to intracellular killing (Jones and Hirsch, 1972; Jones et al., 1972; Wilson et al., 1980).

The difference in the fate of Tox. gondii-containing vacuoles between those formed when the parasite actively penetrates cells (failure to fuse with lysosomes), and those arising from phagocytosis of antibody-coated Toxoplasma (fusion with lysosomes), must be related to the mechanism of vacuole formation, and hence to the mode of interaction of the parasite with the host cell. This point has been clearly illustrated by studies using Fc receptor-transfected CHO cells (Joiner et al., 1990). In these cells, antibody-coated parasites are internalized following recognition by the Fc receptor and are delivered to an LAM-1 positive (lysosomal) compartment, whereas uncoated parasites that actively penetrate cells remain sequestered in non-fusogenic vacuoles.

(b) *The parasitophorous vacuole fails to acidify.* Tox gondii is very sensitive to acid (Sibley et al., 1985) and it appears highly significant for parasite survival that, when mouse macrophages were infected with live parasites, the parasitophorous vacuole failed to acidify (Sibley et al., 1985). Of interest, uptake of heat-killed or antibody-coated Tox. gondii was accompanied by a drop in phagosomal pH, indicating that blocking vacuolar acidification requires an intact, metabolically competent parasite. The fact that macrophages, which are endowed with phagocytic activity, were used in these experiments somewhat confused the issue, as it is now established that live Tox. gondii actively invades cells, whether phagocytic or not (see Section 2.2.2). It is probable that, due to its abnormal composition and fusion incompetence (see above), the modified membrane of the parasitophorous vacuole formed when live parasites penetrate macrophages fails to mature normally and to acquire the H^+-ATPase. When parasites are damaged or covered with antibody, they are no longer able to invade macrophages, and their internalization is then mediated by the constitutive phagocytic activity of the host cell, whereupon phagosomal acidification proceeds normally.

(c) *The membrane of the* Toxoplasma gondii-*containing parasitophorous vacuole.* Live Tox. gondii actively penetrate cells. Although a model for this infection process has been briefly described above (Section 2.2.2), it is important to note that many of its features remain unclear, as the swiftness and complexity of the event hinder its analysis (Joiner and Dubremetz, 1993). A key issue is that of the composition of the membrane of the parasitophorous vacuole, which renders it unable to fuse with lysosomes (for a more thorough discussion of this point, see Sibley, 1993). Soon after invasion, the bilayer is essentially devoid of intramembranous particles, and thus presumably lacks recognition structures required for fusion with other organelles. The mechanisms that underly the absence of host cell plasmalemmal proteins from the membrane of the parasitophorous vacuole are unknown. It has been proposed that the junction created between host cell and parasite might restrict the flow of membrane proteins in the growing parasitophorous vacuole during cell invasion, eventually resulting in exclusion of host cell membrane proteins from the mature vacuole (de Carvalho and de Souza, 1989). Confocal microscope studies indicated, however, that major cell surface proteins (Na/K ATPase, CD44, Mac-1) were internalized during parasite entry. Following penetration, these markers disappeared within minutes from the vacuole, perhaps by moving to other endosomal compartments. Subsequently, the parasitophorous vacuole and its membrane continued to undergo extensive modifications. The vacuolar space became filled with a tubular network of parasite origin (Sibley et al., 1986, 1995). Within minutes of penetration, exocytosis of electron-dense storage granules was observed, the content of which associated with both the intravesicular network (proteins GRA 1, 2, 4 and 6) and the vacuolar membrane (proteins GRA 3 and 5) (Cesbron-Delauw et al., 1989; Achbarou et al., 1991; Dubremetz et al., 1993; Lecordier et al., 1993; Mercier et al., 1993; Cesbron-Delauw, 1994; Ossorio et al., 1994). Antibody to one of the dense granule proteins (GRA 3) was also found to label strands extending from the parasitophorous vacuole into the host cell cytoplasm (Dubremetz et al., 1993). A rhoptry protein (ROP 2) has also been shown to insert in the vacuolar membrane, and to be exposed on the cytoplasmic side of the organelle (Beckers et al., 1994). No function has yet been ascribed to these various molecules and, indeed, a relationship between these events and the reported fusion incompetence of *Toxoplasma*-induced parasitophorous vacuoles remains to be established.

Another element of the parasitophorous vacuole membrane structure may explain its lack of fusion competence. Rhoptries are rich in cholesterol and phosphatidylcholine (Foussard et al., 1991), which are presumably released during host cell invasion. As discussed elsewhere (Joiner, 1991), insertion of these lipids into the vacuole membrane might alter its ability to fuse with lysosomes. Whether this proposition is tenable will have to wait a further analysis of the lipid composition of the *Toxoplasma*-induced parasitophorous vacuole.

3.3. Interference with the Respiratory Burst of Phagocytes

In view of the high sensitivity of certain microorganisms to the toxic action of oxygen metabolites generated during the respiratory burst, it is not surprising that intracellular parasites which interact with phagocytes have developed means of destroying or otherwise avoiding these compounds. From an operational point of view, two types of mechanisms can be distinguished: (i) the neutralization of toxic oxygen derivatives and (ii) the capacity to be internalized without eliciting a respiratory burst, and to dampen further generation of oxidant metabolites once intracellular.

3.3.1. *Neutralization of Oxygen Metabolites*

Different species of intracellular parasites display widely different levels of resistance against oxygen metabolites, and the intracellular fate of parasites invading phagocytes may depend in part on their capacity to detoxify such molecules. Thus promastigotes of *L. donovani*, which contain little catalase (Murray, 1981a), appear less resistant to oxidant-mediated damage (Channon and Blackwell, 1985a) and to intracellular killing (Pearson *et al.*, 1983) than amastigotes, which exhibit significant catalase and superoxide dismutase activity (Channon and Blackwell, 1985b). It is noteworthy that *Tox. gondii* is much more resistant than *L. donovani* promastigotes to hydrogen peroxide-induced toxicity, presumably because *Toxoplasma* contains higher levels of catalase and glutathione peroxidase (Murray, 1981a).

(a) *Antioxidant molecules: glutathione, trypanothione and ovothiol.* Early experiments studying antioxidant defence mechanisms in *Leishmania* indicated that these parasites were poorly endowed with glutathione peroxidase (Murray, 1981a). As glutathione, which is found in millimolar concentrations in most cells, plays a critical role in oxidant detoxification as a cofactor in peroxide reductions (Meister and Anderson, 1983; Kehrer and Lund, 1994), the exquisite sensitivity of *Leishmania* to killing by reagent H_2O_2 (Murray, 1981a) could be ascribed to lack of this enzyme. It was later observed, however, that *Leishmania* (and related trypanosomatids such as *Tryp. cruzi*, *Tryp. congolense*, *Tryp. brucei* and *Crithidia fasciculata*) contained a novel antioxidant composed of two molecules of glutathione conjugated with spermidine, termed trypanothione (Fairlamb *et al.*, 1985), and thought to play a similar role to that of glutathione in other cells. Another conjugate of glutathione with aminopropylcadaverine, homotrypanothione, has also recently been described in *Tryp. cruzi* (Hunter *et al.*, 1994). The relative specificity of glutathione and trypanothione reductases for their respective substrates might explain the difficulty in detecting glutathione reductase in *Leishmania*. Oxidized trypanothione is reduced by trypanothione reductase in an NADPH-dependent reaction very similar to that catalysed by glutathione reductase (Walsh *et al.*, 1991). Reduced trypanothione might therefore be expected to provide a degree of antioxidant protection. When trypanothione reductase was overexpressed in *L. donovani*, the level of reduced trypanothione did not increase in the transfectants relative to control *Leishmania*, owing to the fact that most of the trypanothione is normally already in the reduced state in the microorganism. Accordingly, the transfectants were as sensitive as non-transfectants to reagent H_2O_2 or to drugs known to induce an oxidative stress (nifurtimox, nitrofurazone, gentian violet), suggesting that the ability to express high levels of the enzyme was not a determining protective factor (Kelly *et al.*, 1993).

Recently, a novel antioxidant molecule, ovothiol A, has been identified in *L. donovani* by Spies and Steenkamp (1994). Ovothiol A belongs to a family of 4-mercaptohistidines, also found in the egg cells of several marine organisms, which are endowed with very high antioxidant and radical scavenging properties. Elucidation of the possible role of this thiol in protection of *Leishmania* against oxidant damage is awaited with interest.

(b) *LPG as an antioxidant molecule.* The ability of *Leishmania* promastigotes to establish infection within the hostile environment of macrophage phagolysosomes may be due to the protection afforded by specialized surface molecules such as LPG. In addition to its putative effects on lysosomal enzymes (Section 3.2.1), protection afforded by LPG may be explained by direct scavenging of oxygen intermediates, or by an indirect effect on the respiratory burst through modulation of host cell signalling pathways. Treatment of phorbol myristate acetate (PMA)-triggered macrophages with LPG-coated beads induced an instantaneous drop in chemiluminescence (McNeely and Turco, 1990), compatible with a scavenging effect of LPG on oxygen metabolites. Indeed, LPG belongs to a class of microbial glycolipids which are highly effective in capturing hydroxyl radicals and superoxide anions, as shown by electron-spin resonance spectroscopy and spin-trapping experiments (Chan *et al.*, 1989). In LPG, this activity appears to be a property of the repeating oxidizable phosphorylated disaccharide units found in this molecule.

3.3.2. Inhibition of Macrophage Respiratory Burst Activity

(a) Leishmania *and the macrophage respiratory burst.* The critical importance for *Leishmania* survival of the capacity to avoid triggering the host cell respiratory burst was suggested by experiments using a murine macrophage cell line (J774) which, in its non-activated state, is deficient in respiratory burst activity (Murray, 1981b). J774 cells ingested, but were unable to kill, promastigotes of *L. donovani*. However, when the cells were activated by lymphokine, they became capable of developing a strong respiratory burst, and phagocytosis of *L. donovani* then resulted in parasite destruction. It has to be stressed that the role of NO as a microbicidal agent was not evaluated in these experiments.

Differences have been reported in the relative capacity of amastigotes and promastigotes of *Leishmania* to elicit a respiratory burst in macrophages, the former developmental stage being less active (Haidaris and Bonventre, 1982; Pearson *et al.*, 1983; Channon *et al.*, 1984). This may explain in part the lesser susceptibility of amastigotes to intracellular killing noted by some investigators (Pearson *et al.*, 1983). The fact that failure to elicit respiratory burst activity constitutes an essential survival mechanism is also illustrated by studies on the interaction of *Tox. gondii* with murine and human macrophages. As mentioned in Section 2.2.2, entry of live parasites into the latter cells (now known to involve active penetration rather than phagocytosis) occurs without a respiratory burst and no killing ensues (Wilson *et al.*, 1980). This phenomenon clearly depends on the mechanism of parasite entry; allowing normal phagocytosis to occur, by coating *Tox. gondii* with specific antibody, activates the respiratory burst (Wilson *et al.*, 1980; Murray, 1981a) and induces parasite destruction. Under these conditions, the microorganisms presumably engage the macrophage Fc receptor, which is known to trigger the oxygen burst (see also Section 3.2.3 for a discussion of vacuolar acidification following Fc receptor-mediated parasite entry in macrophages).

After completion of the internalization process, presence of the intracellular parasite may further alter the response of these cells to subsequent stimulation of the respiratory burst, as demonstrated in different experimental models. Human monocyte-derived macrophages infected with *L. donovani* transiently exhibited a lower chemiluminescence response to further challenge with opsonized zymosan compared with non-infected controls (Pearson *et al.*, 1982), and decreased release of O_2^- and H_2O_2 when exposed to the bacterial peptide formyl-methionyl-leucyl-phenylalanine (Olivier *et al.*, 1992). Mouse peritoneal macrophages infected with *L. enriettii* or *L. major* were similarly less responsive to lipopolysaccharide or macrophage activating factors (Buchmüller-Rouiller and Mauël, 1987), as measured by their capacity to generate O_2^- and H_2O_2 upon triggering with opsonized zymosan or PMA. This inhibition was found to correlate with the number of intracellular organisms and was absent when macrophages were allowed to ingest dead rather than living *Leishmania* organisms.

(b) *Protein kinase C, the* Leishmania *lipophosphoglycan and intracellular signalling pathways.* The mechanisms of the effects described above are not fully understood, but several lines of evidence suggest that they may depend on alterations of signalling pathways in the infected cells. Attention has focused particularly on the role of protein kinase C (PKC) in these processes. In macrophages, activation of the respiratory burst correlates with translocation of PKC from the cytosolic to the membrane fraction of the cells (Myers *et al.*, 1985), and is accompanied by PKC-dependent phosphorylation of specific cellular proteins, including p47-*phox*, a component of the NADPH-oxidase (Nauseef *et al.*, 1991). PKC is therefore thought to be critically involved in the signalling pathways leading to the respiratory burst, as is also indicated by the stimulation of the burst induced by PKC agonists such as PMA and its inhibition by PKC inhibitors (Baggiolini and Wymann, 1990; Thelen *et al.*, 1993). Also consistent with a role of PKC in macrophage antileishmanial activity is the observation that depletion of phorbol ester-sensitive PKC isoforms, as obtained by prolonged incubation with PMA, abolished the killing of *L. donovani* in normal or lymphokine-activated murine macrophages (Murray, 1982).

Several lines of evidence indicate that the activity of macrophage PKC is affected by intracellular *Leishmania*. Depression of respiratory burst activity in human monocytes infected with *L. donovani* was accompanied by decreased protein phosphorylation, correlating with weaker translocation of PKC to the membrane (Olivier *et al.*, 1992). In murine macrophage infected with the same parasite, c-*fos* gene expression mediated by PKC was impaired, under conditions where protein kinase A-mediated c-*fos* gene expression remained unaffected (Moore *et al.*, 1993). These findings suggest that *Leishmania* interferes with signal transduction in macrophages, particularly with the PKC-dependent pathways (for a more thorough review of these points, see Reiner, 1994).

What are the parasite molecules responsible for these effects? Several observations again implicate LPG as a mediator of the decreased responsiveness of *Leishmania*-infected macrophages (for a review, see Descoteaux and Turco, 1993). Upon ingestion of *L. donovani* by murine macrophages, LPG becomes detectable on the surface of the infected cells within minutes (Tolson *et al.*, 1990) and is maximally expressed within the first post-infection day, after which it gradually disappears. Purified LPG was able to induce the same effects in non-infected cells as were observed in infected ones and, significantly, LPG inhibited the respiratory burst of human monocytes induced by zymosan uptake (McNeely and Turco, 1990). This finding may be correlated with the observation that LPG inhibits purified PKC (McNeely and Turco, 1987), through competitive inhibition with respect to diacylglycerol (DAG) and noncompetitive inhibition of phosphatidylserine. These two PKC agonists bind to the regulatory subunit of the enzyme, suggesting that LPG also interacts with this moiety. Both the 1-O-alkylglycerol and the phosphoglycan portions of LPG inhibit PKC, whereas the phosphorylated disaccharide units are ineffective (McNeely *et al.*, 1989). When the effect of LPG on the activation process of intact mouse bone marrow-derived macrophages was examined, LPG inhibited the DAG-stimulated phosphorylation of both a specific PKC-dependent peptide and of MARCKS, an endogenous PKC substrate, clearly pointing to defective protein phosphorylation as one of the mechanisms through which LPG can exercise its effects

(Descoteaux *et al.*, 1992). LPG-mediated impaired signal transduction in macrophages resulted in decreased LPS-induced c-*fos* expression (Descoteaux *et al.*, 1991), as was also shown using *L. donovani*-infected cells (Moore *et al.*, 1993).

The above observations are compatible with an LPG-mediated down-regulation of the respiratory burst activity in *Leishmania*-infected macrophages, through inhibition of PKC. However, the precise role of LPG as a protective molecule remains to be fully clarified. As discussed by Reiner (1994), the concentration of LPG required to inhibit PKC may not be achieved *in vivo*. Moreover, whereas LPG might dampen the oxidative response of macrophages upon phagocytosis of promastigotes, this molecule is unlikely to play a similar role in established infections. Indeed amastigotes, which make little or no LPG, elicit a lesser respiratory burst than promastigotes (Pearson *et al.*, 1983). They do, however, synthesize other classes of GIPLs (see Section 2.2.1) which might be endowed with similar activities. Significantly, GIPLs also inhibit PKC activity (McNeely *et al.*, 1989).

It should be emphasized that, although PKC inhibition may promote intracellular survival of *Leishmania*, other mechanisms must also contribute to parasite protection. Indeed, an LPG-deficient *L. donovani* mutant did inhibit PKC-dependent c-*fos* expression in macrophages, yet was readily destroyed intracellularly (McNeely and Turco, 1990).

Molecules other than LPG that have been reported to inhibit the respiratory burst activity of phagocytes include the *Leishmania* acid phosphatase (Remaley *et al.*, 1985b) and the gp63 promastigote surface protease (Sørensen *et al.*, 1994), which are both localized on the parasite surface. The relative contribution of these different factors to the overall inhibitory activities described above remains to be determined.

3.4. Inhibition of Nitric Oxide Synthesis

Murine macrophages activated by exposure to IFN-γ and other agonists express high levels of nitric oxide synthase. The flux of NO generated during the conversion of arginine into citrulline is sufficient to kill intracellular microbes, including *Leishmania* spp., *Tox. gondii* and *Tryp. cruzi.* (cf. Section 3.1.3). As reported recently by Proudfoot *et al.* (1995), *Leishmania* GIPLs can inhibit the synthesis of NO, thereby markedly reducing the leishmanicidal potency of activated macrophages. Strikingly, under the same conditions, LPG failed to affect NO production. This indicates that stimulation of the NADPH-oxidase and nitric oxide synthase are activated through different transduction pathways.

3.5. Other Inhibitory Effects

In different experimental models, infection by *Leishmania* has been shown to affect macrophage physiology in ways that might indirectly promote intracellular survival. Of particular interest is the recent observation that infection of murine bone marrow-derived macrophages by *L. donovani* inhibits apoptosis of the latter cells, presumably by a mechanism involving the stimulation of granulocyte–macrophage colony-stimulating factor (GM-CSF) and TNF-α production (Moore and Matlashewski, 1994). Preventing death of the host cell would be expected to favour parasite survival and spread within the host organism. *L. donovani* also markedly suppressed the enhanced expression of class I and II major histocompatibility (MHC) antigens induced by IFN-γ stimulation of murine macrophages (Reiner *et al.*, 1987). In a detailed study of MHC gene products in phagolysosomes of IFN-γ-activated macrophages which contained *L. amazonensis*, Lang *et al.* (1994) reported the conspicuous absence of MHC class I molecules from parasite-harbouring vacuoles. MHC class II molecules were present, being almost entirely clustered at the site of parasite anchorage to the vacuolar membrane. These observations may be correlated with the fact that *L. major* infection induced defective stimulation of antigen-specific T-cells *in vitro* (Fruth *et al.*, 1993); the same phenomenon was reported for *L. amazonensis* (Prina *et al.*, 1993). In the latter case, no decreased surface expression of Ia molecules could be detected, suggesting that deficient antigen expression by the infected cells could be due to inaccessibility of MHC class II molecules for processed moieties of the exogenous antigens in the parasite-containing vacuole, or to competition of parasite molecules and endocytosed antigen for the same MHC class II structures.

Other effects of infection of murine macrophages by *Leishmania in vitro* include the release of prostaglandin E_2 (Reiner and Malemud, 1985) and of transforming growth factor β (Barral *et al.*, 1993), both of which are regarded as inhibitors of the immune response. Downregulation of TNF-α receptor expression in LPG-treated macrophages (Descoteaux *et al.*, 1991) might also favour intracellular survival by inhibiting macrophage activation by TNF-α (Green *et al.*, 1990; Corradin *et al.*, 1991). A further property of LPG that may promote parasite survival is its capacity to inhibit neutrophil and monocyte chemotaxis (Frankenburg *et al.*, 1990); a similar phenomenon was observed when chemotaxis was assayed in the presence of *L. major*-derived gp63 surface protease (Sørensen *et al.*, 1994). As these different phenomena would only indirectly contribute to parasite survival, they fall outside of the scope of this review and will not be further considered here.

Exposure to *Tox. gondii* tachyzoites markedly alters the profile of eicosanoids that are released by human mononuclear phagocytes *in vitro*

(Yong *et al.*, 1994). In particular, the host cells fail to form 5-lipoxygenase products (leukotrienes (LT) B4 and C4), while retaining the capacity to synthesize prostaglandin E2 and thromboxane B2. LTB4 appears to be very toxic for both extracellular and intracellular *Tox. gondii*. Killing of the parasite in human macrophages activated by IFN-γ was inhibited by a selective 5-lipoxygenase inhibitor, further pointing to a role for 5-lipoxygenase products in the anti-*Toxoplasma* activity of activated human macrophages. Inhibition of LTB4 synthesis seems, therefore, to constitute a protective mechanism that contributes to intracellular survival of this particular parasite.

4. MEETING NUTRITIONAL REQUIREMENTS

4.1. Different Strategies for Different Parasites

Having succeeded in circumventing host cell defences, the parasite must find within its intracellular location all the nutriments required to fulfil its energy requirements, as well as the molecular blocks necessary for synthetic processes. Switching to intracellular life entails major metabolic changes required for adaptation of the organisms to their new physicochemical environment (temperature, pH, osmolarity, oxygen tension, ionic configuration, etc.). Variations in carbon and energy sources, as well as the need to maintain internal homeostasis in the face of drastic environmental fluctuations, requires greater metabolic flexibility from intracellular parasites than is necessary for parasites that dwell in the more steady extracellular milieu. These differences in adaptability between intracellular and extracellular parasites are reflected in clear differences of metabolic regulation. For instance, as discussed by ter Kuile (1993), the metabolic strategy of *L. donovani* is geared to maintaining constant internal conditions in highly different environments. This can be achieved only at the price of considerable energy expenditure and therefore requires great metabolic compliance. On the contrary, *Tryp. brucei*, which lives in the steady environment of the blood stream and therefore has ready access to nutriments, seeks to achieve maximum energy efficiency even at the cost of short-term flexibility. Wide gaps still remain in our understanding of how intracellular parasites acquire and metabolize nutriments in such a way as to preserve their homeostasis. Furthermore, for practical reasons, experiments have often been performed on the free-living forms of the parasites, and extrapolation to the intracellular organisms requires some caution.

It can be expected that meeting nutritional requirements will necessitate different strategies, depending on the intracellular location of the micro-organism. As described in earlier sections, *Leishmania* thrives in phago-lysosomes. Intracellular *Leishmania* amastigotes nevertheless remain in contact with the extracellular milieu, as shown by the flow of electron-dense markers delivered to the parasitophorous vacuole when added to culture fluids around infected macrophages *in vitro* (Chang and Dwyer, 1976) or *in vivo* (Berman *et al.*, 1981). Particles in large phagocytic organelles are also transferred to *Leishmania*-containing phagolysosomes (Veras *et al.*, 1992). Biotinylated β-glucuronidase or dextran could be further followed microscopically as they entered the flagellar pocket of intracellular *L. mexicana* amastigotes and reached the lysosomal compart-ment of the parasite (Russell *et al.*, 1992). The flagellar pocket of trypa-nosomatids, including *Leishmania*, presumably functions as a hydrolytic compartment that helps degrade large compounds into smaller molecules ready for further transmembrane transport, but the precise role of this organelle in the overall process of nutrient acquisition remains poorly understood (Webster and Russell, 1993).

Tox. gondii lives in non-fusogenic organelles that presumably remain sequestered from the extracellular environment. Yet, as indicated below, molecules have also been shown to flow from the extracellular milieu or from the host cell cytosol to within the parasite body. This traffic might be helped by a recently described molecular sieve in the membrane of the parasitophorous vacuole (Schwab *et al.*, 1994).

As *Tryp. cruzi* dwells in the host cell cytosol, it may have easy and direct access to all life-sustaining metabolites and presumably obtains its nutri-ments from the host cell's own cytosolic pool.

4.2. Metabolic Adaptation to Intracellular Life

Metabolic changes have been recorded upon passage of *Leishmania* from the free-living promastigote stage to that of intracellular amastigote. A drastic reduction in respiration rate was noted by Janovy (1967), suggesting a switch to anaerobic metabolism. Interestingly, however, the glycolytic pathway appeared also to be reduced in amastigotes relative to promasti-gotes, and fatty acids were found to become the predominant energy source (Coombs *et al.*, 1982; Hart and Coombs, 1982).

Transfer of trypanosomatids from the insect vector to the vertebrate host entails exposure of the parasite to increased temperature, variations in external pH, and oxidant stress. These changes presumably trigger adap-tive responses, and several investigators have studied the effects of acid exposure, temperature increase and oxidant treatment on the morphology, metabolism and stress responses of *Leishmania* and *Tryp. cruzi* in culture.

Two hours' exposure of *Tryp. cruzi* trypomastigotes to pH 5 was sufficient to trigger their transformation into forms indistinguishable from intracellular-derived amastigotes on morphological, biochemical and immunological criteria (Tomlinson *et al.*, 1995). Similarly, shifting the temperature of *L. braziliensis* promastigote cultures from 26 °C to 34 °C resulted in morphological changes resembling transformation to amastigotes (Stinson *et al.*, 1989); similar results have been reported for other *Leishmania* species such as *L. pifanoi* (see Pan, 1984). This morphological transformation was accompanied by increased oxidation of medium and long-chain fatty acids, reminiscent of the metabolic changes observed in amastigotes (Blum, 1987).

In addition, transferring promastigotes to 37 °C led to the synthesis of several heat shock proteins (HSP) (Lawrence and Robert-Gero, 1985; Van der Ploeg *et al.*, 1985). The precise functional significance of this observation remains unclear, but it is noteworthy that heat treatment of *L. braziliensis* increased the infectivity of the microorganism (Smejkal *et al.*, 1988), suggesting that heat induces the expression of genes whose function may be important for adaptation to intracellular life. Incubation of *L. chagasi* with sublethal concentrations of H_2O_2 or menadione also induced the synthesis of HSP 70, rendered the parasites more resistant to H_2O_2 toxicity, and increased their virulence in a Balb/C mouse model of infection (Wilson *et al.*, 1994). A search for genes specifically expressed in *L. major* amastigotes, and which might be essential to maintenance of the parasite in the intracellular environment, has led to the identification of a histone H1-like gene which is upregulated in the amastigote relative to the promastigote (Fasel *et al.*, 1994), consistent with the role of H1 histones in gene expression.

4.3. Acquisition of Nutrients

Nutriments must somehow cross the parasite plasma membrane; much effort has therefore been devoted to characterizing structures involved in the inward transport of small molecules such as glucose and amino acids. The subject of nutrient transport across plasma membranes of kinetoplastids has been reviewed by Zilberstein (1993).

4.3.1. *Glucose Uptake*

Early experiments suggested that *L. donovani* promastigotes were able to concentrate glucose (Zilberstein and Dwyer, 1984b), implying active transport. This process was shown to be driven by a proton gradient across the parasite membrane (Zilberstein and Dwyer, 1985), resulting both from lysosomal acidification through host cell metabolism (in the case of intracellular amastigotes) and from the activity of a proton-translocating ATPase also identified in the parasite membrane (Zilberstein and Dwyer, 1988). More recent findings have been claimed to be compatible with facilitated diffusion as the mechanism of glucose uptake in *Leishmania* (ter Kuile and Opperdoes, 1993; ter Kuile, 1993). This conclusion is also borne out by the identification of two membrane transport proteins (Pro-1 and D2) that exhibit a high degree of sequence homology with the *Tryp. brucei* and human erythrocyte glucose transporters (Langford *et al.*, 1992; Bringaud and Baltz, 1992; ter Kuile, 1993), which function as facilitated diffusion carriers. Genes for these two transporters are developmentally regulated and expressed primarily in the insect stage of the parasite (Langford *et al.*, 1992). They are therefore unlikely to play a role in glucose uptake by *Leishmania* amastigotes. However, a third transporter, D1, whose expression is not stage regulated and which is structurally different from both Pro-1 and D2, has also been identified (Langford *et al.*, 1992). Information on the functional regulation of this molecule is awaited with interest.

4.3.2. *Uptake of Amino Acids*

L-Proline is an important energy source for the insect stage of *Leishmania* (Krassner and Flory, 1972; Mukkada *et al.*, 1974), wherein it is mostly metabolized to CO_2 and only marginally incorporated into proteins. As with glucose, active, proton-driven transport was strongly suggested by earlier work (Bonay and Cohen, 1983; Zilberstein and Dwyer, 1985; Glaser and Mukkada, 1992), a conclusion recently challenged on the basis that these experiments failed to evaluate the non-metabolized proline pool within the cells (ter Kuile, 1993). In spite of this unresolved issue, it is highly significant that, in *L. donovani*, expression and activity of the L-proline transporter is strongly regulated by the external pH. Amastigotes, whose intracellular location is within macrophage phagolysosomes, are acidophilic and exhibit their highest metabolic activity, including proline uptake, at acid pH (Glaser and Mukkada, 1992; see also Section 3.2.1). On the contrary, promastigotes thrive best at pH 7.0. However, on a per-weight basis, the overall transport of L-proline is lower in amastigotes than in promastigotes (Glaser and Mukkada, 1992). Using acid-adapted promastigotes as amastigote substitutes, Zilberstein and Gepstein (1993) demonstrated a 10-fold drop in V_{max} for L-proline transport in acid-adapted parasites relative to control organisms, whereas the K_m decreased by one half. This is in contrast with transport of D-glucose, which exhibits pH-independent activity. It is expected that the pH-dependent regulation of proline transport is of physiological importance to the parasite, although the nature of the metabolic modifications thus induced and their significance so far escape our understanding.

Transporters for other amino acids (e.g., alanine, phenylalanine and methionine) have also been described in *Leishmania* (see Simon and Mukkada, 1977; Bonay and Cohen, 1983). Amino acid transport can be blocked *in vitro* with tricyclic antidepressants, which have been suggested as potential antileishmanial drugs. These compounds also induce the collapse of $\Delta\mu H^+$ and rapid parasite death (Zilberstein and Dwyer, 1984a). Other toxic effects of these drugs include a decrease in cellular ATP, probably involving uncoupling of the mitochondrion, and a general increase in membrane permeability (Zilberstein *et al.*, 1990).

4.3.3. *Uptake of Nucleosides and Nucleotides*

Several species of protozoan parasites (including *Toxoplasma* and the trypanosomatids) are unable to synthesize the purine ring, and therefore depend on the host for acquisition of this most essential metabolite. *L. donovani* has been claimed to synthesize deoxyribonucleic acid (DNA) from precursors of ribonucleic acid (RNA) derived from host macrophages (Bhattacharya and Janovy, 1975). Both a $3'$-nucleotidase–nuclease and a $5'$-nucleotidase (Gottlieb and Dwyer, 1983; Dwyer and Gottlieb, 1984) have been identified on the surface of *L. donovani* promastigotes and amastigotes. The combined activities of these enzymes are sufficient to process nucleic acids to inorganic phosphate and purine and pyrimidine nucleosides, which would then be carried into the cell by their respective transporters. In this connection, two high affinity nucleoside transporters have been described in *L. donovani*, one for inosine and guanosine, the other for adenosine and the pyrimidine nucleosides (Aronow *et al.*, 1987). The optimum pH of the nuclease activity of the $3'$-nucleotidase–nuclease is 6.0 and that of the nucleotidase activity of the same enzyme is 8.5, whereas that of the $5'$-nucleotidase is in the range of 6.8–8.0. These differences raise the question of the precise role played by these molecules in the acidic environment of the phagolysosome.

In a series of elegant studies, Pfefferkorn and co-workers took advantage of mutant cell lines to analyse the interaction of *Tox. gondii* with its host cells. They demonstrated that *Tox. gondii* in the parasitophorous vacuoles of fibroblasts from patients with the Lesch–Nyhan syndrome (who lack the purine salvage enzyme hypoxanthine guanine–phosphoribosyl transferase) would readily incorporate hypoxanthine and guanine into RNA and DNA, whereas the host cells would not (Pfefferkorn and Pfefferkorn, 1977). This implies that extracellular nucleotides must flow to the intracellular parasite through the host cell. That purines are also supplied to the parasite from the host cell's own nucleotide pool (including ATP), or from its nucleic acids, is nevertheless evident from results showing that *Tox. gondii* grows normally within cells in the absence of extraneously added purines (Schwartzman and Pfefferkorn, 1982; Pfefferkorn *et al.*, 1983). No such transcellular traffic could be demonstrated in the case of pyrimidine nucleotides, providing circumstantial evidence that the parasite is capable of pyrimidine synthesis *de novo* (Schwartzman and Pfefferkorn, 1981). Acquisition of purines by *Tox. gondii* probably involves enzymatic activities necessary to generate the required compounds from host cell donor molecules. In this respect, it is interesting that a potent nucleoside triphosphate hydrolase has been identified recently in *Toxoplasma* by Sibley *et al.* (1994). The enzyme is stored in dense granules, and is secreted into the parasitophorous vacuole following host cell invasion.

Based on the use of fluorescent markers of varying sizes, Schwab *et al.* (1994) provided indirect evidence for the presence of transport systems on the surface of *Tox. gondii*-containing organelles, which would allow the passage of solutes of $M_r < 1300$ from the host cell cytosol into the parasitophorous vacuole. Whether this represents a genuine mechanism enabling the parasite to gain access to host cell-derived metabolites remains to be established.

4.3.4. *Other Activities*

Various transporters that allow the active (against a concentration gradient) or passive (along a concentration gradient) uptake of small molecules have been identified in *Leishmania* (see reviews by Schneider *et al.*, 1992a and Zilberstein, 1993). These molecules help the cell to capture essential metabolites such as folate from the external milieu. Substrates for these (as well as for the previously described) transporters are presumably generated in part by the enzymatic degradation of larger molecules (polysaccharides, proteins, nucleic acids) in the lumen of host macrophage phagolysosomes. As mentioned above, the parasite may itself participate in the production of small metabolites from larger host constituents by enzyme activities associated with its surface membrane. Prominent among the surface molecules of *Leishmania* promastigotes is the gp63 surface metalloprotease. As discussed by Schneider *et al.* (1992a), although the protease occurs on promastigotes of all species of *Leishmania* so far examined, efforts to demonstrate the active molecule at the surface of amastigotes have been essentially unsuccessful. The specific substrates of the gp63 protease, if any, have not been identified. However, the presence of the molecule also on non-infective strains suggests that it is essential to promastigote survival, perhaps by helping amino acid acquisition in the alimentary tract of the vector, through digestion of proteins of the blood meal.

Three membrane acid phosphatases have been described on the surface of *L. donovani* by Remaley *et al.* (1985a). They differ in size, isoelectric point and substrate specificity. The combined activities of the three molecules allow the hydrolysis of a number of phosphorylated substrates, including phosphotyrosine, fructose 1,6-bisphosphate, adenosine mono- and diphosphate, different inositol phosphates and various phosphoproteins (Remaley *et al.*, 1985a; Das *et al.*, 1986). The pH optima of these different phosphatases are in the range 5.0–6.0. This is compatible with an activity in the phagolysosomal environment, but the role of the phosphatases in parasite survival remains unresolved. As indicated above (Section 3.3.2), one of the leishmanial acid phosphatases was shown to block neutrophil respiratory burst by inhibition of the NADPH oxidase. This effect was, however, independent of the phosphatase activity of the molecule (Das *et al.*, 1986).

5. CONCLUDING REMARKS

Twenty years have elapsed since the essential features of intracellular survival of *Leishmania*, *Tox. gondii* and *Tryp. cruzi* were established. These early studies were essentially concerned with morphological descriptions of the host cell–parasite interaction. The intracellular location of these parasites was identified, which provided the first clues to the strategies evolved to live within cells, e.g. survival inside lysosomes for *Leishmania*, inclusion in non-fusogenic vacuoles for *Toxoplasma*, and escape to the host cell cytosol for *Tryp. cruzi*.

Subsequently, the availability of more refined biochemical and immunological tools allowed researchers to begin defining molecular interactions governing the invasion process, and to uncover some of the tricks evolved by these parasites for circumventing host defences and capturing nutrients required for homeostasis and growth. However, despite our expanding knowledge, many aspects of the host cell–parasite relationship remain obscure. In particular, little is known of the nature and regulation of the genes whose expression is triggered when parasites become established within a cell. A rapid evolution should now be witnessed concerning the application of molecular biology to the study of parasitism. There is little doubt that the coming years will bring a wealth of new information, which will have a profound impact on our understanding of the host–parasite interaction. These new approaches will allow the further identification of molecules that are critical for intracellular survival; only then can new generations of vaccines, diagnostic tools and targeted drugs be developed to combat these pathogens.

ACKNOWLEDGEMENTS

I thank Drs Sally Betz Corradin, Jacques Bouvier, Theresa Glaser, and Pascal Schneider for their many helpful suggestions, and Dr Jeremy Mottram for allowing the use of unpublished results. Research by the author is supported by grant no. 31-40712.94 from the Swiss National Fund for Scientific Research.

REFERENCES

Achabarou, A., Mercereau-Puijalon, O. and Sadak, A. (1991). Differential targeting of dense granule proteins in the parasitophorous vacuole of *Toxoplasma gondii*. *Parasitology* **103**, 321–329.

Adams, L.B., Hibbs, J.B.J., Taintor, R.R. and Krahenbuhl, J.L. (1990). Microbiostatic effect of murine-activated macrophages for *Toxoplasma gondii*. Role for synthesis of inorganic nitrogen oxides from L-arginine. *Journal of Immunology* **144**, 2725–2729.

Alexander, J. and Russell, D.G. (1992). The interaction of *Leishmania* species with macrophages. *Advances in Parasitology* **31**, 175–254.

Alexander, J. and Vickerman, K. (1975). Fusion of host cell secondary lysosomes with parasitophorous vacuoles of *Leishmania mexicana*-infected macrophages. *Journal of Protozoology* **22**, 502–508.

Andrews, N.W. and Whitlow, M.B. (1989). Secretion by *Trypanosoma cruzi* of a hemolysin active at low pH. *Molecular and Biochemical Parasitology* **33**, 249–256.

Andrews, N.W., Abrams, C.K., Slatin, S.L. and Griffiths, G. (1990). *T. cruzi*-secreted protein immunologically related to the complement component C9: evidence for membrane pore forming activity at low pH. *Cell* **61**, 1277–1287.

Antoine, J.C., Prina, E., Jouanne, C. and Bongrand, P. (1990). Parasitophorous vacuoles of *Leishmania*-infected macrophages maintain an acidic pH. *Infection and Immunity* **58**, 779–787.

Aronow, B., Kaur, K., McCartan, K. and Ullman, B. (1987). Two high affinity nucleoside transporters in *Leishmania donovani*. *Molecular and Biochemical Parasitology* **22**, 29–37.

Aunis, D. and Bader, M.F. (1988). The cytoskeleton as a barrier to exocytosis in secretory cells. *Journal of Experimental Biology* **139**, 253–266.

Avila, J.L. and Convit, J. (1976). Physicochemical characteristics of the glycosaminoglycan–lysosomal enzyme interactions *in vitro*. *Biochemical Journal* **160**, 129–136.

Baggiolini, M. and Wymann, M.P. (1990). Turning on the respiratory burst. *Trends in Biochemical Sciences* **15**, 69–72.

Barral, A., Barral-Netto, M., Young, E.C., Brownell, C.E., Twardzik, D.R. and Reed, S.G. (1993). Transforming growth factor-β as a virulence mechanism for *Leishmania braziliensis*. *Proceedings of the National Academy of Sciences of the USA* **90**, 3442–3446.

Beckers, C.J., Dubremetz, J.F., Mercereau-Puijalon, O. and Joiner, K.A. (1994).

The *Toxoplasma gondii* rhoptry protein ROP 2 is inserted into the parasitophorous vacuole membrane, surrounding the intracellular parasite, and is exposed to the host cell cytoplasm. *Journal of Cell Biology* **127**, 947–961.

Berman, J.D., Fioretti, T.B. and Dwyer, D.M. (1981). *In vivo* and *in vitro* localization of *Leishmania* within macrophage phagolysosomes: use of colloidal gold as a lysosomal label. *Journal of Protozoology* **28**, 239–242.

Bhattacharya, A. and Janovy, J. (1975). *Leishmania donovani*: autoradiographic evidence for molecular exchanges between parasites and host cells. *Experimental Parasitology* **37**, 353–360.

Blackwell, J.M., Ezekowitz, R.A.B., Roberts, M.B., Channon, J.Y., Sim, R.B. and Gordon, S. (1985). Macrophage complement receptors and lectin-like receptors bind *Leishmania* in the absence of serum. *Journal of Experimental Medicine* **162**, 324–331.

Blum, J.J. (1987). Oxidation of fatty acids by *Leishmania braziliensis panamensis*. *Journal of Protozoology* **34**, 169–174.

Bonay, P. and Cohen, E. (1983). Neutral amino-acid transport in *Leishmania* promastigotes. *Biochimica et Biophysica Acta* **731**, 222–228.

Bonay, P. and Fresno, M. (1995). Characterization of carbohydrate binding proteins in *Trypanosoma cruzi*. *Journal of Biological Chemistry* **270**, 11062–11070.

Bordier, C. (1987). The promastigote surface protease of *Leishmania*. *Parasitology Today* **3**, 151–156.

Bouvier, J., Schneider, P., Etges, R. and Bordier, C. (1990). Peptide bond specificity of the membrane-bound protease of *Leishmania*. *Biochemistry* **29**, 10113–10119.

Bouvier, J., Schneider, P. and Etges, R. (1995). Leishmanolysin: surface metalloproteinase of *Leishmania*. *Methods in Enzymology* **248**, 313–332.

Bretscher, A. (1991). Microfilament structure and function in the cortical cytoskeleton. *Annual Review of Cell Biology* **7**, 337–374.

Bringaud, F. and Baltz, T. (1992). A potential hexose transporter gene expressed predominantly in the blood stream form of *Trypanosoma brucei*. *Molecular and Biochemical Parasitology* **52**, 111–122.

Brown, E.J. (1992). Complement receptors, adhesion, and phagocytosis. *Infectious Agents and Disease* **1**, 63–70.

Buchmüller-Rouiller, Y. and Mauël, J. (1987). Impairment of the oxidative metabolism of mouse peritoneal macrophages by intracellular *Leishmania* spp. *Infection and Immunity* **55**, 587–593.

Cesbron-Delauw, M.F. (1994). Dense-granule organelles of *Toxoplasma gondii*: their role in the host–parasite relationship. *Parasitology Today* **10**, 293–296.

Cesbron-Delauw, M.F., Guy, B., Torpier, G., Pierce, R.J., Lenzen, G., Cesbron, J.Y., Charif, H., Lepage, P., Darcy, F., Lecocq, J.P. and Capron, A. (1989). Molecular characterization of a 23-kilodalton major antigen secreted by *Toxoplasma gondii*. *Proceedings of the National Academy of Sciences of the USA* **86**, 7537–7541.

Chan, J., Fujira, T., Brennan, B., McNeil, M., Turco, S.J., Sibille, J.C., Snapper, M., Aisen, P. and Bloom, B.R. (1989). Microbial glycolipids: possible virulence factors that scavenge oxygen radicals. *Proceedings of the National Academy of Sciences of the USA* **86**, 2553–2557.

Chang, K.P. and Dwyer, D.M. (1976). Multiplication of a human parasite (*Leishmania donovani*) in phagolysosomes of hamster macrophages *in vitro*. *Science* **193**, 678–680.

Channon, J.Y. and Blackwell, J.M. (1985a). A study of the sensitivity of

Leishmania donovani promastigotes and amastigotes to hydrogen peroxide. I. Differences in sensitivity correlate with parasite-mediated removal of hydrogen peroxide. *Parasitology* **91**, 197–206.

Channon, J.Y. and Blackwell, J.M. (1985b). A study of the sensitivity of *Leishmania donovani* promastigotes and amastigotes to hydrogen peroxide. II. Possible mechanisms involved in protective H_2O_2 scavenging. *Parasitology* **91**, 207–217.

Channon, J.Y., Roberts, M.B. and Blackwell, J.M. (1984). A study of the differential respiratory burst activity elicited by promastigotes and amastigotes of *Leishmania donovani* in murine resident peritoneal macrophages. *Immunology* **53**, 345–355.

Chaudhuri, G. and Chang, K.P. (1988). Acid protease activity of a major surface membrane glycoprotein (gp63) from *Leishmania mexicana* promastigotes. *Molecular and Biochemical Parsitology* **27**, 43–52.

Chaudhuri, G., Chaudhuri, M., Pan, A. and Chang, K.P. (1989). Surface acid proteinase (gp63) of *Leishmania mexicana*. A metalloenzyme capable of protecting liposome-encapsulated proteins from phagolysosomal degradation by macrophages. *Journal of Biological Chemistry* **264**, 7483–7489.

Chuenkova, M. and Pereira, M.E. (1995). *Trypanosoma cruzi* trans-sialidase: enhancement of virulence in a murine model of Chagas' disease. *Journal of Experimental Medicine* **181**, 1693–1703.

Coombs, G.H., Craft, J.A. and Hart, D.T. (1982). A comparative study of *Leishmania mexicana* amastigotes and promastigotes. Enzyme activities and subcellular locations. *Molecular and Biochemical Parasitology* **5**, 199–211.

Cooper, A., Rosen, H. and Blackwell, J.M. (1988). Monoclonal antibodies that recognize distinct epitopes of macrophage type three complement receptor differ in their ability to inhibit binding of *Leishmania* promastigotes harvested at different phases of their growth cycle. *Immunology* **65**, 511–514.

Corradin, S.B., Buchmüller-Rouiller, Y. and Mauël, J. (1991). Phagocytosis enhances murine macrophage activation by interferon-γ and tumor necrosis factor-α. *European Journal of Immunology* **21**, 2553–2558.

da Silva, R.P., Hall, R.F., Joiner, K.A. and Sacks, D.J. (1989). The C3b receptor mediates binding of infective *Leishmania major* metacyclic promastigotes to human macrophages. *Journal of Immunology* **143**, 617–622.

Das, S., Saha, A.K., Remaley, A.T., Glew, R.H., Dowling, J.N., Kajiyoshi, M. and Gottlieb, M. (1986). Hydrolysis of phosphoproteins and inositol phosphates by cell surface phosphatase of *Leishmania donovani*. *Molecular and Biochemical Parasitology* **20**, 143–153.

de Carvalho, L. and de Souza, W. (1989). Cytochemical localization of plasma membrane enzyme markers during internalization of the tachyzoites of *Toxoplasma gondii* by macrophages. *Journal of Protozoology* **36**, 164–170.

de Melo, E.J., de Carvalho, T.U. and de Souza, W. (1992). Penetration of *Toxoplasma gondii* into host cells induces changes in the distribution of the mitochondria and the endopalsmic reticulum. *Cell Structure and Function* **15**, 311–317.

Descoteaux, A. and Turco, S.J. (1993). The lipophosphoglycan of *Leishmania* and macrophage protein kinase C. *Parasitology Today* **9**, 468–471.

Descoteaux, A., Turco, S.J., Sacks, D.L. and Matlashewski, G. (1991). *Leishmania donovani* lipophosphoglycan selectively inhibits signal transduction in macrophages. *Journal of Immunology* **146**, 2747–2753.

Descoteaux, A., Matlashewski, G. and Turco, S.J. (1992). Inhibition of macrophage protein kinase C-mediated protein phosphorylation by *Leishmania donovani* lipophosphoglycan. *Journal of Immunology* **149**, 3008–3015.

Dubremetz, J.F., Achbarou, A., Bermudes, D. and Joiner, K.A. (1993). Kinetics and pattern of organelle exocytosis during *Toxoplasma gondii*/host cell interaction. *Parasitology Research* **79**, 402–408.

Dwyer, D.M. and Gottlieb, M. (1984). Surface membrane localization of 3′- and 5′-nucleotidase activities in *Leishmania donovani* promastigotes. *Molecular and Biochemical Parasitology* **10**, 139–150.

Eilam, Y., El-On, J. and Spira, D.T. (1985). *Leishmania major*: excreted factor, calcium ions, and the survival of amastigotes. *Experimental Parasitology* **59**, 161–168.

El-On, J., Bradley, D.J. and Freeman, J.C. (1980). *Leishmania donovani*: action of excreted factor on hydrolytic enzyme activity of macrophages from mice with genetically different resistance to infection. *Experimental Parasitology* **49**, 167–74.

Etges, R. (1992). Identification of a surface metalloproteinase on 13 species of *Leishmania* isolated from humans, *Crithidia fasciculata* and *Herpetomonas samuelpessoai*. *Acta Tropica* **50**, 205–217.

Etges, R., Bouvier, J. and Bordier, C. (1986). The major surface protein of *Leishmania* promastigotes is a protease. *Journal of Biological Chemistry* **261**, 9098–9101.

Evans, T.G., Thai, L., Granger, D.L. and Hibbs, J.B.J. (1993). Effect of *in vivo* inhibition of nitric oxide production in murine leishmaniasis. *Journal of Immunology* **151**, 907–915.

Fairlamb, A.H., Blackburn, P., Ulrich, P., Chait, B.T. and Cerami, A. (1985). Trypanothione: a novel bis(glutathionyl)spermidine cofactor for glutathione reductase in trypanosomatids. *Science* **227**, 1485–1487.

Fasel, N.J., Robyr, D.C., Mauël, J. and Glaser, T.A. (1994). Identification of a histone H1-like gene expressed in *Leishmania major*. *Molecular and Biochemical Parasitology* **62**, 321–324.

Fogac, M., Cantley, L., Wiedenmann, B., Altstiel, L. and Branton, D. (1983). Clathrin-coated vesicles contain an ATP-dependent proton pump. *Proceedings of the National Academy of Sciences of the USA* **80**, 1300–1303.

Foussard, F., Leriche, M.A. and Dubremetz, J.F. (1991). Characterization of the lipid content of *Toxoplasma gondii* rhoptries. *Parasitology* **102**, 367–370.

Franke de Cazzulo, B.M., Martinez, J., North, M.J., Coombs, G.H. and Cazzulo, J.J. (1994). Effects of proteinase inhibitors on the growth and differentiation of *Trypanosoma cruzi*. *FEMS Microbiology Letters* **124**, 81–86.

Frankenburg, S., Leibovici, V., Mansbach, N., Turco, S.J. and Rosen, G. (1990). Effect of glycolipids of *Leishmania* parasites on human monocyte activity. Inhibition by lipophosphoglycans. *Journal of Immunology* **145**, 4284–4289.

Frevert, U., Schenkman, S. and Nussenzweig, V. (1992). Stage-specific expression and intracellular shedding of the cell surface trans-sialidase of *Trypanosoma cruzi*. *Infection and Immunity* **60**, 2349–2360.

Fruth, U., Solioz, N. and Louis, J.A. (1993). *Leishmania major* interferes with antigen presentation by infected macrophages. *Journal of Immunology* **150**, 1857–1864.

Furtado, G.C., Slowick, M., Kleinman, H.K. and Joiner, K.A. (1992a). Laminin enhances binding of *Toxoplasma gondii* tachyzoites to J774 murine macrophage cells. *Infection and Immunity* **60**, 2337–2342.

Furtado, G.C., Cao, Y. and Joiner, K.A. (1992b). Laminin on *Toxoplasma gondii* mediates parasite binding to the integrin receptor α6/β1 on human foreskin fibroblasts and Chinese hamster ovary cells. *Infection and Immunity* **60**, 4925–4931.

Galloway, C.J., Dean, J.E., Marsh, M., Rudwick, G. and Mellman, I. (1983). Acidification of macrophage and fibroblast endocytic vesicles *in vitro*. *Proceedings of the National Academy of Sciences of the USA* **80**, 3334–3338.

Gazzinelli, R.T., Oswald, I.P., Hieny, S., James, S.L. and Sher, A. (1992). The microbicidal activity of interferon-γ-treated macrophages against *Trypanosoma cruzi* involves an L-arginine dependent, nitrogen oxide-mediated mechanism inhibitable by interleukin-10 and transforming growth factor-β. *European Journal of Immunology* **22**, 2501–2506.

Geisow, M.J., D'Arcy Hart, P. and Young, M.R. (1981). Temporal changes of lysosome and phagosome pH during phagolysosome formation in macrophages: studies by fluorescence spectroscopy. *Journal of Cell Biology* **89**, 645–652.

Giordano, R., Chammas, R., Veiga, S.S., Colli, W. and Alves, M.J. (1994). *Trypanosoma cruzi* binds to laminin in a carbohydrate-independent way. *Brazilian Journal of Medical and Biological Research* **27**, 2315–2318.

Glaser, T.A. and Mukkada, A.J. (1992). Proline transport in *Leishmania donovani* amastigotes: dependence on pH gradients and membrane potential. *Molecular and Biochemical Parasitology* **51**, 1–8.

Glaser, T.A., Baatz, J.E., Kreishman, G.P. and Mukkada, A.J. (1988). pH homeostasis in *Leishmania donovani* amastigotes and promastigotes. *Proceedings of the National Academy of Sciences of the USA* **85**, 7602–7606.

Glaser, T.A., Utz, G.L. and Mukkada, A.J. (1992). The plasma membrane electrical gradient (membrane potential) in *Leishmania donovani* promastigotes and amastigotes. *Molecular and Biochemical Parasitology* **51**, 9–15.

Gottlieb, M. and Dwyer, D.M. (1983). Evidence for distinct 5′- and 3′-nucleotidase activities in the surface membrane fraction of *Leishmania donovani* promastigotes. *Molecular and Biochemical Parasitology* **7**, 303–317.

Green, P.J., Feizi, T., Stoll, M.S., Thiel, S., Prescott, A. and McConville, M.J. (1994). Recognition of the major cell surface glycoconjugates of *Leishmania* parasites by the human serum mannan-binding protein. *Molecular and Biochemical Parasitology* **66**, 319–328

Green, S.J., Crawford, R.M., Hockmeyer, J.T., Meltzer, M.S. and Nacy, C.A. (1990). *Leishmania major* amastigotes initiate the L-arginine-dependent killing mechanism in IFN-γ-stimulated macrophages by induction of tumor necrosis factor-α. *Journal of Immunology* **145**, 4290–4297.

Grimwood, J. and Smith, J.E. (1992). *Toxoplasma gondii*: the role of a 30-kDa surface protein in host cell invasion. *Experimental Parasitology* **74**, 106–111.

Haidaris, C.G. and Bonventre, P.F. (1982). A role for oxygen-dependent mechanisms in killing of *Leishmania donovani* tissue forms by activated macrophages. *Journal of Immunology* **129**, 850–855.

Hall, B.F. (1993). *Trypanosoma cruzi*: mechanisms for entry into host cells. *Seminars in Cell Biology* **4**, 323–333.

Hall, B.F., Furtado, G.C. and Joiner, K.A. (1991). Characterization of host cell vacuolar membrane proteins surrounding different intracellular forms of *Trypanosoma cruzi* in J774 cells: evidence for receptor sorting during the early stages of parasite entry. *Journal of Immunology* **147**, 4313–4321.

Hall, B.F., Webster, P.M.A., Joiner, K.A. and Andrews, N.W. (1992). Desialyation of lysosomal membrane glycoproteins in *Trypanosoma cruzi*: a role for the surface neuraminidase in facilitating parasite entry into host cells. *Journal of Experimental Medicine* **176**, 313–325.

Halonen, S.K. and Weidner, E. (1994). Overcoating of *Toxoplasma gondii* parasitophorous vacuoles with host cell vimentin type intermediate filaments. *Journal of Eukaryotic Microbiology* **41**, 65–71.

Handman, E. and Greenblatt, C.L. (1977). Promotion of leishmanial infection in non-permissive host-macrophages by conditioned medium. *Zeitschrift für Parasitenkunde* **53**, 143–147.

Handman, E., Schnur, L.F., Spithill, T.W. and Mitchell, G.F. (1986). Passive transfer of *Leishmania* lipopolysaccharide confers parasite survival in macrophages. *Journal of Immunology* **137**, 3608–3613.

Hart, D.T. and Coombs, G.H. (1982). *Leishmania mexicana*: energy metabolism of amastigotes and promastigotes. *Experimental Parasitology* **54**, 397–409.

Holtzman, E. (1989). *Lysosomes*. New York: Plenum Press.

Hughes, H.P.A. (1988). Oxidative killing of intracellular parasites mediated by macrophages. *Parasitology Today* **4**, 340–347.

Hulett, M. and Hogarth, P. (1994). Molecular basis of Fc receptor function. *Advances in Immunology* **57**, 1–127.

Hunter, K.J., Le Quesne, S.A. and Fairlamb, A.H. (1994). Identification and biosynthesis of N1,N9-bis(glutathionyl)aminopropylcadaverine (homotrypanothione) in *Trypanosoma cruzi*. *European Journal of Biochemistry* **226**, 1019–1027.

Hynes, R.O. (1992). Integrins: versatility, modulation and signalling in cell adhesion. *Cell* **69**, 11–25.

Ilg, T., Harbecke, D. and Overath, P. (1993). The lysosomal gp63-related protein in *Leishmania mexicana* amastigotes is a soluble metalloproteinase with an acidic pH optimum. *FEBS Letters* **327**, 103–107.

Ilg, T., Stierhof, Y.D., Wiese, M., McConville, M.J. and Overath, P. (1994). Characterization of phosphoglycan-containing secretory products of *Leishmania*. *Parasitology* **108**, supplement, S63–S71.

Janovy, J.J. (1967). Respiratory changes accompanying leishmania to leptomonad transformation in *Leishmania donovani*. *Experimental Parasitology* **20**, 51–55.

Jiang, S., Ojcius, D.M., Persechini, P.M. and Young, J.D.-E. (1990). Resistance of cytolytic lymphocytes to perforin-mediated killing. Inhibition of perforin binding activity by surface membrane proteins. *Journal of Immunology* **144**, 998–1003.

Johansson, A., Jesaitis, A.J., Lundqvist, H., Magnusson, K.E., Sjolin, C., Karlsson, A. and Dahlgren, C. (1995). Different subcellular localization of cytochrome b and the dormant NADPH-oxidase in neutrophils and macrophages: effect on the production of reactive oxygen species during phagocytosis. *Cellular Immunology* **161**, 161–171.

Johnston, R.B. (1978). Oxygen metabolism and the microbicidal activity of macrophages. *Federation Proceedings* **37**, 2759–2764.

Joiner, K.A. (1991). Rhoptry lipids and parasitophorous vacuole formation: a slippery issue. *Parasitology Today* **7**, 226–227.

Joiner, K.A. and Dubremetz, J.F. (1993). *Toxoplasma gondii*: a protozoan for the nineties. *Infection and Immunity* **61**, 1169–1172.

Joiner, K.A., Fuhrman, S.A., Miettinen, H.M., Kasoer, L.H. and Mellman, I. (1990). *Toxoplasma gondii*: fusion competence of parasitophorous vacuoles in Fc receptor-transfected fibroblasts. *Science* **249**, 641–646.

Jones, T.C. and Hirsch, J.G. (1972). The interaction between *Toxoplasma gondii* and mammalian cells II. The absence of lysosomal fusion with phagocytic vacuoles containing living parasites. *Journal of Experimental Medicine* **136**, 1173–1194.

Jones, T.C., Yeh, S. and Hirsch, J.G. (1972). The interaction between *Toxoplasma gondii* and mammalian cells. I. Mechanism of entry and intracellular fate of the parasite. *Journal of Experimental Medicine* **136**, 1157–1172.

Kaku, M., Yagawa, K., Nagao, S. and Tanaka, A. (1983). Enhanced superoxide anion release from phagocytes by muramyl dipeptide and lipopolysaccharide. *Infection and Immunity* **39**, 559–564.

Kasper, L.H. and Mineo, J.R. (1994). Attachment and invasion of host cells by *Toxoplasma gondii*. *Parasitology Today* **10**, 184–188.

Kehrer, J.P. and Lund, L.G. (1994). Cellular reducing equivalents and oxidative stress. *Free Radicals in Biology and Medicine* **17**, 65–75.

Kelleher, M., Bacic, A. and Handman, E. (1992). Identification of a macrophage-binding determinant on lipophosphoglycan from *Leishmania major* promastigotes. *Proceedings of the National Academy of Sciences of the USA* **89**, 6–10.

Kelly, J.M., Taylor, M.C., Smith, K., Hunter, K.J. and Fairlamb, A.H. (1993). Phenotype of recombinant *Leishmania donovani* and *Trypanosoma cruzi* which over-express trypanothione reductase. Sensitivity towards agents that are thought to induce oxidative stress. *European Journal of Biochemistry* **218**, 29–37.

Krassner, S.M. and Flory, B. (1972). Proline metabolism in *Leishmania donovani* promastigotes. *Journal of Protozoology* **19**, 682–685.

Kress, Y., Bloom, B.R., Wittner, M., Rowen, A. and Tanowitz, H. (1975). Resistance of *Trypanosoma cruzi* to killing by macrophages. *Nature* **257**, 394–396.

Krych, M., Atkinson, J.P. and Holers, V.M. (1992). Complement receptors. *Current Opinion in Immunology* **4**, 8–13.

Kutish, G.F. and Janovy, J.J. (1981). Inhibition of *in vitro* macrophage digestion capacity by infection with *Leishmania donovani* (Protozoa: Kinetoplastida). *Journal of Parasitology* **67**, 457–462.

Lang, T., de Chastellier, C., Frehel, C., Hellio, R., Metezeau, P., de Souza Leao, S. and Antoine, J.C. (1994). Distribution of MHC class I and of MHC class II molecules in macrophages infected with *Leishmania amazonensis*. *Journal of Cell Science* **107**, 69–82.

Langford, C.K., Ewbank, S.A., Hanson, S.S., Ullman, B. and Landfear, S.M. (1992). Molecular characterization of two genes encoding members of the glucose transporter superfamily in the parasitic protozoan *Leishmania donovani*. *Molecular and Biochemical Parasitology* **55**, 51–64.

Lawrence, F. and Robert-Gero, M. (1985). Induction of heat shock and stress proteins in promastigotes of three *Leishmania* species. *Proceedings of the National Academy of Sciences of the USA* **82**, 4414–4417.

Lecordier, L., Mercier, C., Torpier, G., Tourvieille, B., Darcy, F., Liu, J.L., Maes, L., Tartar, A., Capron, A. and Cesbron-Delauw, M.F. (1993). Molecular structure of a *Toxoplasma gondii* dense granule antigen (GRA 5) associated with the parasitophorous vacuole membrane. *Molecular and Biochemical Parasitology* **59**, 143–153.

Ley, V., Robbins, E.S., Nussenzweig, V. and Andrews, N.W. (1990). The exit of *Trypanosoma cruzi* from the phagosome is inhibited by raising the pH of acidic compartments. *Journal of Experimental Medicine* **171**, 401–413.

Love, D.C., Esko, J.D. and Mosser, D.M. (1993). A heparin-binding activity on *Leishmania* amastigotes which mediates adhesion to cellular proteoglycans. *Journal of Cell Biology* **123**, 759–766.

Lukacs, G.L., Rotstein, O.D. and Grinstein, S. (1990). Phagosomal acidification is mediated by a vacuolar-type H⁺-ATPase in murine macrophages. *Journal of Biological Chemistry* **265**, 21099–21107.

Lukacs, G.L., Rotstein, O.D. and Grinstein, S. (1991). Determinant of the phagosomal pH in macrophages. In situ assessment of vacuolar H⁺-ATPase activity, counterion conductance, and H⁺ "leak". *Journal of Biological Chemistry* **266**, 24540–24548.

Marletta, M.A. (1993). Nitric oxide synthase structure and mechanism. *Journal of Biological Chemistry* **268**, 12231–12234.

Mauël, J., Ransijn, A. and Buchmüller-Rouiller, Y. (1991). Killing of *Leishmania* parasites in activated murine macrophages is based on an L-arginine-dependent process that produces nitrogen derivatives. *Journal of Leukocyte Biology* **49**, 73–82.

McConville, M.J. and Blackwell, J.M. (1991). Developmental changes in the glycosylated phosphatidylinositols of *Leishmania donovani*. *Journal of Biological Chemistry* **266**, 15170–15179.

McConville, M.J. and Ferguson, M.A.J. (1993). The structure, biosynthesis and function of glycosylated phosphatidylinositols in the parasitic protozoa and other eukaryotes. *Biochemical Journal* **294**, 305–324.

McConville, M.J., Homans, S.W., Thomas-Oates, J.E., Dell, A. and Bacic, A. (1990). Structures of the glycoinositol phospholipids of *Leishmania major*: a family of novel galactosylfuranose containing glycolipids. *Journal of Biological Chemistry* **265**, 7385–7394.

McConville, M.J., Turco, S.J., Ferguson, M.A.J. and Sacks, D.L. (1992). Developmental modification of lipophosphoglycan during the differentiation of *Leishmania major* promastigotes to an infectious stage. *EMBO Journal* **11**, 3593–3690.

McNeely, T.B. and Turco, S.J. (1987). Inhibition of protein kinase C activity by the *Leishmania donovani* lipophosphoglycan. *Biochemical and Biophysical Research Communications* **148**, 653–657.

McNeely, T.B. and Turco, S.J. (1990). Requirement of lipophosphoglycan for intracellular survival of *Leishmania donovani* within human monocytes. *Journal of Immunology* **144**, 2745–2750.

McNeely, T.B., Rosen, G., Londner, V. and Turco, S.J. (1989). Inhibitory effects on protein kinase C by lipophosphoglycan and glycosyl-phosphatidylinositol antigens of *Leishmania*. *Biochemical Journal* **259**, 601–604.

McWilliam, A.S., Tree, P. and Gordon, S. (1992). Carbohydrate recognition receptors of the macrophage and their regulation. In: *Mononuclear Phagocytes — Biology of Monocytes and Macrophages* (R. van Furth, ed.), pp. 224–232. Dordrecht, Boston, London: Kluwer Academic Publishers.

Meade, J.C., Shaw, J., Lemaster, S., Gallagher, G. and Stringer, J.R. (1987). Structure and expression of a tandem gene pair in *Leishmania donovani* that encodes a protein structurally homologous to eucaryotic cation-transporting ATPases. *Molecular and Cellular Biology* **7**, 3937–3946.

Meade, J.C., Hudson, K.M., Stringer, S.L. and Stringer, J.R. (1989). A tandem pair of *Leishmania donovani* cation transporting ATPase genes encode isoforms that are differentially expressed. *Molecular and Biochemical Parasitology* **33**, 81–91.

Meirelles, M.N.L. and de Souza, W. (1986). The fate of plasma membrane macrophage enzyme markers during endocytosis of *Trypanosoma cruzi*. *Journal of Submicroscopic Cytology* **18**, 99–107.

Meirelles, M.N.L., Martinez-Palomo, A., Souto-Padron, T. and de Souza, W.

(1983). Participation of concanavalin A binding sites in the interaction between *Trypanosoma cruzi* and macrophages. *Journal of Cell Science* **62**, 287–299.

Meirelles, M.N.L., Souto-Padron, T. and de Souza, W. (1984). Participation of cell surface anionic sites in the interaction between *Trypanosoma cruzi* and macrophages. *Journal of Submicroscopic Cytology* **16**, 533–545.

Meirelles, M.N.L., Araujo-Gorge, T.C., Miranda, C.F., de Souza, W. and Barbosa, H.S. (1986). Interaction of *Trypanosoma cruzi* with heart muscle cells: ultrastructural and cytochemical analysis of endocytic vacuole formation and effect upon myogenesis *in vitro*. *European Journal of Cell Biology* **41**, 198–206.

Meister, A. and Anderson, M.E. (1983). Glutathione. *Annual Review of Biochemistry* **52**, 711–760.

Mercier, C., Lecordier, L., Darcy, F., Deslee, D., Murray, A., Tourvieille, B., Maes, P., Capron, A. and Cesbron-Delauw, M.F. (1993). Molecular characterization of a dense granule antigen (Gra 2) associated with the network of the parasitophorous vacuole in *Toxoplasma gondii*. *Molecular and Biochemical Parasitology* **58**, 71–82.

Miller, R.A., Reed, S.G. and Parsons, M. (1990). *Leishmania* gp63 molecule implicated in cellular adhesion lacks an Arg-Gly-Asp sequence. *Molecular and Biochemical Parasitology* **39**, 267–274.

Mineo, J.R. and Kasper, L.H. (1994). Attachment of *Toxoplasma gondii* to host cells involves major surface protein, SAG-1 (P30). *Experimental Parasitology* **79**, 11–20.

Ming, M., Chuenkova, M., Ortega-Barria, E. and Pereira, M.E. (1993). Mediation of *Trypanosoma cruzi* invasion by sialic acid on the host cell and trans-sialidase on the trypanosome. *Molecular and Biochemical Parasitology* **59**, 243–252.

Moncada, S. and Higgs, A. (1993). The L-arginine–nitric oxide pathway. *New England Journal of Medicine* **329**, 2002–2012.

Moody, S.F., Handman, E. and Bacic, A. (1991). Structure and antigenicity of the lipophosphoglycan from *Leishmania major* amastigotes. *Glycobiology* **1**, 419–424.

Moody, S.F., Handman, E., McConville, M.J. and Bacic, A. (1993). The structure of *Leishmania major* amastigote lipophosphoglycan. *Journal of Biological Chemistry* **268**, 18457–18466.

Moore, K.J. and Matlashewski, G. (1994). Intracellular infection by *Leishmania donovani* inhibits macrophage apoptosis. *Journal of Immunology* **152**, 2930–2937.

Moore, K.J., Labrecque, S. and Matlashewski, G. (1993). Alteration of *Leishmania donovani* infection levels by selective impairment of macrophage signal transduction. *Journal of Immunology* **150**, 4457–4465.

Moreno, S.N., Silva, J., Vercesi, A.E. and Docampo, R. (1994). Cytosolic-free calcium elevation in *Trypanosoma cruzi* is required for cell invasion. *Journal of Experimental Medicine* **180**, 1535–1540.

Mosser, D.M. (1994). Receptors on phagocytic cells involved in microbial recognition. *Immunology Series* **60**, 99–114.

Mosser, D. and Edelson, P. (1985). The mouse macrophage receptor for C3bi (CR3) is a major mechanism in the phagocytosis of *Leishmania* promastigotes. *Journal of Immunology* **135**, 2785–2789.

Mosser, D.M. and Edelson, P.J. (1987). The third component of complement (C3) is responsible for the intracellular survival of *Leishmania major*. *Nature* **327**, 329–331.

Mosser, D.M. and Rosenthal, L.A. (1993). *Leishmania*–macrophage interactions: multiple receptors, multiple ligands and diverse cellular responses. *Seminars in Cell Biology* **4**, 315–322.

Mosser, D.M., Vlassara, H., Edelson, P.J. and Cerami, A. (1987). *Leishmania* promastigotes are recognized by the macrophage receptor for advanced glycosylation endproducts. *Journal of Experimental Medicine* **165**, 140–145.

Mosser, D.M., Springer, T.A. and Diamond, M.S. (1992). *Leishmania* promastigotes require opsonic complement to bind to the human leukocyte integrin Mac-1 (CD 11b/CD18). *Journal of Cell Biology* **116**, 511–520.

Mukkada, A.J., Schaefer, F.W.I., Simon, M.W. and Nue, C. (1974). Delayed *in vitro* utilization of glucose by *Leishmania donovani* promastigotes. *Journal of Protozoology* **21**, 393–397.

Mukkada, A.J., Maede, J.C., Glaser, T. and Bonventre, P.V. (1985). Enhanced metabolism of *Leishmania donovani* amastigotes at acid pH: an adaptation for intracellular growth. *Science* **229**, 1099–1101.

Munoz-Fernandez, M.A., Fernandez, M.A. and Fresno, M. (1992). Synergism between tumor necrosis factor-α and interferon-γ on macrophage activation for the killing of intracellular *Trypanosoma cruzi* through a nitric oxide-dependent mechanism. *European Journal of Immunology* **22**, 301–307.

Murray, H.W. (1981a). Susceptibility of *Leishmania* to oxygen intermediates and killing by normal macrophages. *Journal of Experimental Medicine* **153**, 1302–1315.

Murray, H.W. (1981b). Interaction of *Leishmania* with a macrophage cell line. Correlation between intracellular killing and the generation of oxygen intermediates. *Journal of Experimental Medicine* **153**, 1690–1695.

Murray, H.W. (1982). Pretreatment with phorbol myristate acetate inhibits macrophage activity against intracellular protozoa. *Journal of the Reticuloendothelial Society* **31**, 479–487.

Murray, H.W. and Cohn, Z.A. (1980). Macrophage oxygen-dependent antimicrobial activity. III. Enhanced oxidative metabolism as an expression of macrophage activation. *Journal of Experimental Medicine* **152**, 1596–1609.

Murray, H.W. and Teitelbaum, R.F. (1992). L-Arginine-dependent reactive mitrogen intermediates and the antimicrobial effect of activated human macrophages. *Journal of Infectious Diseases* **165**, 513–517.

Murray, H.W., Spitalny, G.L. and Nathan, C.F. (1985). Activation of murine macrophages by interferon-γ *in vitro* and *in vivo*. *Journal of Immunology* **134**, 1619–1622.

Myers, M.A., McPhail, L.C. and Snyderman, R. (1985). Redistribution of protein kinase C activity in human monocytes: correlation with activation of the respiratory burst. *Journal of Immunology* **135**, 3411–3416.

Nathan, C.F. and Gabay, J. (1992). Antimicrobial mechanisms of macrophages. In: *Mononuclear Phagocytes — Biology of Monocytes and Macrophages* (R. van Furth, ed.), pp. 259–267. Dordrecht, Boston, London: Kluwer Academic Publishers.

Nathan, C.F. and Hibbs, J.B. (1991). Role of nitric oxide synthesis in macrophage antimicrobial activity. *Current Opinion in Immunology* **3**, 65–70.

Nathan, C.F., Murray, H.W., Wiebe, M.E. and Rubin, B.Y. (1983). Identification of interferon-γ as the lymphokine that activates human macrophage oxidative metabolism and antimicrobial activity. *Journal of Experimental Medicine* **158**, 670–689.

Nauseef, W.M., Volpp, B.D., McCormick, S., Leidal, K.G. and Clark, R.A. (1991).

Assembly of the neutrophil respiratory burst oxidase. Protein kinase C promotes cytoskeletal and membrane association of cytosolic oxidase components. *Journal of Biological Chemistry* **266**, 5911–5917.

Nogueira, N. and Cohn, Z. (1976). *Trypanosoma cruzi*: mechanisms of entry and intracellular fate in mammalian cells. *Journal of Experimental Medicine* **143**, 1402–1420.

Noisin, E.L. and Villalta, F. (1989). Fibronectin increases *Trypanosoma cruzi* amastigote binding to and uptake by murine macrophages and human monocytes. *Infection and Immunity* **57**, 1030–1034.

Ohta, M., Okada, M., Yamashina, I. and Kawasaki, T. (1990). The mechanism of carbohydrate-mediated complement activation by the serum mannan-binding protein. *Journal of Biological Chemistry* **265**, 1980–1984.

Olivier, M., Brownsey, R.W. and Reiner, N.E. (1992). Defective stimulus–response coupling in human monocytes infected with *Leishmania donovani* is associated with altered activation and translocation of protein kinase C. *Proceedings of the National Academy of Sciences of the USA* **89**, 7481–7485.

Ortega-Barria, E. and Pereira, M.E.A. (1991). A novel *T. cruzi* heparin-binding protein promotes fibroblast adhesion and penetration of engineered bacteria and trypanosomes into mammalian cells. *Cell* **67**, 411–421.

Ossorio, P.N., Schwarzman, J.D. and Boothroyd, J.C. (1992). A *Toxoplasma gondii* rhoptry protein associated wtih host cell penetration has an unusual charge asymmetry. *Molecular and Biochemical Parasitology* **50**, 1–16.

Ossorio, P.N., Dubremetz, J.F. and Joiner, K.A. (1994). A soluble secretory protein of the intracellular parasite *Toxoplasma gondii* associates with the parasitophorous membrane through hydrophobic interactions. *Journal of Biological Chemistry* **269**, 15350–15357.

Ouaissi, M.A., Afchain, D., Capron, A. and Grimaud, J.A. (1984). Fibronectin receptors on *Trypanosoma cruzi* trypomastigotes and their biological function. *Nature* **308**, 380–382.

Pan, A.A. (1984). *Leishmania mexicana*: serial cultivation of intracellular stages in a cell-free medium. *Experimental Parasitology* **58**, 72–80.

Pearson, R.D., Harcus, J.L., Symes, P.H., Romito, R. and Donowitz, G.R. (1982). Failure of the phagocytic oxidative response to protect human monocyte-derived macrophages from infection by *Leishmania donovani*. *Journal of Immunology* **129**, 1282–1286.

Pearson, R.D., Haraus, J.L., Roberts, D. and Donowitz, G.R. (1983). Differential survival of *Leishmania donovani* amastigotes in human monocytes. *Journal of Immunology* **131**, 1994–1999.

Petray, P., Rottenberg, M.E., Grinstein, S. and Orn, A. (1994). Release of nitric oxide during the experimental infection with *Trypanosoma cruzi*. *Parasite Immunology* **16**, 193–199.

Peyrol, S., Ouaissi, M.A., Capron, A. and Grimaud, J.A. (1987). *Trypanosoma cruzi*: ultrastructural visualization of fibronectin bound to culture forms. *Experimental Parasitology* **63**, 112–114.

Pfefferkorn, E.R. and Pfefferkorn, L.C. (1977). *Toxoplasma gondii*: specific labeling of nucleic acids of intracellular parasites in Lesch–Nyhan cells. *Experimental Parasitology* **41**, 95–104.

Pfefferkorn, E.R., Schwartzman, J.D. and Kasper, L.H. (1983). *Toxoplasma gondii*: use of mutants to study the host–parasite relationship. *Ciba Foundation Symposium* **99**, 74–91.

Pitt, A., Mayorga, L.S., Stahl, P.D. and Schwartz, A.L. (1992). Alterations in the protein composition of maturing phagosomes. *Journal of Clinical Investigation* **90**, 1978–1983.

Previato, J.Q., Andrade, A.F., Pessolani, M.C.V. and Mendonça-Previato, L. (1985). Incorporation of sialic acid into *Trypanosoma cruzi* macromolecules. A proposal for a new metabolic route. *Molecular and Biochemical Parasitology* **16**, 85–96.

Prina, E., Antoine, J.C., Wiederanders, B. and Kirschke, H. (1990). Localization and activity of various lysosomal proteases in *Leishmania amazonensis*-infected macrophages. *Infection and Immunity* **58**, 1730–1737.

Prina, E., Jouanne, C., de Souza, L.S., Szabo, A., Guillet, J.G. and Antoine, J.C. (1993). Antigen presentation capacity of murine macrophages infected with *Leishmania amazonensis* amastigotes. *Journal of Immunology* **151**, 2050–2061.

Proudfoot, L., O'Donnell, C.A. and Liew, F.Y. (1995). Glycoinositolphospholipids of *Leishmania major* inhibit nitric oxide synthesis and reduce leishmanicidal activity in murine macrophages. *European Journal of Immunology* **25**, 745–750.

Puentes, S.M., Sacks, D.L., da Silva, R.P. and Joiner, K.A. (1988). Complement binding by two developmental stages of *Leishmania major* promastigotes varying in expression of a surface lipophosphoglycan. *Journal of Experimental Medicine* **167**, 887–902.

Puentes, S.M., da Silva, R.P., Sacks, D.L., Hammer, C.H. and Joiner, K.A. (1990). Serum resistance of metacyclic stage *Leishmania major* promastigotes is due to the release of C5b-9. *Journal of Immunology* **145**, 4311–4316.

Pupkis, M.F., Tetley, L. and Coombs, G.H. (1986). *Leishmania mexicana*: amastigote hydrolases in unusual lysosomes. *Experimental Parasitology* **62**, 29–39.

Reiner, N.E. (1994). Altered cell signaling and mononuclear phagocyte deactivation during intracellular infection. *Immunology Today* **15**, 374–381.

Reiner, N.E. and Malemud, C.J. (1985). Arachidonic acid metabolism by murine peritoneal macrophages infected with *Leishmania donovani*: in vitro evidence for parasite-induced alterations in cyclooxygenase and lipoxygenase pathways. *Journal of Immunology* **134**, 556–563.

Reiner, N.E., Ng, W. and McMaster, W.R. (1987). Parasite accessory cell interactions in murine leishmaniasis. II. *Leishmania donovani* suppresses macrophage expression of class I and class II major histocompatibility complex gene products. *Journal of Immunology* **138**, 1926–1932.

Remaley, A.T., Das, S., Campbell, P.I., LaRocca, G.M., Pope, M.T. and Glew, R.H. (1985a). Characterization of *Leishmania donovani* acid phosphatases. *Journal of Biological Chemistry* **260**, 880–886.

Remaley, A.T., Glew, R.H., Kuhns, D.B., Basford, R.E., Waggoner, A.S., Ernst, L.A. and Pope, M. (1985b). *Leishmania donovani*: surface membrane acid phosphatase blocks neutrophil oxidative metabolite production. *Experimental Parasitology* **60**, 331–341.

Rizvi, F.S., Ouaissi, M.A., Marty, B., Santoro, F. and Capron, A. (1988). The major surface protein of *Leishmania* promastigotes is a fibronectin-like molecule. *European Journal of Immunology* **18**, 473–476.

Robert, R., Leynia de la Jarrige, P., Mahaza, C., Cottin, J., Marot-Leblond, A. and Senet, J.-M. (1991). Specific binding of neoglycoproteins to *Toxoplasma gondii* tachyzoites. *Infection and Immunity* **59**, 4670–4673.

Roberts, S.C., Wilson, M.E. and Donelson, J.E. (1995). Developmentally regulated expression of a novel 59-kDa product of the major surface protease (Msp or gp63) gene family of *Leishmania chagasi*. *Journal of Biological Chemistry* **270**, 8884–8892.

Robinson, J.M. and Badwey, J.A. (1994). Production of active oxygen species by phagocytic leukocytes. *Immunology Series* **60**, 159–178.

Rodriguez, A., Rioult, M.G., Ora, A. and Andrews, N.W. (1995). A trypanosome soluble factor induces IP3 formation, intracellular Ca^{2+} mobilization and microfilament rearrangement in host cells. *Journal of Cell Biology* **129**, 1263–1273.

Rosen, G.M., Pou, S., Ramos, C.L., Cohen, M.S. and Britigan, B.E. (1995). Free radicals and phagocytic cells. *FASEB Journal* **9**, 200–209.

Rosenberg, I.A., Prioli, R.P., Mejia, J.S. and Pereira, M.E. (1991). Differential expression of *Trypanosoma cruzi* neuraminidase in intra- and extracellular trypomastigotes. *Infection and Immunity* **59**, 464–466.

Russell, D.G. (1987). The macrophage-attachment glycoprotein gp63 is the predominant C3-acceptor site on *Leishmania mexicana* promastigotes. *European Journal of Biochemistry* **164**, 213–221.

Russell, D.G. and Wright, S.D. (1988). Complement receptor type 3 (CR3) binds to an Arg-Gly-Asp-containing region of the major surface glycoprotein, gp63 of *Leishmania* promastigotes. *Journal of Experimental Medicine* **168**, 279–292.

Russell, D.G., Xu, S. and Chakraborty, P. (1992). Intracellular trafficking and the parasitophorous vacuole of *Leishmania mexicana*-infected macrophages. *Journal of Cell Science* **103**, 1193–1210.

Sacks, D.L. and da Silva, R.P. (1987). The generation of infective stage *Leishmania major* promastigotes is associated with the cell surface expression and release of developmentally regulated glycolipid. *Journal of Immunology* **139**, 3099–3106.

Sacks, D.L. and Perkins, P.V. (1984). Identification of an infective stage of *Leishmania* promastigotes. *Science* **223**, 1417–1419.

Sacks, D.L., Hieny, S. and Sher, A. (1985). Identification of cell surface carbohydrate and antigenic changes between noninfective and infective developmental stages of *Leishmania major* promastigotes. *Journal of Immunology* **135**, 564–569.

Sacks, D.L., Brodin, T.N. and Turco, S.J. (1990). Developmental modification of the lipophosphoglycan from *Leishmania major* promastigotes during metacyclogenesis. *Molecular and Biochemical Parasitology* **42**, 225–234.

Saffer, L.D., Mercereau-Peuijalon, O., Dubremetz, J.F. and Schwartzman, J.D. (1992). Localization of a *Toxoplasma gondii* rhoptry protein by immunoelectronmicroscopy during and after host cell penetration. *Journal of Protozoology* **39**, 526–530.

Schauer, R., Reuter, G., Mühlpbordt, H., Andrade, A.F. and Pereira, M.E. (1983). The occurrence of N-acetyl- and N-glycoloylneuraminic acid in *Trypanosoma cruzi*. *Hoppe-Seylers Zeitschrift für Physiologische Chemie* **364**, 1053–1057.

Schenkman, R.P., Vandekerckhove, F. and Schenkman, S. (1993). Mammalian sialic acid enhances invasion by *Trypanosoma cruzi*. *Infection and Immunity* **61**, 898–902.

Schenkman, S. and Eichinger, D. (1993). *Trypanosoma cruzi* trans-sialidase and cell invasion. *Parasitology Today* **9**, 218–222.

Schenkman, S., Jiang, M.S., Hart, G.W. and Nussenzweig, V. (1991a). A novel cell surface trans-sialidase of *Trypanosoma cruzi* generates a stage-specific epitope required for invasion of mammalian cells. *Cell* **65**, 1117–1125.

Schenkman, S., Robbins, E.S. and Nussenzweig, V. (1991b). Attachment of

Trypanosoma cruzi to mammalian cells requires parasite energy, and invasion can be independent of the target cell cytoskeleton. *Infection and Immunity* **59**, 645–654.

Schenkman, S., Eichinger, D., Pereira, M.E. and Nussenzweig, V. (1994). Structural and functional properties of *Trypanosoma cruzi* trans-sialidase. *Annual Review of Microbiology* **48**, 499–523.

Schneemann, M., Schoedon, G., Hofer, S., Blau, N., Guerrero, L. and Schaffner, A. (1993). Nitric oxide synthase is not a constituent of the antimicrobial armature of human mononuclear phagocytes. *Journal of Infectious Diseases* **167**, 1358–1363.

Schneider, P., Bordier, C. and Etges, R. (1992a). Membrane proteins and enzymes of *Leishmania*. *Sub-Cellular Biochemistry* **18**, 39–72.

Schneider, P., Rosat, J.P., Bouvier, J., Louis, J. and Bordier, C. (1992b). *Leishmania major*: differential regulation of the surface metalloprotease in amastigote and promastigote stages. *Experimental Parasitology* **75**, 196–206.

Schneider, P., Rosat, J.P., Ransijn, A., Ferguson, M.A.J. and McConville, M.J. (1993). Characterization of glycoinositol phospholipids in the amastigote stage of the protozoan parasite *Leishmania major*. *Biochemical Journal* **295**, 555–564.

Schreiber, A.D., Rossman, M.D. and Levinson, A.I. (1992). The immunobiology of human Fcγ receptors on hematopoietic cells and tissue macrophages. *Clinical Immunology and Immunopathology* **62**, S66–72.

Schwab, J.C., Beckers, C.J. and Joiner, K.A. (1994). The parasitophorous vacuole membrane surrounding intracellular *Toxoplasma gondii* functions as a molecular sieve. *Proceedings of the National Academy of Sciences of the USA* **91**, 509–513.

Schwartzman, J.D. (1986). Inhibition of a penetration-enhancing factor of *Toxoplasma gondii* by monoclonal antibodies specific for rhoptries. *Infection and Immunity* **51**, 760–764.

Schwartzman, J.D. and Krug, E.C. (1989). *Toxoplasma gondii*: characterization of monoclonal antibodies that recognize rhoptries. *Experimental Parasitology* **68**, 74–82.

Schwartzman, J.D. and Pfefferkorn, E.R. (1981). Pyrimidine synthesis by intracellular *Toxoplasma gondii*. *Journal of Parasitology* **67**, 150–158.

Schwartzman, J.D. and Pfefferkorn, E.R. (1982). *Toxoplasma gondii*: purine synthesis and salvage in mutant host cells and parasites. *Experimental Parasitology* **53**, 77–86.

Segal, A.W. and Abo, A. (1993). The biochemical basis of the NADPH oxidase of phagocytes. *Trends in Biochemical Sciences* **18**, 43–47.

Sibley, L.D. (1993). Interaction between *Toxoplasma gondii* and its mammalian host cells. *Seminars in Cell Biology* **4**, 335–344.

Sibley, L.D., Weidner, E. and Krahenbuhl, J.L. (1985). Phagosome acidification blocked by intracellular *Toxoplasma gondii*. *Nature* **315**, 416–419.

Sibley, L.D., Krahenbuhl, J.L., Adams, G.M.W. and Weidner, E. (1986). *Toxoplasma* modifies macrophage phagosomes by secretion of a vesicular network rich in surface proteins. *Journal of Cell Biology* **103**, 867–874.

Sibley, L.D., Niesman, I.R., Asai, T. and Takeuchi, T. (1994). *Toxoplasma gondii*: secretion of a potent nucleoside triphosphate hydrolase into the parasitophorous vacuole. *Experimental Parasitology* **79**, 301–311.

Sibley, L.D., Niesman, I.R., Parmley, S.F. and Cesbron-Delauw, M.F. (1995). Regulated secretion of multi-lamellar vesicles leads to formation of a tubulovesicular network in host-cell vacuoles occupied by *Toxoplasma gondii*. *Journal of Cell Science* **108**, 1669–1677.

Simon, M.W. and Mukkada, A.J. (1977). *Leishmania tropica*: regulation of specificity of the methionine transport system in promastigotes. *Experimental Parasitology* **42**, 97–105.

Smejkal, R.M., Wolff, R. and Olenick, J.G. (1988). *Leishmania braziliensis panamensis*: increased infectivity resulting from heat shock. *Experimental Parasitology* **65**, 1–9.

Sørensen, A.L., Hey, A.S. and Kharazmi, A. (1994). *Leishmania major* surface protease Gp63 interferes with the function of human monocytes and neutrophils *in vitro*. *Acta Pathologica, Microbiologica et Immunologica Scandinavica (APMIS)* **102**, 265–271.

Soteriadou, K.P., Remoundos, M.S., Katsikas, M.C., Tzinia, A.K., Tsikaris, V., Sakarellos, C. and Tzartas, S.J. (1992). The Ser-Arg-Tyr-Asp region of the major surface glycoprotein of *Leishmania* mimics the Arg-Gly-Asp-Ser cell attachment region of fibronectin. *Journal of Biological Chemistry* **267**, 13980–13985.

Spies, H.S. and Steenkamp, D.J. (1994). Thiols of intracellular pathogens. Identification of ovothiol A in *Leishmania donovani* and structural analysis of a novel thiol from *Mycobacterium bovis*. *European Journal of Biochemistry* **224**, 203–213.

Stahl, P.D. (1992). The mannose receptor and other macrophage lectins. *Current Opinion in Immunology* **4**, 49–52.

Stinson, S., Sommer, J.R. and Blum, J.J. (1989). Morphology of *Leishmania braziliensis*: changes during reversible heat-induced transformation from promastigote to an ellipsoidal form. *Journal of Parasitology* **75**, 431–440.

Strauss, A.H., Levery, S.B., Jasiulionis, M.G., Salyan, M.E.K., Steele, S.J., Travassos, L.R., Hakomori, S. and Takahashi, H.K. (1993). Stage-specific glycosphingolipids from amastigote forms of *Leishmania amazonensis*. *Journal of Biological Chemistry* **268**, 13723–13730.

Talamas-Rohana, B., Wright, S.D., Lennartz, M.R. and Russell, D.G. (1990). Lipophosphoglycan from *Leishmania mexicana* promastigotes binds to members of the CR3, p150.95 and LFA-1 family of leukocyte integrins. *Journal of Immunology* **144**, 4817–4827.

Tanowitz, H., Wittner, M., Kress, Y. and Bloom, B. (1975). Studies of *in vitro* infection by *Trypanosoma cruzi*. I. Ultrastructural studies on the invasion of macrophages and L-cells. *American Journal of Tropical Medicine and Hygiene* **24**, 25–33.

Tardieux, I., Webster, P., Ravesloot, J., Boron, W., Lunn, J.A., Heuser, J.E. and Andrews, N.W. (1992). Lysosome recruitment and fusion are early events required for trypanosome invasion of mammalian cells. *Cell* **71**, 1117–1130.

Tardieux, I., Nathanson, M.H. and Andrews, N.W. (1994). Role in host cell invasion of *Trypanosoma cruzi*-induced cytosolic-free Ca^{2+} transients. *Journal of Experimental Medicine* **179**, 1017–1022.

ter Kuile, B.H. (1993). Glucose and proline transport in kinetoplastids. *Parasitology Today* **9**, 206–210.

ter Kuile, B.H. and Opperdoes, F.R. (1993). Uptake and turnover of glucose in *Leishmania donovani*. *Molecular and Biochemical Parasitology* **60**, 313–322.

Thelen, M., Dewald, B. and Baggiolini, M. (1993). Neutrophil signal transduction and activation of the respiratory burst. *Physiological Reviews*, 797–821.

Titus, R.G., Theodos, C.M., Shankar, A. and Hall, L. (1994). Interactions between *Leishmania major* and macrophages. *Immunology Series* **60**, 437–459.

Tolson, D.L., Turco, S.J. and Pearson, T.W. (1990). Expression of repeating phosphorylated disaccharide lipophosphoglycan epitope on the surface of macrophages infected with *Leishmania donovani. Infection and Immunity* **58**, 3500–3507.

Tomlinson, S., Vandekerckhove, F., Frevert, U. and Nussenzweig, V. (1995). The induction of *Trypanosoma cruzi* trypomastigote to amastigote transformation by low pH. *Parasitology* **110**, 547–554.

Turco, S.J. and Descoteaux, A. (1992). The lipophosphoglycan of *Leishmania* parasites. *Annual Review of Microbiology* **46**, 65–94.

Turco, S.J. and Sacks, D.L. (1991). Expression of a stage-specific lipophosphoglycan in *Leishmania major. Molecular and Biochemical Parasitology* **45**, 91–100.

Tzinia, A.K. and Soteriadou, K.P. (1991). Substrate-dependent pH optima of gp63 purified from seven strains of *Leishmania. Molecular and Biochemical Parasitology* **47**, 83–90.

Van der Ploeg, L.H., Giannini, S.H. and Cantor, C.R. (1985). Heat shock genes: regulatory role for differentiation in parasitic protozoa. *Science* **228**, 1443–1446.

Veras, P.S., de Chastellier, C. and Rabinovitch, M. (1992). Transfer of zymosan (yeast cell walls) to the parasitophorous vacuoles of macrophages infected with *Leishmania amazonensis. Journal of Experimental Medicine* **176**, 639–646.

Von Kreuter, B.F., Walton, B.L. and Santos-Buch, C.A. (1995). Attenuation of parasite cAMP levels in *T. cruzi*–host cell membrane interactions *in vitro. Journal of Eukaryotic Microbiology* **42**, 20–26.

Vouldoukis, I., Riveros-Moreno, V., Dugas, B., Ouaaz, F., Bécherel, P., Debré, P., Moncada, S. and Mossalayi, M.D. (1995). The killing of *Leishmania major* by human macrophages is mediated by nitric oxide induced after ligation of the FcεRII/CD23 surface antigen. *Proceedings of the National Academy of Sciences of the USA* **92**, 7804–7808.

Walsh, C., Bradley, M. and Nadeau, K. (1991). Molecular studies on trypanothione reductase, a target for antiparasitic drugs. *Trends in Biochemical Sciences* **16**, 305–309.

Webster, P. and Russell, D.G. (1993). The flagellar pocket of trypanosomatids. *Parasitology Today* **9**, 201–206.

Wei, X.Q., Charles, I.G., Smith, A., Ure, J., Feng, G.J., Huang, F.P., Xu, D., Muller, W., Moncada, S. and Liew, F.Y. (1995). Altered immune responses in mice lacking inducible nitric oxide synthase. *Nature* **375**, 408–411.

Weinberg, J.B., Misukonis, M.A., Shami, P.J., Mason, S.N., Sauls, D.L., Dittman, W.A., Wood, E.R., Smith, G.K., McDonald, B., Bachus, K.E., Haney, A.F. and Granger, D.L. (1995). Human mononuclear phagocyte inducible nitric oxide synthase (iNOS): analysis of iNOS mRNA, iNOS protein, biopterin, and nitric oxide production by blood monocytes and peritoneal macrophages. *Blood* **86**, 1184–1195.

Wilson, C.B., Tsai, V. and Remington, J.S. (1980). Failure to trigger the oxidative metabolic burst by normal macrophages. Possible mechanism for survival of intracellular pathogens. *Journal of Experimental Medicine* **151**, 328–346.

Wilson, M.E. and Pearson, R.D. (1986). Evidence that *Leishmania donovani* utilizes a mannose receptor on human mononuclear phagocytes to establish intracellular parasitism. *Journal of Immunology* **136**, 4681–4688.

Wilson, M.E. and Pearson, R.D. (1988). Roles of CR3 and mannose receptors in the attachment and ingestion of *Leishmania donovani* by human mononuclear phagocytes. *Infection and Immunity* **56**, 363–369.

Wilson, M.E., Andersen, K.A. and Britigan, B.E. (1994). Response of *Leishmania chagasi* promastigotes to oxidant stress. *Infection and Immunity* **62**, 5133–5141.

Wozencraft, A.O., Sayers, G. and Blackwell, J.M. (1986). Macrophage type 3 complement receptors mediate serum-independent binding of *Leishmania dono-vani*. Detection of macrophage-derived complement on the parasite surface by immunoelectron microscopy. *Journal of Experimental Medicine* **164**, 1332–1337.

Wright, S.D. and Silverstein, S.C. (1983). Receptors for C3b and C3bi promote phagocytosis but not release of toxic oxygen from human phagocytes. *Journal of Experimental Medicine* **158**, 2016–2023.

Wyler, D.J., Sypek, J.P. and McDonald, J.A. (1985). *In vitro* parasite–monocyte interactions in human leishmaniasis: possible role of fibronectin in parasite attachment. *Infection and Immunity* **49**, 305–311.

Yong, E.C., Chi, E.Y. and Henderson, W.R.J. (1994). *Toxoplasma gondii* alters eicosanoid release by human mononuclear phagocytes: role of leukotrienes in IFN-γ-induced antitoxoplasma activity. *Journal of Experimental Medicine* **180**, 1637–1648.

Zilberstein, D. (1993). Transport of nutrients and ions across membranes of trypanosomatid parasites. *Advances in Parasitology* **32**, 261–291.

Zilberstein, D. and Dwyer, D.M. (1984a). Antidepressants cause lethal disruption of membrane function in the human protozoan parasite *Leishmania*. *Science* **226**, 977–979.

Zilberstein, D. and Dwyer, D.M. (1984b). Glucose transport in *Leishmania dono-vani* promastigotes. *Molecular and Biochemical Parasitology* **12**, 327–336.

Zilberstein, D. and Dwyer, D.M. (1985). Protonmotive force-driven active trans-port of D-glucose and L-proline in the protozoan parasite *Leishmania donovani*. *Proceedings of the National Academy of Sciences of the USA* **82**, 1716–1720.

Zilberstein, D. and Dwyer, D.M. (1988). Identification of a surface membrane proton-translocating ATPase in promastigotes of the parasitic protozoan *Leishmania donovani*. *Biochemical Journal* **256**, 13–21.

Zilberstein, D. and Gepstein, A. (1993). Regulation of L-proline transport in *Leishmania donovani* by extracellular pH. *Molecular and Biochemical Para-sitology* **61**, 197–205.

Zilberstein, D., Liveanu, V. and Gepstein, A. (1990). Tricyclic drugs reduce proton motive force in *Leishmania donovani* promastigotes. *Biochemical Pharmacology* **39**, 935–940.

Zingales, B., Carniol, C., de Lederkremer, R.M. and Colli, W. (1987). Direct sialic acid transfer from a protein donor to glycolipids of trypomastigote forms of *Trypanosoma cruzi*. *Molecular and Biochemical Parasitology* **26**, 135–144.

Regulation of Infectivity of *Plasmodium* to the Mosquito Vector

R.E. Sinden, G.A. Butcher, O. Billker and S.L. Fleck

Molecular and Cellular Parasitology Research Group, Infection and Immunity Section, Department of Biology, Imperial College of Science, Technology and Medicine, Prince Consort Road, South Kensington, London, SW7 2BB, UK

1. Introduction 54
2. Gametocytogenesis 55
3. Biology of the Mature Gametocyte 58
   3.1. Distribution of gametocytes in blood 59
   3.2. Gametocyte cell biology 61
   3.3. Host factors modulating infection 64
4. Gametogenesis 79
   4.1. Cell biology 79
   4.2. Induction mechanisms 84
5. Fertilization 92
   5.1. The mosquito blood-meal 92
   5.2. Efficiency of fertilization 93
6. Post-fertilization Development 95
   6.1. Development of the zygote 95
   6.2. Fertilization as a developmental trigger 95
   6.3. Ookinete formation and biology 96
Acknowledgements 99
References 99

ADVANCES IN PARASITOLOGY VOL 38
ISBN 0–12–031738–9

1. INTRODUCTION

The malarial parasite is required to alternate between its vertebrate and invertebrate hosts to complete its life cycle. In the case of mammalian parasites the vectors are anopheline mosquitoes, whereas the avian and saurian malarial parasites are transmitted by as many as five genera of mosquitoes including the culicines, *Aedes* and *Culex*. There are reports that saurian parasites are also transmitted by phlebotomine sandflies (Ayala, 1973).

Transmission of all malarial parasites from the vertebrate host to the invertebrate host is mediated exclusively by the mature gametocytes, cells that are arrested in G_0 of the cell cycle until taken up in the blood-meal of the engorging female insect. It is to the production of those few parasites that are successfully transferred to the mosquito that the entire vertebrate phase of the malarial life cycle is directed, yet we observe very different patterns of gametocyte production and longevity in the different species of parasites. It is only now, as we begin to understand the complex factors that regulate the infectivity of the gametocyte to the mosquito vector, that we can understand how these different patterns of parasite biology may be individually advantageous to particular host–parasite combinations.

We have chosen to subdivide this review of published and unpublished data into five interdependent subjects.

1. Gametocytogenesis, including the factors that regulate the differentiation of the erythrocytic stage parasites into sexual or asexual parasites, and the development of the gametocytes into either male or female cells.
2. The biology of the mature gametocyte, including its cell biology, distribution and longevity/viability in the peripheral bloodstream; and the biology of the gametocyte-infected host and the regulation of mosquito infection by host factors that interact either directly with the parasite or indirectly upon the parasite following its ingestion by the mosquito vector.
3. Gametogenesis, its molecular regulation with respect to the inducers required to trigger the gametocyte to leave its arrested state, the secondary signal pathways that regulate the constituent events of microgametogenesis (exflagellation), and the impact of external regulatory molecules (usually inhibitors) on gamete formation.
4. Fertilization, its efficiency and role as a developmental regulator.
5. Post-fertilization development, including the mechanisms that are known to regulate the survival or destruction of the parasite before its successful establishment under the midgut basal lamina where the oocyst then forms.

2. GAMETOCYTOGENESIS

Following infection of the vertebrate host by the malarial sporozoite, the parasite grows initially in an asexual phase of pre-erythrocytic merogony (schizogony). In those parasites considered ancestral to *Plasmodium*, the subsequent erythrocytic phase is entirely sexual (Garnham, 1966) and the coccidian parasite *Eimeria* initiates its sexual differentiation following a fixed number of asexual generations (Klimes *et al.*, 1972). In *Plasmodium*, sexual stage parasites are produced directly from the merozoites emerging from the pre-erythrocytic meront (schizont) (Killick-Kendrick and Warren, 1968) and at each subsequent round of merogony. Using *in vitro* culture techniques to initiate erythrocytic cultures of *Plasmodium berghei* from pre-erythrocytic stages, Suhrbier *et al.* (1987) showed that the highest proportion of sexual stages were committed directly from the pre-erythro-cytic merozoites, and that progressively lower proportions were generated with each successive round of asexual erythrocytic merogony. Recognizing the plasticity of the malarial genome (Frontali, 1994), these results are consistent with the observations of Birago *et al.* (1982) and Janse *et al.* (1991) that the ability to generate sexual stage parasites is critically dependent on the presence of an intact genome. Certain chromosomes in particular have been identified as having a high number of genes encoding molecules expressed in the sexual stage parasite, e.g. chromosome 5 in *P. berghei* (see Janse *et al.*, 1989, 1991; Janse, 1993). Current studies recognize a critical correlation with the integrity of the region internal to the subtelomeric sequences on the left arm of this chromosome. There are, however, apparent discrepancies in data suggesting a similar role for chromosome 9 in *P. falciparum* (see Day *et al.*, 1993; Chaiyaroj *et al.*, 1994), although more recent data support a correlation between the integrity of this chromosome and the ability of cloned lines to express sexual stage proteins and to generate morphologically identifiable gametocytes.

Ponnudurai *et al.* (1982a,b), Graves *et al.* (1984) and Ponnudurai (1987) made the further important observation that different clones of *P. falciparum* have different abilities to produce gametocytes *in vitro*. Cultures have been reported that produce exclusively gametocytes (though only once), those that never produce gametocytes (the majority of long-term cloned lines, in which it is assumed chromosomal deletions have accumulated in the absence of selection for sexual competence), and those that produce fairly consistent proportions of gametocytes for predictable periods of time (see Figure 1). The proportion of gametocytes produced is a stable phenotype for each clone (Graves *et al.*, 1984). Similarly the sex ratio of gametocytes has been shown to vary between different cloned lines of parasites (Read, D. *et al.*, 1994) and interesting hypotheses have been

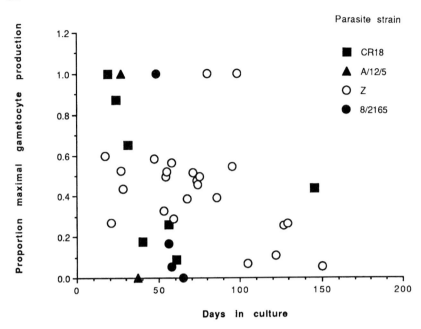

Figure 1 Clonal variation in *P. falciparum* in the ability to sustain gametocyte production *in vitro*. (Data taken from Carter, 1983.)

proposed on the importance of the sex ratios to the population dynamics of parasite breeding structure (Read, F.W. *et al.*, in press). Noden *et al.* (1994) noted, however, that the sex ratio of a given clone/isolate of *P. falciparum* had no measurable effect on the infectivity of cultured gametocytes in the laboratory.

Although the above data might suggest that sexual development is regulated entirely by innate mechanisms, other evidence suggests that environmental factors are important. Observations on *P. falciparum* gametocyte development in human subjects suggested that numerous physiological responses in the host could trigger (induce) gametocytogenesis; among these were the immune status, the pH of the blood and spleen, and the impact of chemotherapy (reviewed by Sinden, 1983a,b). Subsequently, Mons (1986) noted that mice infected from previously cryopreserved aliquots of blood containing *P. berghei* routinely had higher proportionate gametocytaemias than those infected with parasites passaged from an infected host. Additionally, Motard *et al.* (1995), studying *P. yoelii in vivo*, noted that immunization of hosts with recombinant HSP70 in Freund's adjuvant, followed by sporozoite infection, produced blood infections that had significantly higher numbers of gametocytes on days 3 and 4 than on day 5. No such increase was noted if the infection was blood-induced. The authors noted that the immunity induced was primarily directed against the pre-erythrocytic parasite and was mediated by interleukin-6 and interferon-γ, which are known to interact to produce reactive nitrogen intermediates.

Some of these environmental factors have been evaluated in culture experiments: e.g., unspecified parasite-induced stress (Carter and Miller, 1979) including blood cell lysis (Schneweis *et al.*, 1991); chemotherapy (Trager and Gill, 1989; Ono and Nakabayashi, 1990; Ono *et al.*, 1993); and immune pressure (Smalley and Brown, 1981; Ono *et al.*, 1986). Often the studies on *P. falciparum* failed to take into account one important factor, namely that a change in the proportion of sexual to asexual parasites does not indicate a switch between sexual and asexual development unless the absolute number of parasites is unchanged: i.e., an increase in the proportion of sexual parasites could indicate merely an increased death rate among the asexual parasites due to the treatment given. This is particularly important in the case of *P. falciparum* because it is well established that the more mature gametocytes (stages III–V) are more resistant to many stresses than either the immature gametocytes (stages I–II) or asexual trophozoites (Sinden and Smalley, 1979; Sinden, 1983a,b; see also Section 3.3.1(b)). Thus drug treatment of a culture containing mixed asexual and sexual parasites will almost invariably give rise to an increase in the proportion of mature sexual stages surviving in the treated culture. Suggestions that drugs may introduce gametocytogenesis in *P. falciparum in vivo* have been challenged by studies in Mozambique on patients treated with chloroquine (Field and Shute, 1955) or Fansidar (pyrimethamine + sulfadoxine) (Høgh *et al.*, 1995). These studies compared the difference in the rates of loss of chloroquine-sensitive parasites from both the asexual and sexual populations, and reaffirmed the observation that gametocyte "enriched" (or more accurately asexual parasite depleted) populations persisted for 28 days after treatment. The profile of gametocyte numbers was compared to a simple model calculation based on the knowledge of the times at which the young gametocyte is sensitive to the drugs and the period of sequestration in the deep tissues. The observed and predicted profiles showed high correlation, consistent with the hypothesis that the drugs did not induce a switch from asexual to sexual development.

On the evidence presented to date it is probable that the switch from asexual to sexual development is both innately and environmentally regulated. No study of the molecular regulation of sex ratios in gametocytes has yet been published. The stability of the different sex ratios expressed by individual clones suggests genetic regulation. These factors presumably act subsequent to any commitment of the parasite to the asexual or sexual pathway. When is the parasite committed to the sexual pathway? Is it in the preceding meront or in the merozoite following invasion of the red blood cell? — i.e., are the merozoites from a single asexual meront (which must include pre-erythrocytic meronts) committed wholly or partly to either the sexual or asexual pathway? Early studies on the coccidian parasite *Eimeria* showed clearly that second generation asexual meronts were wholly committed to one or other developmental route (Klimes *et al.*, 1972). Carter and Miller (1979) suggested that, in *Plasmodium*, commitment to sexual development could occur either in the meront or in the erythrocytic merozoite immediately after invasion. Subsequent studies (Holloway *et al.*, 1994) suggested that individual meronts produced a mixed progeny of asexual and sexual parasites. Mons (1986), using a mixed system *in vivo* and *in vitro* to grow *P. berghei*, concluded that sexual commitment was determined 8 hours after erythrocyte invasion by the merozoite. Subsequently Bruce *et al.* (1990) concluded that, within the limits of their study, meronts of *P. falciparum*, like those of *Eimeria*, produced either gametocytes or asexual parasites, suggesting that the meronts were pre-committed to one or other developmental pathway. In view of the wide evolutionary distance between these two apicomplexan parasites, it may be suggested that all *Plasmodium* species are similarly regulated, although the evidence cannot be considered absolutely convincing.

In the vast majority of species of *Plasmodium*, maturation of the asexual and sexual stages occurs concurrently in erythrocytes freely circulating in the peripheral bloodstream. In marked contrast, two species (*P. falciparum* and *P. reichenowi*) have very extended periods of gametocytogenesis compared to their short asexual cycle (10–12 days vs. 2 days), and for most of this period the immature gametocytes are sequestered, possibly in the bone marrow (Smalley *et al.*, 1981). Current data suggesting that the very immature forms (stages I and II) are cytoadherent due to the expression of the protein Pfemp1 (K.P. Day, personal communication) cannot offer the whole explanation, first because the older stages (II–V) are also sequestered but do not express Pfemp1 on the erythrocyte surface, and secondly because there is at present no evidence for the co-sequestration of young gametocytes and erythrocytic meronts (which also express this cytoadherent ligand) in the capillaries of the brain. The present availability of reagents (e.g., anti Pfs27 antibody) that could specifically identify young gametocytes in autopsy material may soon resolve this question.

3. BIOLOGY OF THE MATURE GAMETOCYTE

Malarial gametocytes, when mature, circulate in the peripheral bloodstream. They fulfil two very different cellular functions, both of which are essential to the survival of the parasite. First, they provide an effective mechanism for transmission from the vertebrate host to the mosquito vector. Second, they are responsible for the sexual events of the life cycle which will be completed only within the blood-meal of the mosquito. Despite clearly flawed reports as recently as 1994 (Hummert, 1994) describing the alleged presence of ookinetes of *P. vivax* in peripheral blood samples, it is unquestionable that transfer to the mosquito vector is a prerequisite for the natural completion of the sexual cycle.

3.1 Distribution of Gametocytes in Blood

Mature gametocytes, arrested in the G_0 phase of the cell cycle, persist in the bloodstream for differing, species-specific periods of time. Among the longest lived are those of *P. falciparum* and *P. reichenowi* whose half-life is 2.5 days, giving a gametocyte "wave" that lasts typically 10–30 days (Smalley and Sinden, 1977). If the resulting sustained availability of the parasite in the venous bloodstream also applies to the capillaries in the skin — which is strongly suggested by the infectivity studies of Eyles *et al.* (1948) amongst others — it extends the potential "infectious window", and enhances the probability that any female anopheline mosquito will become infected when taking a blood-meal. Notwithstanding these observations, evidence has been presented that distribution of haemosporidean gameto-cytes in the vertebrate host changes in a circadian pattern (Hawking *et al.*, 1968a,b, 1971) and that the highest densities in the peripheral blood coincide with the peak time of mosquito blood-feeding, i.e., late eve-ning/early morning. Hawking *et al.* (1968a) argued that this was achieved by a circadian pattern in the preceding rounds of synchronous asexual erythrocytic merogony which therefore determined the availability of mature gametocytes (whose infectivity, it was assumed, lasted less than a day) (Hawking *et al.*, 1971). This attractive theory, although not supported by subsequent observations on gametocyte infectivity in *P. falciparum* (Bray *et al.*, 1976; Githeko *et al.*, 1993), may perhaps be relevant in other species where gametocyte maturation and survival is much shorter, e.g., the avian malarial parasites (Gambrell, 1937), the rodent parasites in which maturation takes 26–29 hours and infectivity persists for only 24 hours (Janse *et al.*, 1984, 1989; Mons, 1986), and *P. cynomolgi* (see Garnham and Powers, 1974). Even for these species the persistence of viability of the gametocyte exceeds the infectivity "window" described by Hawking *et al.* (1968a,b, 1971). In a series of elegant experiments on *Leucocytozoon smithi*, Gore *et al.* (1982) showed that gametocytes cycled between the deep tissues and the peripheral circulation, and that this was controlled by the host response to the day–night cycle through the pineal gland and could therefore be related to melatonin production. At present it must be acknowledged that conclusive evidence for circadian changes in the distribution of gametocytes of *Plasmodium* spp. is still lacking (Sinden, 1983b). It should be noted in any future experimental work involving murine malaria that, under laboratory light cycles (12 h light : 12 h dark), only 5 of 36 inbred mouse strains examined had detectable melatonin in the pineal gland (Goto *et al.*, 1989).

It is perhaps surprising that parasite transmission from a gametocyte-infected host to the mosquito vector does not correlate well with gametocyte numbers in the peripheral blood (Figure 2). This suggests that the gametocytes are of variable infectivity/viability (Carter and Graves, 1988; Dearsly *et al.*, 1990), or that factors in the blood of the vertebrate host or in the mosquito vector naturally regulate infectivity (Mendis *et al.*, 1987a,b; Carter *et al.*, 1988; Weiss, 1990; Sinden, 1991; Mendis and Carter, 1992; Fleck *et al.*, 1994) or that both mechanisms may operate. Studies on *P. berghei* were unable to demonstrate changes in the innate viability of gametocytes in the course of natural infection (Dearsly *et al.*, 1990), suggesting that inhibitory host factors may be important.

Circadian patterns of production of potentially inhibitory molecules which could result in the patterns of infectivity reported by Hawking *et al.* (1968a)

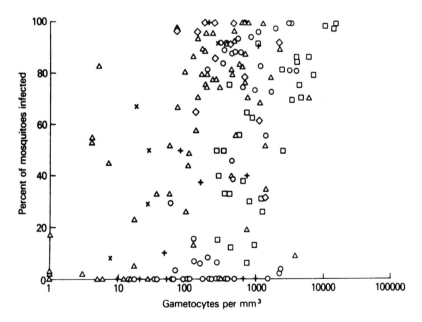

Figure 2 The relationship between gametocytaemia in patients naturally infected with *P. falciparum* and the prevalence of oocyst infection in *Anopheles gambiae*. (Figure taken from Carter and Graves, 1988; data derived from various authors.)

could be sought. In contrast to the well-documented effects of transmission blocking antibodies, the killing mechanisms of other non-specific inhibitors are poorly understood. Gametogenesis, fertilization, zygote development and ookinete to oocyst transformation are just some of the processes in the parasite life cycle that we suspect may be regulated by inhibitory factors ingested from the vertebrate host in the blood-meal (Figure 3).

3.2. Gametocyte Cell Biology

3.2.1. *Sexual Development*

Gametocytes enact the sexual phase of the parasite's life cycle in the mosquito blood-meal, i.e. gametogenesis and fertilization, leading to the production of a diploid zygote which then undergoes immediate meiotic division to re-establish the haploid genome (Sinden and Hartley, 1985; Sinden *et al.*, 1985; Ranford-Cartwright *et al.*, 1993). The zygote then differentiates into the ookinete. As the mosquito will begin to secrete lethal proteases into the blood-meal as early as 5 hours after feeding, reaching peak levels between 18 and 30 hours (Briegel and Lea, 1975), sexual development must be completed quickly to avoid digestion of the parasite by the mosquito. It is not therefore surprising that the structure of the mature macro- and microgametocytes is heavily precommitted to the rapid production of female and male gametes in the mosquito blood-meal.

3.2.2. *The Macrogametocyte*

The mature macrogametocyte is comparable to the vertebrate oocyte. It has an extensive endoplasmic reticulum and many ribosomes which are required for rapid *de novo* protein synthesis which follows the induction of gametogenesis. Within the nucleus (notably of the avian/saurian species and *P. falciparum*) lies an electron-dense structure morphologically indistinguishable from a nucleolus (Aikawa *et al.*, 1969; Sinden, 1978). However, to date there is no definitive evidence that this structure contains ribosomal RNA (J. Thompson, unpublished observations). *Plasmodium* is unusual in that it switches from the A to the C or O form of small subunit ribosomal ribonucleic acid (RNA) during its migration from the vertebrate to the mosquito host (Waters *et al.*, 1989; Li *et al.*, 1994). The complexity of this change is only now coming to light (Li *et al.*, 1994; McCutchan *et al.*, 1995).

The mature macrogametocyte nucleus reportedly contains 50% more deoxyribonucleic acid (DNA) (1.5C) than the haploid merozoite (1.0C) (Janse *et al.*, 1986a,b). The genome does not replicate further until meiosis

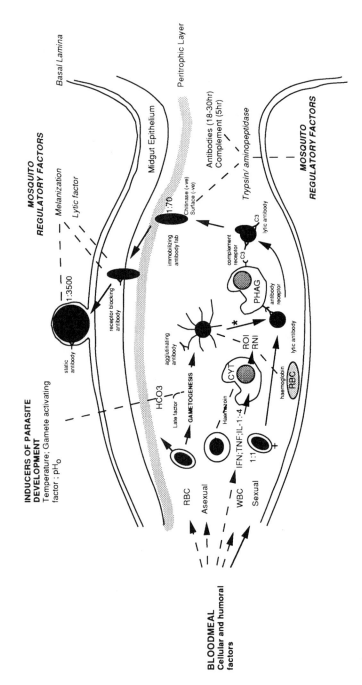

Figure 3 Summary diagram describing the reported mechanisms by which malaria development in the mosquito midgut may be regulated. Abbreviations: C_3, complement; fab, single chain variable region of immunoglobulin G; IFN, interferon-γ; IL-1, interleukin 1; IL-4, interleukin 4; PHAG, phagocyte; pH_o, extracellular pH; RBC, red blood cell; RNI, reactive nitrogen intermediates; ROI, reactive oxygen intermediates; TNF, tumour necrosis factor-α; WBC, white blood cell. Asterisk (*) indicates fertilization blocking/enhancing antibody.

in the fertilized zygote, 2–3 hours after the induction of gamete formation (Janse *et al.*, 1986b). An unusual intranuclear spindle-like structure is present in the macrogametocyte and macrogamete of *P. falciparum*. In view of the fact that the DNA of macrogametocytes never reaches 2C, and since the normal restriction points in the eukaryotic cell cycle lie either in G_1 (start) or in G_2, it is necessary to conclude either that the macrogametocyte is held at "start" and that the observations on DNA replication and mitosis are technical artefacts, or that these cells are atypically arrested mid-way through the S phase, as recently reported for *Saccharomyces pombe* by Murakami (1995). The fact that the macrogametocyte neither resumes synthesis to a 2C state nor completes a mitotic division before fertilization leads one to conclude that the former explanation is the more logical. The significance of the single spindle-pole and microtubule array found in the young macrogametocyte and gamete remains unclear (Sinden *et al.*, 1978).

The macro- and microgametocytes are morphologically distinct both from each other and from the asexual stages. Not surprisingly therefore they contain gametocyte specific proteins, e.g. Pfg27 (Carter *et al.*, 1989a), and sex-limited molecules, e.g. α-tubulin in the male (Rawlings *et al.*, 1992) or messenger (m) RNA encoding Pbs21 and Pfs25 in the female (Kaslow *et al.*, 1988; Paton *et al.*, 1993). Proteins, expressed rapidly following induction of gametogenesis, may be pre-synthesized in the macrogametocyte in active or inactive form, or encoded by stable mRNA whose processing is under translational or post-translational control, e.g., Pbs21 (Paton *et al.*, 1993). Complementary DNA studies suggesting that gametocyte and asexual mRNAs encoding the same protein may differ in relative molecular mass (M_r) deserve careful evaluation. Pre-protein activation, if proved, in early development of the gametes and zygotes may be expected to be dependent, *inter alia*, upon enzymatic modification, pH shift (Bullough *et al.*, 1994), or chaperonin/heat shock protein-based mechanisms (Georgopoulos and Welch, 1993).

3.2.3. *The Microgametocyte*

The mature microgametocyte also contains 1.5C DNA, but in a much larger nucleus. The kinetochores of the 14 chromosomes (the haploid number) are attached to a large amorphous organelle found on the nuclear envelope. It extends through a nuclear pore and is contiguous with the cytoplasmic microtubule organizing centre (MTOC) (Aikawa *et al.*, 1969; Sinden, 1978). The mature microgametocyte is readily distinguished from the female as the cytoplasm contains very few ribosomes and sparse, predominantly smooth endoplasmic reticulum. This morphology suggests that the very extensive organelle assembly (e.g., axonemes, plasma membrane, mitotic spindles) accompanying microgametogenesis may be based largely on the redistribution or activation of pre-synthesized macromolecules. Synthesis *de novo* of DNA (Janse *et al.*, 1986a,b), RNA (Toyé *et al.*, 1977; see Figure 12) and protein (Toyé *et al.*, 1977) is none the less required for exflagellation.

Both macro- and microgametocytes have small (0.1–0.2 μm diameter) osmiophilic membrane-bound vesicles in their cytoplasm, the female having 4–5 times as many. Secretion of these vesicles into the parasitophorous vacuole precedes the escape of the parasite from the enclosing erythrocyte (Sinden *et al.*, 1976).

3.3. Host Factors Modulating Infection

3.3.1. *Naturally Acquired Immunity to Sexual Stages*

Interest in immune responses to the sexual stages of malaria parasites stemmed from the first immunization experiments in which the infectivity to *Aedes* of *P. gallinaceum* was blocked by prior immunization with formalin-treated blood (Huff, 1957). These results were confirmed and extended by R.W. Gwadz, R. Carter and colleagues (reviewed by Carter and Graves, 1988; Carter *et al.*, 1988; Mendis *et al.*, 1990), demonstrating that transmission to the mosquito vector was vulnerable to immune attack in a variety of animal hosts and that this immunity was independent of immunity to the asexual stages. In this early work, antibody responses to gametocytes, gametes, zygotes and oocysts in immunized animals were examined and some of the mechanisms of antibody action characterized, eventually leading to the identification of important transmission-blocking antigens. Detailed reviews of sexual stage antigens, including properties of experimental antibodies raised to them, have been published elsewhere (Alano, 1991; Kaslow, 1993; Sinden, 1994) and will be included here only when relevant. Subsequently, naturally occurring responses came under scrutiny as the possibility of developing a practical transmission blocking (TB) vaccine approached. The mechanisms by which these and many other factors may regulate parasite development in the mosquito phases of development are illustrated in Figure 3.

   (a) *Antibody-mediated immunity.* The intracellular location of the gametocyte may largely protect it from direct antibody action but the succeeding extracellular parasite stages in the mosquito midgut are exquisitely vulnerable. In the early experiments, sera from gamete-immunized chickens that blocked infectivity were found to agglutinate gametes of *P. gallinaceum.* Subsequent experiments with a wide range of host–parasite combinations have revealed other possibilities of antibody action, including inhibition of formation of ookinetes, damage to the ookinete surface, inhibition of invasion of the mosquito gut wall, and damage to the developing oocyst with a reduction in the number of sporozoites developing (see Carter *et al.*, 1988; Ranawaka *et al.*, 1993, 1994a,b). These effector mechanisms may in certain circumstances be enhanced by the cellular arm of the immune system, by phagocytic activity and/or release of toxic molecules (Ranawaka *et al.*, 1993, 1994a,b; Lensen *et al.*, 1996 (in press)) (Figure 3).

Recognition of sexual stage antigens of *P. falciparum* in naturally infected adults appears to occur early in the immune response following a single or few infections (Graves *et al.*, 1990; Ong *et al.*, 1990). Malarial T cell epitopes on sexual stage antigens may be shared by other previously encountered infectious organisms (as well as other asexual stage antigens) (Good *et al.*, 1987; Fell *et al.*, 1994) which could, in theory, contribute to this early recognition. However, gametocyte prevalence in *P. falciparum* infections declines with age in endemic areas (Graves *et al.*, 1988b), presumably reflecting generally increasing immunity and declining parasite load.

In immune or semi-immune subjects in Papua New Guinea (PNG), over 50% of sera showed some inhibition of infectivity of *P. falciparum* compared to controls in membrane feeding experiments; 22% of sera reduced infectivity by 95% (Graves *et al.*, 1988a) and inhibition correlated with recognition of the protein Pfs 230. There was no correlation of transmission blockade with antibody to the protein Pfs48/45, which is coprecipitated with Pfs230. There is, however, marked variability in the response to Pfs230 in these populations (Carter *et al.*, 1989b). In a later study employing the same sera, and additional samples from Thailand, over one-third recognized Pfs45/48 and it was noted that gametocyte carriers potentially capable of infecting mosquitoes were not infective if their sera recognized an epitope (IIa) on this antigen (Graves *et al.*, 1992). Over one-third of PNG sera recognized four epitopes on Pfs45/48, but the response was biphasic, the highest responses being in the 5–9 and over 20 years age groups. The Pfs25 zygote/ookinete antigen which, on current evidence, is not expressed in the human host was not recognized by PNG sera (Carter *et al.*, 1989b).

In sera from non-immune subjects from Sri Lanka (where *P. falciparum* is less common than in PNG) blockade of infectivity did not correlate with immunoprecipitation of Pfs230. When sera from Sri Lanka and PNG were compared in the same experiment, fewer Sri Lankan sera than PNG sera blocked transmission (56% vs. 75% respectively) (Premawansa *et al.*, 1994). Over 70% of sera from Cameroon blocked infectivity of autologous isolates of *P. falciparum* to some degree (Mulder *et al.*, 1994), but in a more recent report transmission blocking immunity did not correlate with immunoprecipitation of Pfs230 (Roeffen *et al.*, 1995b). Forty per cent of Gambians failed to make antibody to Pfs230 (Riley *et al.*, 1994). In the same study, reactivity was thought not to be under direct genetic control but may have reflected environmental factors (including exposure to other immunogenic molecules) because both mono- and dizygous twins showed similar responses (Riley *et al.*, 1994).

A series of studies involving competitive inhibition of immune sera and a panel of monoclonal antibodies to Pfs230 and Pfs45/48 further demonstrated the complexity of responses to these antigens. Although it is clear that naturally occurring antibodies frequently recognize the same antigens as monoclonal antibodies (Graves *et al.*, 1988c), four patterns of response were observed to four different epitopes in Pfs45/48 when sera were taken during and following a malaria attack (Graves *et al.*, 1991). Such variability of response to each epitope makes it less surprising that there is no strong correlation between the numbers of gametocytes present in patients' blood and their ability to infect mosquitoes (see Figure 2). Roeffen *et al.* (1995a) recorded only "fair-to-moderate" agreement between a competitive enzyme-linked immunosorbent assay for Pfs48/45 and a transmission-blocking assay.

The possibility of antigenic diversity is always an important question with regard to malarial immunity. Variation between *P. falciparum* isolates was detected in early work with monoclonal antibodies to gamete surface antigens (Graves *et al.*, 1985). With Pfs48/45, some epitope variation occurs between isolates (Foo *et al.*, 1991; Riley *et al.*, 1990) but the relevant gene was described as minimally variable (Kocken *et al.*, 1995). Pfs230 is thought to be antigenically non-variable (see Read *et al.*, 1994), in spite of some gene sequence variation (Williamson and Kaslow, 1993). *P. vivax* antigens (M_r = 20, 24 and 37/42) expressed by the parasite while in the vertebrate host, and against which transmission-blocking antibodies can be generated, show considerable polymorphism (Premawansa *et al.*, 1990).

Post-fertilization antigens (zygote/ookinete), not exposed to the mammalian host and not under immune pressure, are likely to exhibit minimal antigen variability (Kocken *et al.*, 1995). Pfs25, described as an invariant antigen (Mendis *et al.*, 1990), shows variation in gene sequence at such a low level that only single amino acid changes are predicted (Kaslow *et al.*, 1989a,b). Pfs25 induces potent transmission-blocking immunity. A disadvantage of such antigens as potential vaccine candidates is the lack of the natural boosting which occurs with blood-stage and pre-fertilization antigens; this is, however, heavily outweighed by the lack of polymorphism in the target protein.

On current evidence, antibody-mediated immunity to *P. vivax* (and *P. cynomolgi*) exhibits some marked differences from that to *P. falciparum*. Most significantly, sera from exposed subjects often enhance infectivity (compared to the controls), as opposed to blocking it. Sera from toque monkeys (*Macaca sinica*) infected with *P. cynomolgi* enhanced infectivity 3–4 months following infection (Naotunne *et al.*, 1990) and enhancing activity was observed in sera following *P. vivax* infections (Mendis *et al.*, 1987b; Peiris *et al.*, 1988; Gamage-Mendis *et al.*, 1992). Enhancement (which has also been occasionally observed with merozoite invasion of red blood cells) could, for example, arise from low levels of antibodies possibly bringing micro- and macrogametes into contact more efficiently by cross-linking, though other explanations are possible, including the use of inappropriate controls. An unexpected observation, that infection with one species of malaria followed by a second can enhance infectivity of the latter (Collins, W.E. *et al.*, 1975; Arrada-Meyer *et al.*, 1979), is difficult to explain, but could perhaps also result from cross-reactive antibody. If this applies in natural infections it would be of considerable advantage to the various parasite species (probably the majority) that frequently share hosts (Boyd, 1949).

Acute sera from *P. vivax* patients block infectivity through complement-mediated action (Mendis *et al.*, 1987a). However, in *P. vivax* there is no rise in inhibitory activity with increasing age (unlike *P. falciparum*) and, although more frequent attacks increase inhibition, the immunological memory lasts for only four months in the absence of boosting (Gamage-Mendis *et al.*, 1992; Ranawaka *et al.*, 1988). This led to suggestions of T cell-independent immunity (Mendis and Carter, 1991). Nevertheless, in PNG gametocyte prevalence declines with age in *P. vivax* (as in *P. falciparum* — see above) (Graves *et al.*, 1988b), presumably because of increasing immunity to asexual stages. However, infectivity can revive with the declining immunity that may result from successful mosquito control (Metselaar, 1960).

Few studies on human leucocyte antigen (HLA) restriction of immune responses to sexual stage antigens have been reported. No correlation was observed between any particular genotype and antibody to Pfs230 and Pfs45/48 in PNG (Graves *et al.*, 1989). Similarly, when T cell responses by West Africans to Pfs45/48 were assessed by proliferation and interferon-γ production, no relationship between HLA type and antigen recognition was observed, although only 40% of samples responded positively (Riley *et al.*, 1990).

(b) *Cell-mediated immunity.* Gametocytes appearing early in infections inevitably come into contact with mediators involved in the developing resistance to asexual parasites. This problem may vary in its importance between the majority of species where gametocyte and asexual maturation is of equal duration (e.g., *P. vivax, P. gallinaceum, P. berghei*) and the minority where gametocyte maturation is so long that gametocytes often appear only after the asexual stage has passed (e.g., *P. falciparum*). Of particular importance are macrophages and neutrophils, at least in those hosts which are able to make a non-specific response. First infections of parasites in some experimental models such as *P. knowlesi* in rhesus monkeys (*M. mulatta*), and *P. berghei* given as a high infective dose of blood-stage parasites to TO mice, appear not to induce a significant protective non-specific response and the parasites multiply at their maximal rate until the host dies of anaemia. In *P. berghei*, infectivity declines as parasitaemias rise (Figure 4), for reasons not necessarily associated with immune reactions (see below), and this has caused some confusion in the past (Carter and Graves, 1988). Where there is significant macrophage/neutrophil activation, triggered by release of lipid malarial toxins (Bate *et al.*, 1989) and perhaps haemazoin (Pichyangkul *et al.*, 1994), a wide variety of parasite inhibitory mediators may be released (Butcher, 1989; Playfair *et al.*, 1989) that are likely to have an effect on gametocytes, depending on the susceptibility of the species (Taverne *et al.*, 1982) and the concentrations of the various mediators.

It has often been observed that gametocytes rapidly become non-infective at about the time at which the asexual parasites come under the host's

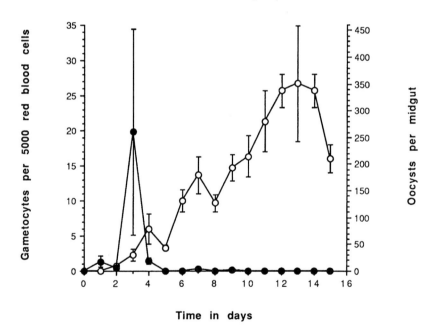

Time in days

Figure 4 Comparison of *P. berghei* gametocyte density in the peripheral circulation (○) with the success of infection (oocyst number, ●) in *Anopheles stephensi*. Error bars are ± SEM. Data are pooled from three mice that had been infected between 1 and 15 days. (Data modified from Dearsly *et al.*, 1990.)

control (Cantrell and Jordan, 1946; Eyles, 1952; Petit *et al.*, 1982; Bastien *et al.*, 1987; Naotunne *et al.*, 1990; Fleck, 1993). The asexual stages may exhibit highly abnormal morphology within the red blood cell and are referred to as crisis forms, which were first reported by Taliaferro and Taliaferro (1944) in *P. brasilianum*. Crisis forms are not, however, seen *in vivo* in all malaria infections, including those of humans. When they do occur it appears likely that the gametocytes are also affected by whatever cytotoxic host factors are produced in response to the high asexual parasitaemia (see above). Subtle changes in gametocyte morphology associated with lack of infectivity may occur that are less dramatic than of the "classical" asexual crisis forms and may be reversible (Landau *et al.*, 1979). Crisis forms can be produced in infections in which they may not have been initially apparent by the expedient of injecting lipopolysaccharide (Clark, 1978) or very large numbers of parasites (Butcher *et al.*, 1978), and the non-specific nature of the response becomes clear when two different species of parasite exhibit damage at the same time (Butcher *et al.*, 1978). Under certain circumstances *in vitro, P. falciparum* gametocytes may not be affected, at least morphologically, when asexual stages show severe damage. This was apparent when mouse serum containing tumour necrosis factor-α (TNF) was cultured with *P. falciparum in vitro* in the absence of hypoxanthine (Butcher *et al.*, 1991), and simply reflects the fact that this apparently highly toxic serum interferes with hypoxanthine uptake mechanisms in the rapidly metabolizing asexual cell, presumably through modulation of the nucleoside transporter in the parasite plasmalemma. These gametocytes are also resistant to sorbitol lysis, and may either lack some of the anion pathways of asexual stages (Saul *et al.*, 1990) or simply be less metabolically active and therefore less susceptible.

It is currently thought that the early non-specific response reflects activation of Th_1 cells and that this gradually gives way to a Th_2 driven response (Langhorne, 1989) that suppresses the over-production of cytokines such as TNF, reportedly responsible for much of malarial pathology (Clark, 1987). Nevertheless, it is important to keep in mind that infectivity may decline in spite of rising parasite numbers in those situations where crisis forms do not appear and where there is little apparent effective immunological response, e.g., *P. berghei* infections in laboratory mice (see above).

The first indication that cellular mechanisms may be important in limiting infectivity arose from the observations of Rutledge *et al.* (1969) and Sinden and Smalley (1976) that blood leucocytes were able to phagocytose or kill extracellular gametes and gametocytes. In gamete-immunized animals infected with *P. yoelii nigeriensis* there was evidence for the involvement of antibody-independent T cells in inhibiting infectivity (Harte *et al.*, 1985). Further investigations into non-specific effector mechanisms that could block transmission followed the general approach of similar work on asexual stages, involving stimulation of leucocytes by lipopolysaccharide, parasites, or parasite material and cytokines (reviewed by Mendis and Carter, 1992). Within the first few days of *P. cynomolgi* infections in *M. sinica*, morphologically abnormal gametocytes appear at the same time as asexual crisis forms, and although parasitaemias rise to only about 3%, infectivity declines over this period. Evidence that TNF and interferon-γ stimulation of peripheral blood mononuclear cells (PBMC) may damage gametocytes was reported (Naotunne *et al.*, 1991). Surprisingly, a very low dose (5 ng/kg) of lipopolysaccharide injected into a normal monkey was said to generate inhibitory serum, equivalent to the "crisis serum" in its blocking effect; an appropriate control was not, however, included. Incidentally, if such a low dose was indeed effective it suggests a dramatically high sensitivity of the macrophages of these animals to this stimulus, compared to other species (see, for example, Clark, 1978).

P. vivax infection in humans is characterized by low parasitaemias and less severe symptoms than *P. falciparum*, but may induce very high plasma TNF concentrations even in semi-immune subjects who are not seriously ill (Butcher *et al.*, 1990). During paroxysms there is a temporary loss of infectivity associated with the rapid rise and decline of TNF (Karunaweera *et al.*, 1992a,b,c). Serum taken during paroxysms inhibited the infectivity of membrane-fed gametocytes, as did supernatants from PBMC cultures stimulated with *P. vivax* antigen; the effect was reversed by the addition of anti-TNF antibodies (Karunaweera *et al.*, 1992a). The authors also invoked the presence of an additional "complementary factor" in the paroxysm (crisis) serum (as they did in the *P. cynomolgi* serum experiments outlined above), but the inevitable presence of blood leucocytes in the gametocyte feeds complicates the interpretation of these experiments. It does seem likely, however, that PBMC stimulated by plasma TNF generated in response to the synchronous bursting of many *P. vivax* schizonts (see Kwiatkowski *et al.*, 1989) may have cytotoxic effects on asexual and sexual parasites, causing a loss of infectivity. *In vitro*, morphological damage of parasites was reported (Karunaweera *et al.*, 1992a), but this may not be apparent *in vivo* even though functional loss of activity could occur. In this study, sera from semi-immune patients contained less TNF than those of non-immune subjects and had no blocking effect, suggesting that these patients may have developed antibody to the TNF-inducing parasite toxin (Bate *et al.*, 1990; Bate and Kwiatkowski, 1994), although there are other regulatory factors acting on TNF generation.

K.N. Mendis and colleagues extended their studies on PBMC to *P. falciparum*, using a similar approach to that with *P. vivax* (see Naotunne *et al.*, 1991, 1993). Supernatants of antigen-stimulated PBMC incubated with gametocytes in the presence of fresh white blood cells were observed to block infectivity. Following reports that nitric oxide (NO) or its downstream products could kill asexual parasites *in vitro* (Rockett *et al.*, 1991), they attempted to show that the effector molecule in this case was also NO by including an inhibitor of NO production in some cultures (Naotunne *et al.*, 1993). Although this appeared to remove the blockade of gametocyte infectivity, the result has to be viewed cautiously as does the claim for NO-mediated killing of asexual *P. falciparum* by human monocytes (Gyan *et al.*, 1994). There has been much disagreement about the ability of human PBMC to generate NO (Crawford *et al.*, 1994) and workers in other fields, particularly in that of tuberculosis, have been unable to detect NO in these cells (see, for example, Rook, 1994). In experiments of this nature it is vital to measure NO output rather than rely on the use of inhibitors alone to demonstrate an effect (Archer and Hampl, 1992). Recent data, however, indicate that macrophages derived from blood monocytes can be induced to generate NO under appropriate conditions involving interleukin-4 stimulation and cross-linking of CD69 (De Maria *et al.*, 1994) or CD23 receptors (Mossalayi *et al.*, 1994).

There has been no disagreement concerning the ability of mouse macrophages to make NO. Synchronous infections of *P. vinckei petteri*, that exhibit a temporary decline in infectivity at merogony, were made more infectious by injecting the mice with an L-arginine analogue (Nω-nitro-L-arginine) (Motard *et al.*, 1993). Even the infectivity of *P. berghei*, which in most mice infected with high doses of parasites (10^6) induces virtually no effective immunity, can be raised slightly by administration of an L-arginine analogue at the normal infectivity peak (Fleck, 1993), but this analogue does not prevent the continuous decline in infectivity thereafter (Fleck, 1993; I.W. Jones *et al.*, unpublished data). Indeed, with non-lethal *P. vinckei petteri* and *P. berghei* infections we could find no evidence for significant activation of the inducible form of nitric oxide synthase in any organ of the TO mouse (I.W. Jones *et al.*, unpublished data), except for a small, brief rise in the kidney. Even if it can be shown conclusively that NO *per se* inhibits infectivity or growth of asexual parasites in experimental situations, as has been claimed (Taylor-Robinson *et al.*, 1993; Gyan *et al.*, 1994), the potent capacity of haemoglobin to scavenge NO would appear to make it an unlikely cytotoxin for malaria parasites constantly surrounded by haemoglobin *in vivo*.

Although *P. berghei* in TO mice exhibits the commonly seen fall in infectivity with rising parasitaemias (Figures 4 and 8) (Dearsly *et al.*, 1990; Fleck *et al.*, 1994), we have not been able to demonstrate any immunological factor responsible for this decline (Fleck, 1993). The pattern of infectivity is the same in immunodeficient SCID mice as in the congenic Balb/C strain and the outbred TO strain. Stimulation of the mice by

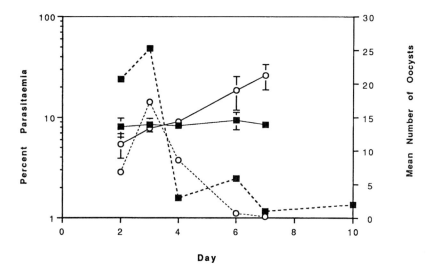

Figure 5 Percentage parasitaemia (——) (each symbol represents the arithmetic mean ± sᴇᴍ for three mice) in, and geometric mean number of oocysts (– – –) per mosquito fed on, splenectomized (■) and sham splenectomized (○) Balb/C mice inoculated intraperitoneally with 10^7 erythrocytes infected with *P. berghei* on day 0.

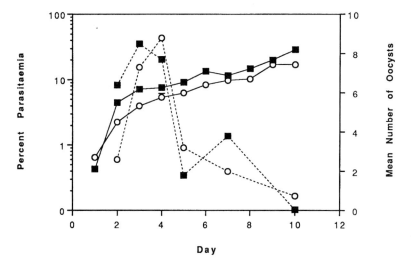

Figure 6 The effects of treating SCID mice, inoculated intraperitoneally on day 0 with 10^7 erythrocytes infected with *P. berghei*, with either control ascitic fluid (○) or rat anti-mouse neutrophil anitbody (■), upon mosquito infectivity (– – –) expressed as the geometric mean number of oocysts in fed mosquitoes and upon parasitaemia (——).

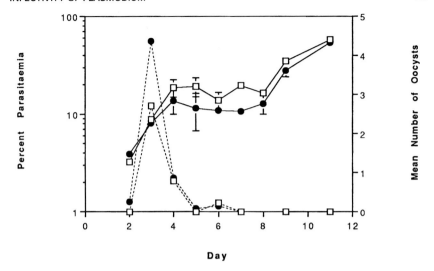

Figure 7 Percentage parasitaemia (——) (each symbol represents the arithmetic mean ± SEM for three mice) in, and geometric mean number of oocysts per mosquito (– – –) fed on, SCID mice inoculated intraperitoneally with 10^7 erythrocytes infected with *P. berghei* on day 0 and treated on days 1 and 7 with rabbit antisera against tumour necrosis factor-α and interferon-γ (□) or normal rabbit immunoglobulin (●).

injections of endotoxin just before mosquito feeding caused no fall in infectivity, nor did splenectomy of the mice (Figure 5), or the presence or absence of complement (Figure 8A). Neutrophil depletion by antibody also failed to arrest the decline in infectivity (Figure 6), as did injections of anti-interferon-γ and anti-TNF antibodies (Figure 7). Infectivity rises for the first five days and thereafter falls to a low level as the parasitaemia continues to rise in both mice (Figures 4, 8; Fleck, 1993) and susceptible young rats (Figure 9). In older rats, in which the parasitaemia does not rise as markedly, infectivity declines more slowly (Figure 9). Figure 8B illustrates that both the duration of the infection and the intensity of the parasitaemia independently correlate with the inhibition of infection. Serum taken after the peak of infectivity ("late serum") profoundly inhibits oocyst formation in mosquitoes fed infective gametocyte (Fleck *et al.*, 1994). It appears, therefore, that one or more physiological factors, independent of both B and T cells, affect transmission in this model, and perhaps in others where a rising gametocyte number is not accompanied by increasing infectivity. Addition of bicarbonate to feeds containing "late serum" partially reinstates infectivity to mosquitoes without significantly changing the pH of the infected blood (G.A. Butcher, unpublished observations). This suggests that bicarbonate levels, not serum pH, may in part be responsible for this inhibition;

A

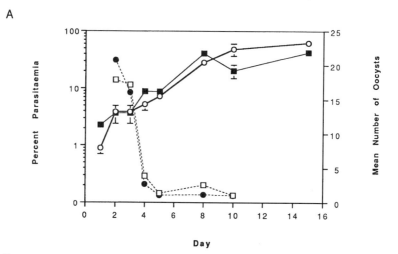

B

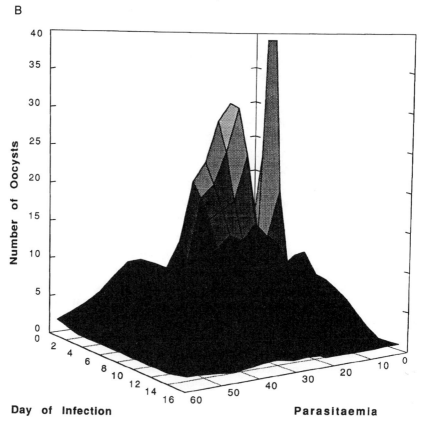

however, it does not fully explain the rapid decline in infectivity that occurs before the pH/bicarbonate levels fall on day 7 (Figure 10).

In this model the inhibitory effect appears to occur before ookinete formation because combining late serum (after day 10) with ookinetes in a membrane feed caused no loss of infectivity (Table 1). We suggest that exflagellation, which is a particularly vulnerable phase of the life cycle, may be the point at which late serum exerts its effect as far fewer free microgametes are present shortly after feeding has commenced in the guts of mosquitoes fed on mice on day 10 of infection compared with day 3 of infection (Table 1).

The delay in the wave of gametocytaemia compared to that of the asexual parasitaemia in *P. falciparum* may offer an unusually effective form of defence against host-generated toxic factors produced in response to the asexual infection, such as the factors outlined above, or free-oxygen radicals (FOR) or their products, which are known to be cytotoxic to asexual stages of *Plasmodium in vitro* (Clark *et al.*, 1987a; Rockett *et al.*, 1988; Buffinton *et al.*, 1991; Hunt *et al.*, 1992) and probably *in vivo* (Eaton *et al.*, 1976; Clark *et al.*, 1987b). It may be pertinent that asexual stages of *P. falciparum* have the potential to make their own glucose-6-phosphate dehydrogenase (O'Brien *et al.*, 1994) but evidence concerning the capacity of asexual stages of other parasites to defend themselves from FOR is contradictory (Hunt *et al.*, 1992). There is no report of research on similar protective defences, if such exist, in sexual stages.

From the epidemiological viewpoint, transmission-blocking immunity by relatively long-lived molecules such as antibodies may be more important than short-term loss of infectivity induced by acute infections. In view of the short period of non-infectiousness in acutely ill patients, it is likely that this can have a significant effect on overall parasite transmission only in epidemic situations. However, several aspects of host resistance to malaria have not been examined with the sexual stages as the focus of the investigation. For example, parasite-induced immunosuppression

Figure 8 **A**. Percentage parasitaemia (——) (each symbol represents the arithmetic mean ± SEM for three mice) in, and geometric mean number of oocysts per mosquito (– – –) fed on, Balb/C (○) or DBA/2 (■) mice inoculated intraperitoneally with 10^7 erythrocytes infected with *P. berghei* on day 0. **B**. The relationship between total parasitaemia, age of infection and infectivity to mosquitoes (mean oocyst number) in *P. berghei* infections of mice fed on directly by *Anopheles stephensi*. Data are presented as a three-dimensional surface plot using a negative exponential interpolation. Individual surface "plates" are shaded according to the minimal oocyst number in each "plate", irrespective of the maximal value, from black (< 5) through dark grey (5–9), mid-grey (10–14), light grey (15–19) and the lightest shade (20–24). (Data from Dearsly *et al.*, 1990; Sinden, 1991; Fleck, 1993; G.A. Butcher, unpublished observations.)

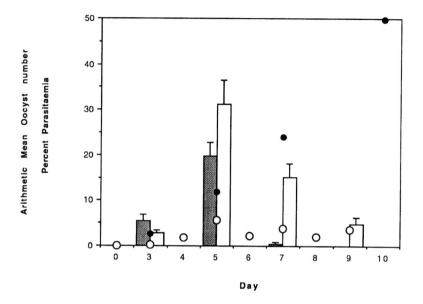

Figure 9 Percentage parasitaemia in young (●) and mature (○) rats (each symbol represents the arithmetic mean for three animals), and geometric mean number (± SEM) of oocysts in mosquitoes fed on young (black bars) and mature (open bars) rats inoculated intraperitoneally with 10^7 erythrocytes infected with *P. berghei* on day 0.

(Goonewardena *et al.*, 1990; Theander *et al.*, 1990) and pregnancy, which reduce resistance to both *P. falciparum* and *P. vivax* (see Mutabingwa, 1994), may contribute to the overall picture and require further investigation. Whether the various red blood cell polymorphisms that are said to give some degree of protection from malaria cause decreased gametocytaemia and infectivity does not appear to have been determined.

3.3.2. *Physiological Factors*

The nutritional status of the host can affect both asexual parasitaemias directly and the effectiveness of the immune response. Although some studies have demonstrated nutritional influences on malaria, in general they have to be rather severe to have any marked effect (McGregor, 1988). Dietary factors can also cause an amelioration of parasitaemia by influencing the antioxidant status of the host (see above). The long maturation time for *P. falciparum* gametocytes could perhaps be subject to nutritional or dietary factors but we know of no investigation into this.

In assessing the contribution of any factor to infectivity (Cantrell and

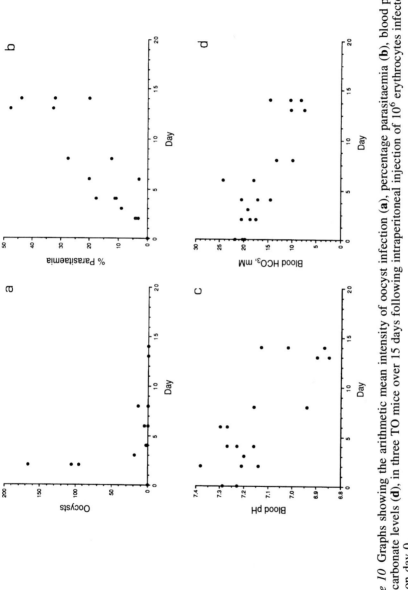

Figure 10 Graphs showing the arithmetic mean intensity of oocyst infection (**a**), percentage parasitaemia (**b**), blood pH (**c**) and blood bicarbonate levels (**d**), in three TO mice over 15 days following intraperitoneal injection of 10^6 erythrocytes infected with *P. berghei* on day 0.

Table 1 Effects of early and late infection serum upon early sporogonic development of *Plasmodium berghei* in the mosquito *Anopheles stephensi*.

Development step examined	Early serum	Late serum
Microgametocyte to exflagellation[a,b]	102	25
Gametocytes to retort-form ookinete[a,c]	++++	+
Gametocytes to ookinete[a,c]	++++	+
Ookinete to oocyst[d]		
Intensity	8.02±2.09	11.28±2.77
Prevalence	42.9	50.00

[a] Replicated experiments *in vivo*.
[b] Number of free microgametes seen per 100 microscope fields in Giemsa-stained films of mosquito gut contents 10 minutes after blood feeding. NB. Total parasite number (and therefore gametocyte number) in late infection was six times greater than in early infection.
[c] Number of retorts/ookinetes in Giemsa-stained films from cultures *in vitro* prepared from mosquito blood-meals taken 10 minutes after feeding (Sinden *et al.*, 1985).
[d] Ookinetes from culture *in vitro* were given to mosquitoes by membrane feeding suspended in early or late serum. The intensity (± SEM) and prevalence of the resulting oocyst infection was recorded 10 days after the feed (see Ranawaka *et al.*, 1994).

Jordan, 1946; Huff *et al.*, 1958), it has to be kept in mind that components of the blood-meal may influence not only the fate of gametocytes but also the mosquito's responses (Galun *et al.*, 1984) and thus indirectly affect infectivity. *Anopheles stephensi*, for example, feeds more readily in the presence of bicarbonate (G.A. Butcher, unpublished data).

Non-dialysable serum components appear to be beneficial but not absolutely essential for the development of normal numbers of oocysts in *P. gallinaceum* (see Cantrell and Jordan, 1946; Eyles, 1952). Rosenberg and colleagues (Rosenberg and Koontz, 1984; Rosenberg *et al.*, 1984) concluded that substances of high molecular mass in the red blood cells are necessary for zygote development and that dialysable factors in the serum enhanced infection but were not essential. More recently, Motard *et al.* (1990) reported that gametocyte feeds in the presence of haemoglobin resulted in reduced infectivity, but in the light of their other data showing that NO inhibits infectivity this is somewhat contradictory, as haemoglobin should scavenge NO. Simply stated, the authors of many of the earlier studies on infectivity were unaware of the numerous independent and interdependent modulating factors. It remains to be clarified what constituents of the blood-meal are individually beneficial or inhibitory to infectivity.

3.3.3. *Host Behaviour*

A series of studies on the impact of malarial and other parasites on both the vertebrate host and the insect vector highlighted the importance of examining how the interactions between parasite and host modulate the success of parasite transmission between host and vector (Rossignol *et al.*, 1984, 1985). Generalized haematopathology was considered to be an important factor and in more severe cases the reduced irritability of the vertebrate host (mice infected with *P. chabaudi*) was shown to reduce the median probing time by at least one minute (usually 2–5 minutes) (Vaughan *et al.*, 1991, 1994). The decreased nuisance caused, combined with the reduced irritability, would clearly enhance the probability of survival of the mosquito by reducing the extent of defensive action taken by infected hosts. (Though not directly relevant to this thesis it is interesting to note the converse — that mosquitoes with a salivary gland sporozoite infection take longer to feed, and probe more frequently, than uninfected mosquitoes.) Both changes in host/vector response can be predicted to have a significant impact on the potential for parasite transmission (Ribiero *et al.*, 1985).

4. GAMETOGENESIS

4.1. Cell Biology

4.1.1. *Escape from the Host Cell*

Macro- and microgametogenesis are simultaneously induced in the bloodmeal of the feeding mosquito. During gametogenesis both female and male parasites escape from the host red blood cell, thus permitting gamete dispersal and fertilization. (The whole process of gametogenesis is diagrammatically summarized in Figure 11.)

Escape from the host cell is similar (though not identical) with both sexes. Within one or two minutes of induction, the gametocytes have swollen into distended spheres, increasing in volume as much as three times (Sinden and Croll, 1975; Sinden *et al.*, 1976, 1978). At the same time, the parasitophorous vacuole membrane is disrupted, followed by the breakdown of the erythrocyte cytoplasm and plasmalemma (Aikawa *et al.*, 1984). Disruption of the parasitophorous vacuole membrane and the erythrocyte coincides with the secretion of the contents of the osmiophilic bodies (OB) into the parasitophorous vacuole (Sinden *et al.*, 1976). It is difficult to avoid the conclusion that there is a causal relationship between these events. In the microgametocyte, which has fewer osmiophilic bodies,

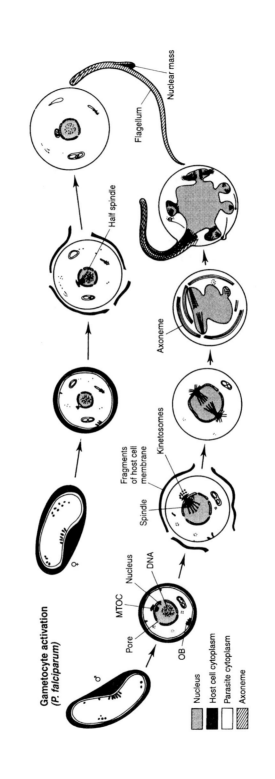

Figure 11 Events of macrogametogenesis (top) and microgametogenesis (bottom) of *P. falciparum*. (Reproduced with the kind permission of The Wellcome Trust, from The Wellcome Trust Tropical Medicine Resource — Image No. R20937.)

final disruption of the disorganized erythrocyte may be facilitated by the lashing movements of microgametes. Suggestions that the antigens RESA (Quakyi *et al.*, 1989) and GLURP (B. Høgh and R.E. Sinden, unpublished observations) are located in the parasitophorous vacuole in addition to Pfs230 (Quakyi *et al.*, 1987) must be viewed with caution in view of the extensive cross reactivity of some monoclonal antibodies. Recent biochemical data (K.C. Williamson, unpublished observations) show that Pfs230 is found in the parasitophorous vacuole of the mature gametocytes as a 360-kDa molecule, but that during gametogenesis the protein is cleaved to 310 kDa. Whether the 360-kDa molecule is a pre-protein which is activated by products from the osmiophilic bodies is unknown.

4.1.2. *Macrogametogenesis*

Having escaped from the host red blood cell, further maturation of the macrogamete may be required before fertilization. Nonetheless the macrogamete can be fertilized within minutes of formation (McCallum, 1897; Sinden and Croll, 1975; Janse *et al.*, 1986a,b). The majority of the molecular events of macrogamete development cannot be detected biochemically until some one or two hours after the induction of gametogenesis. It remains to be distinguished, therefore, which, if any, of these events is (are) critical to gamete maturation and which to zygote development. Molecular events described in "late" macrogamete development include: the translation of major ookinete surface proteins, e.g., Pbs21 (Paton *et al.*, 1993; Thompson and Sinden, 1994); the expression of zygote proteins, e.g., Pgs25/28 (Carter and Kaushal, 1984); the conversion of the parasite 3–6 hours after induction from complement insensitivity to sensitivity (Grotendorst *et al.*, 1986; Grotendorst and Carter, 1987); and the switch from the A to the C (or O) form of small subunit ribosomal RNA (Waters *et al.*, 1989; Li *et al.*, 1994).

4.1.3. *Microgametogenesis*

Microgametogenesis has been examined in detail by both light and electron microscopy (Sinden *et al.*, 1976; Aikawa *et al.*, 1984), and will not be redescribed in detail here, except insofar as is necessary to understand the molecular regulation of the events.

Microgametogenesis can be blocked by inhibitors of DNA, RNA and protein synthesis (Figure 12; Toyé *et al.*, 1977). Nevertheless, simple media will support microgametogenesis. It is therefore assumed that all macromolecular precursors, with the possible exception of glucose, are already available in intracellular pools in the mature gametocyte. During escape from the host cell the microgametocyte undergoes rapid nuclear and

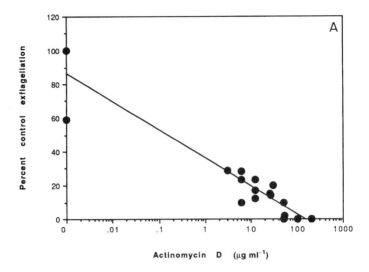

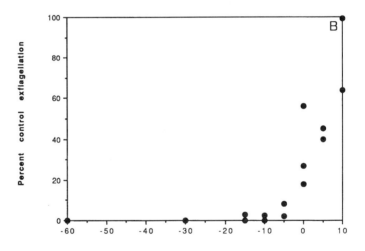

Time from induction in minutes

Figure 12 **A** Dose–response plot of the ability of actinomycin D to block exflagellation of cultured microgametocytes of *P. falciparum in vitro*. All values expressed as a percentage of controls containing culture medium alone (at 22°C and pH 8.0). **B** Plot showing the increasing ability of 50 µg/ml actinomycin D to block exflagellation of *P. falciparum* with increasing periods of preincubation at 37°C. At time 0 the temperature of the gametocyte suspension was dropped from 37°C to 22°C and the pH was raised to 8.0 by the addition of medium (containing the appropriate concentration of drug) at pH 8.3; exflagellation occurred 10 minutes later.

cytoplasmic cytoskeletal changes, the first of which is detectable morphologically only 15 seconds after induction (Sinden *et al.*, 1976). Within the cytoplasm the microtubule organizing centre changes from an amorphous structure into eight highly organized kinetosomes, a complex process involving numerous different proteins, e.g., γ-tubulin, some in quantities that possibly preclude *de novo* synthesis (e.g., α-tubulin — which is also found in the cytoplasm of the microgametocyte (Rawlings *et al.*, 1992)). Over the next 8–10 minutes the kinetosomes "organize" the assembly of the eight axonemes (each 22 mm long) that ultimately propel the microgametes from the cell at exflagellation (Figure 13). In parallel with these changes, and physically linked to them, are the three endomitotic nuclear divisions. These divisions are preceded by very rapid replication of the entire genome to 8C (Janse *et al.*, 1986a). The speed of replication contrasts with that during mitosis in asexual parasites, and suggests that *c.* 1300 replication forks are activated simultaneously in the sexual stage parasites (Janse and Mons, 1987). The induction of so many independent replication forks itself suggests that the induction mechanism is highly amplified within the parasite nucleus.

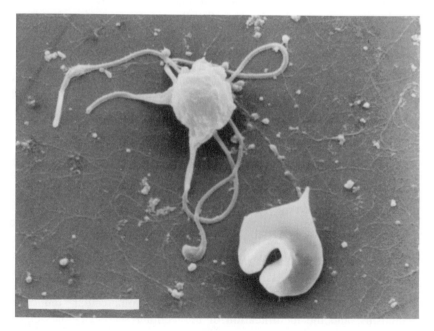

Figure 13 Scanning electron micrograph of exflagellating microgametocyte of *P. yoelii* (scale bar = 5 μm).

4.2. Induction Mechanisms

When considering the possible induction mechanisms for gametogenesis it is necessary to recall four simple overriding constraints. (i) At the time of induction both micro- and macrogametocytes are intracellular, lying within the parasitophorous vacuole of the infected red blood cell. The vexed question whether the parasitophorous vacuole is connected to the extracellular milieu by a parasitophorous duct is of course critical to the question of whether macromolecular inducers in the blood-meal have direct access to the parasite plasmalemma (Pouvelle *et al.*, 1991): it is assumed here that no such access exists. (ii) Complex organelles, e.g., basal bodies and axonemes, are assembled with exceptional speed (15 seconds) in the activated microgametocyte. (iii) Three independent yet parallel events are activated with near synchrony, namely DNA replication/segregation, cytoskeletal assembly, and processes leading to erythrocyte lysis (see Figure 14, C). (iv) Gametogenesis is readily induced *in vitro* in tissue culture medium (in the absence of the mosquito vector) by a reduction in temperature combined with a rise in extracellular pH (Bishop, 1955; Bishop and McConnachie, 1956, 1960; Sinden and Croll, 1975; Carter and Nijhout, 1977; Nijhout and Carter, 1978; Ogwan'g *et al.*, 1993a,b). However, it may be naive to consider that this induction is a natural process, or that the events induced by it are the natural sequence of events.

It was the recognition of the vital events of microgametogenesis and exflagellation that led Laveran (1881) to identify *Plasmodium* as the causative agent of malaria. Since that time studies on the induction of gametogenesis have constantly attracted the attention of students of malaria biology. Their studies fall into two categories. First, those correlating specific environmental conditions with the successful infection of mosquitoes; and secondly, studies *in vitro* on the regulation of exflagellation. Only in the past decade have the possible secondary signalling pathways regulating the individual constituent processes in microgamete formation been examined separately; most of these studies were conducted *in vitro*.

The design and interpretation of studies on induction mechanisms have been heavily influenced by the biological observation that gametocyte induction coincides with transition of the parasite from the homeostatic environment of the vertebrate bloodstream into the poikilothermic milieu of the mosquito gut. Broadly these studies examined four topics: the role of temperature; the role of "external" pH; macromolecular induction mechanisms; and the identification of intracellular secondary-messenger pathways.

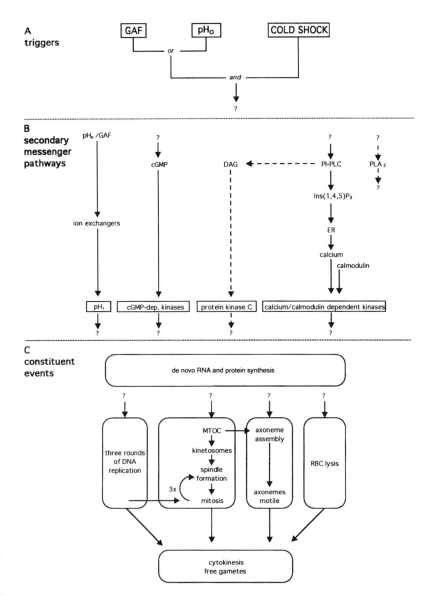

Figure 14 Diagram showing the mechanisms described as regulating the different component events of microgametogenesis of *Plasmodium*. Abbreviations: cGMP, cyclic guanosine monophosphate; DAG, diacyl glycerol; DNA, deoxyribonucleic acid; ER, endoplasmic reticulum; GAF, gamete activating factor; $Ins(4,5)P_2$, inositol bisphosphate; $Ins(1,4,5)P_3$, inositol trisphosphate; MTOC, microtubule organizing centre; PI-PLC, phosphoinositol dependent phospholipase C; pH_i, intracellular pH; pH_o, extracellular pH; PLA_2, phospholipase A_2; RBC, red blood cell; RNA, ribonucleic acid.

4.2.1. *Induction by Temperature Change*

An early study incriminated the fall in temperature that usually accompanies gametocyte entry into the mosquito blood-meal as an inducer of gametogenesis in *Plasmodium* (see Danilewsky, 1889). This idea was later rejected by Ross (1897), and also seemed inconsistent with findings in other Haemosporidia (Haemoproteidae, Leucocytozoidae) in which gamete formation is not inhibited at the body temperature of the host (Marchoux and Chorine, 1932; Roller and Desser, 1973). However, detailed studies *in vitro* on *Plasmodium* by Kliger and Mer (1937), Sinden and Croll (1975) and Sinden (1983b) clearly demonstrated the dynamic relationship between a reduction in temperature and the timing and efficiency of exflagellation *in vitro*. These studies confirmed the earlier observations that a fall of at least 5 °C below the temperature of the vertebrate host was essential to the induction of exflagellation in *Plasmodium*. This obligatory role of a temperature fall has been endorsed recently by Ogwan'g *et al.* (1993b) and by our unpublished observations, the latter studies being conducted both *in vivo* and *in vitro*.

Perhaps the most provocative observations on the role of temperature were made by Kawamoto *et al.* (1991, 1993), who concluded that three of the constituent events of microgametogenesis in *P. berghei* (DNA replication, axoneme assembly, and escape from the host cell) occurred in media with a pH and an ionic composition which is non-permissive for exflagellation and may therefore be triggered by a fall in temperature alone, but that further development leading to exflagellation required an additional stimulus. In recent experiments we were unable to confirm Kawamoto's report (unpublished observations). Earlier observations on the emergence of *P. gallinaceum* gametocytes from the red blood cell *in vitro* also suggested that this process required a drop in temperature but was not independent of other triggers (Bishop and McConnachie, 1960; Nijhout and Carter, 1978). Current evidence therefore suggests that no constituent event of exflagellation investigated so far can be induced by cold shock alone.

4.2.2. *Induction by pH Change*

After initial studies on *Haemoproteus* by Marchoux and Chorine (1932), Chorine (1933) was the first to recognize that an increase in extracellular pH (pH_o) could induce exflagellation of *P. praecox* (= *P. falciparum*) *in vitro*. Bishop (1955) and Bishop *et al.* (1956, 1960), studying *P. gallinaceum*, later substantiated these findings and further noted that the presence of extracellular bicarbonate could be critical for this event. Later studies on *P. gallinaceum* clarified the role of bicarbonate ions and recognized a clear pH optimum (pH 7.9–8.1) for induction *in vitro* (Nijhout and Carter, 1978).

A slightly lower optimum (pH 7.8–8.0) was later reported for the rodent parasite *P. berghei* (Sinden, 1983b). Subsequent studies by Kawamoto *et al.* (1991, 1993) showed that amiloride, a potent inhibitor of Na^+/H^+ exchange, could block exflagellation, suggesting that the drug-induced decrease in intracellular pH, due to the accumulation of hydrogen ions, may be inhibitory. Kawamoto *et al.* (1992) also studied the dependency of exflagellation in *P. berghei* on the extracellular ionic composition, first described by Bishop *et al.* (1956). They proposed the involvement of two amiloride sensitive, electroneutral ion exchangers in the induction process located in the parasite or host cell membranes. Activated by a low pH_o these exchangers could act synergistically and lead to recycling of chloride and bicarbonate, sodium influx and proton efflux, resulting in a rise in pH_i. This hypothesis is also consistent with the identified role of extracellular bicarbonate in the regulation of exflagellation in *P. gallinaceum* and the studies *in vivo* on the inhibition of infectivity in hosts with low serum bicarbonate ion concentrations (Figure 10d).

Although a pH increase at a permissive temperature is frequently used to induce exflagellation *in vitro*, the significance of pH as a stimulus *in vivo* is not at all certain. Micks *et al.* (1948) investigated whether a pH increase in blood ingested by a mosquito can be responsible for the induction of exflagellation *in vivo*. They measured the pH of ingested duck blood from the guts of four mosquito species and also examined the ability of the avian malaria parasite *P. elongatum* to exflagellate in these mosquitoes. No pH increase in the blood was observed upon engorgement by two (permissive) mosquito species in which exflagellation occurred (*C. pipiens* and *C. quinquefasciatus*) or in the non-permissive *Ae. aegypti*, and only a slight increase from 7.65 to 7.75 in the non-permissive *An. quadrimaculatus*. These authors concluded that no correlation existed between exflagellation and pH of the mosquito blood meal *in vivo*. Bishop and McConnachie (1956) found that chicken blood withdrawn directly from a vein had a pH of 7.51 and the pH of *Ae. aegypti* blood-meals ingested from the same chickens was 7.68. In contrast to blood exposed to air, there was no gradual increase of pH in mosquitoes in the hour after engorgement. Using a clinical blood gas analyser, the pH of infectious mouse blood parasitized by *P. berghei*, and infective to mosquitoes was found to be 7.15–7.4 (Figure 10c). Noting that the pH optimum for exflagellation *in vitro* is 7.8–8.0, a substantial and unprecedented pH increase would be required to induce microgamete development *in vivo* in this blood. pH may therefore not be the natural trigger of exflagellation *in vivo*.

4.2.3. Macromolecular Inducers of Exflagellation

Micks *et al.* (1948) were the first to offer evidence that obligate inducers of exflagellation were not demonstrable in all mosquito species. They showed that *P. elongatum* exflagellates in the gut of *Culex* spp. regardless of whether these species were susceptible to infection with the parasite (*C. pipiens*) or not (*C. quinquefasciatus*), but not in the gut of either *Aedes* or *Anopheles*. Subsequently Omar (1968) showed that *P. cynomolgi* did not exflagellate *in vivo* in the midgut of *C. thalassius* but did exflagellate, with different intensities, in *Ae. aegypti* and *Anopheles* spp. Today many laboratories consider that mosquito extracts enhance infection of mosquitoes by *P. falciparum* when using membrane feeders (Naotunne *et al.*, 1993; Vaidya *et al.* 1995) or use them to induce exflagellation *in vitro* (Quakyi *et al.*, 1989). Nijhout (1979) provided solid evidence for the existence of macromolecular inducers of exflagellation by *P. gallinaceum*, but since these were present in the guts of male and female *Aedes* and in *Anopheles* spp. the factor(s) was (were) neither mosquito sex- nor species-specific and may not therefore be responsible for the phenomena described by Micks *et al.* (1948). Nijhout's (1979) data suggested that the inducer termed by her mosquito exflagellation factor (MEF), but which on current evidence is more correctly termed gametocyte activating factor (GAF), was a heat-stable molecule, with molecular mass below 10 kDa. More recent data suggest that this molecule may be as small as 500 Da, organic in nature and containing double-bonded ring structures (Carter and Graves, 1988; Alano and Carter, 1990). It is fundamental to our understanding of the regulation of gametogenesis that the existence and identity of these putative mosquito factors be resolved. However, it is clear that mosquito factors do not transcend the critical role of cold shock in the induction process (unpublished observations), but may act as an essential co-inducer.

When considering a role for the alleged macromolecular inducers of exflagellation, it is essential to recall the intracellular location of the "target" cell. Two classes of inducer could be considered. (i) Small lipophilic molecules that readily pass through the plasmalemma and cytoplasm of the erythrocyte, and the parasitophorous vacuole membrane, and then bind to an intramembranous, cytoplasmic or nuclear receptor in the parasite. This is not inconsistent with the available data, especially if the parasite increases the permeability of the host cell for small anions and uncharged molecules (Gero and Kirk, 1994). (ii) A hydrophilic molecule that binds to a receptor on the parasite plasmalemma; this mechanism, however, would not be possible unless escape from the erythrocyte were independently triggered.

4.2.4. *Intracellular Messenger Systems*

Data collected to date indicate that at least three classical intracellular signalling mechanisms regulate microgametogenesis when induced by temperature shock and elevated pH_o. These are changes in intracellular pH (Kawamoto *et al.*, 1991, 1992, 1993; Kawamoto, 1993), the inositol phosphate/calcium–calmodulin pathway (Kawamoto *et al.*, 1990, 1993; Ogwan'g *et al.*, 1993a; Martin *et al.*, 1994), and protein kinases (Kawamoto *et al.*, 1993). Additional evidence for the roles of the cyclic nucleotides adenosine and guanosine monophosphate (cAMP and cGMP) are less well substantiated (Martin *et al.*, 1978; Kawamoto *et al.*, 1990, 1993). Most data originate from studies investigating the ability of pharmacological agents to induce exflagellation at a non-permissive pH (7.3–7.4) or to exflagellation induced by pH (7.8–8.2) or temperature *in vitro*. The latter are by definition investigating pathways induced by potentially artificial inducers.

In many eukaryotic cells a correlation exists between an increase in cytoplasmic pH (pH_i) and entrance into the M phase of division. This pH change may be mediated by different mechanisms in different organisms (Madshus, 1988). In some metazoan cells cytoplasmic alkalization, which is regulated by the activation of an amiloride-sensitive Na^+/H^+ antiporter by hormone or growth factor, is required for cells to move from G_0 to the G_1/S phases of the cell cycle, and this antiporter depends on a permissive extracellular pH (Pouysségur *et al.*, 1985). The studies of Kawamoto *et al.* (1991, 1992, 1993) are consistent with the regulation of pH_i by both simple and complex amiloride-sensitive ion exchangers in *Plasmodium*. Furthermore the induction of microgametogenesis *in vitro* by a combination of temperature drop and rise in pH_o is entirely consistent with the hypothesis of pH_i being a powerful regulatory element in the induction process. The fact that ammonium, an agent frequently used to increase pH_i independently of pH_o, increases the extent of exflagellation at an otherwise non-permissive pH_o of 7.3 (Kawamoto *et al.*, 1992) also supports this hypothesis. Early embryonic development in invertebrate systems can similarly be induced by an ammonium-mediated increase in pH_i (Shen and Steinhardt, 1978; Madshus, 1988) and it is assumed that induction by ammonium ions bypasses natural signalling pathways (e.g., sperm–receptor mediated induction). Similar artefacts cannot be excluded in the malarial studies described here.

Sequential rises in concentration of intermediates of the inositol pathway are found in activated gametocytes (Ogwan'g *et al.*, 1993a). Compounds that reportedly inhibit inositol triphosphate degradation may stimulate exflagellation (Ogwan'g *et al.*, 1993a), although the methods as presented are not convincing. In *Plasmodium* as in other systems, inositol triphosphate is produced by phosphoinositide-specific phospholipase C (Elabbadi *et al.*, 1994). Inhibitors of phospholipases (which Ogwan'g *et al.* (1993a) suggested could include quinine and chloroquine) reduce exflagellation. In recent experiments we were able to confirm the inhibitory effect of different phospholipase C inhibitors (2-nitro-4-carboxyphenyl-N,N-diphenylcarbonate;

1-{6-[(β-3-methoxyestra-1,3,5(10)-trien-17-yl)amino]hexyl}-1H-pyrrole-2,5-dione) on exflagellation (unpublished observations). These results suggest that inositol triphosphate is a key molecule in the regulation of microgametocyte exflagellation. Inositol triphosphate has a similar key position in the regulation of flagellar excision in *Chlamydomonas* (see Quarmby *et al.*, 1992). Heparin, which has long been known to inhibit exflagellation (R.E. Sinden, unpublished observations), was reported by Ogwan'g *et al.* (1993a) to be able to block the binding of inositol triphosphate to its receptor on the intracellular calcium stores (endoplasmic reticulum and mitochondria), though how such a large molecule could reach this target was not explained. Kawamoto *et al.* (1990, 1992, 1993) none the less showed that inhibition of calcium release from internal stores, or calcium–calmodulin binding, similarly block exflagellation and that extracellular calcium is not required. Based on the well-established regulatory pathways in other eukaryotic cells, it is appropriate to speculate that calcium release from internal stores induced by inositol triphosphate might regulate both exocytosis of the osmiophilic bodies and some of the numerous phosphorylation mechanisms through a range of calmodulin-dependent kinases that in turn regulate cytoskeletal assembly and DNA replication.

Less evidence exists for a possible role of diacyl glycerol, another product of phospholipase C activity and an activator of protein kinase C. Diacyl glycerol levels reportedly increase in activated gametocytes (Ogwan'g *et al.*, 1993a) and its analogue 1-oleoyl-2-acetylglycerol is a weak inducer of exflagellation in *P. falciparum* (see Martin *et al.*, 1994). However, a phorbol ester and potent activator of protein kinase C did not replace the pH_o stimulus in exflagellation of *P. berghei in vitro* (Kawamoto *et al.*, 1990), indicating that protein kinase C activation alone is not sufficient to mediate the process at a permissive temperature. However, the inhibition of exflagellation by staurosporin reported in the same paper suggests that protein kinases (probably protein kinase C) may be required.

Phospholipase A activity of unknown function has been detected in merozoites (Braun-Breton *et al.*, 1992a) and it has been suggested that phospholipase A_2 and arachidonic acid metabolism may be involved in mechanisms leading to exflagellation (Ogwan'g *et al.*, 1993a; Martin *et al.*, 1994). We found that phospholipase A_2 activity may be required, but that arachidonic acid is not involved (O. Billker, unpublished observations).

When discussing the role of phospholipases in exflagellation it should be recalled that phospholipase C activity may also be required in merozoite invasion, another event in the parasite's life cycle that involves modification of the erythrocyte membrane. In this process activation of a glycosyl-phosphatidylinositol(GPI)-specific phospholipase C may be required to solubilize a GPI-anchored serine protease (Braun-Breton *et al.*, 1992a), which can degrade erythrocyte surface proteins and is required for the invasion process (Braun-Breton *et al.*, 1992b). It is tempting to speculate that proteases may also be involved in the host cell lysis required for gametogenesis and that the role of phospholipases may not be restricted to signal transduction but may include the solubilization of parasite proteins and, indirectly, erythrocyte membrane destruction.

The role of cyclic nucleotides in the regulation of exflagellation is far from clear. Early studies by Martin *et al.* (1978), which suggested that cAMP may play a key regulatory role, were questioned by Kawamoto *et al.* (1990), who speculated that the data from both studies were more consistent with regulation mediated by cGMP, and presented evidence for the involvement of cGMP-dependent kinases or protein kinase C. The mechanisms by which either pathway might be induced are at present totally unknown.

The evidence discussed above is summarized in Figure 14. Any mechanism for the primary induction of gametogenesis in malarial parasites must explain the apparent roles of lowered temperature and mosquito species-limited and species-transcending factors, and the lack of dependence on mosquito factors under conditions of raised pH *in vitro* (Figure 14A). The inducers regulate parallel but separate cellular events of RNA and protein synthesis, DNA replication, nuclear division, cytoskeletal assembly, and escape from the host erythrocyte (Figure 14C). Further investigations are required to link any of the secondary intracellular signalling pathways (Figure 14B) both to the specific triggers and to the constituent events of gametogenesis.

Recognizing the intracellular location of the mature gametocyte, it is difficult to envisage induction pathways that are dependent on transmembrane receptors in the parasite plasmalemma. Such receptors are often linked via G proteins or tyrosine kinase activity to phospholipase C and the inositol pathway. A *ras*-like GTP-binding protein and the α subunit of an oligomeric G protein have been reported to exist in asexual stages of *Plasmodium* by Dontfraid and Chakrabarti (1994) and Thelu *et al.* (1994). Whether, and how, gamete activating factor binds *in vivo* to such a receptor needs to be determined.

Both gamete activating factor *in vivo*, and the rise in pH_o *in vitro*, require the co-stimulus of temperature shock, possibly acting through heat shock proteins, some of which are known to be encoded in the malarial genome (Bianco *et al.*, 1988; Kumar *et al.*, 1991; Tsuji *et al.*, 1994). Even in the absence of any data it is tempting to suggest that these may be key regulators of exflagellation (together with products of the inositol pathway or intracellular pH). Heat shock proteins reportedly regulate replication, secretion, cytoskeletal function and protein phosphorylation in many cell systems (Georgopoulos and Welch, 1993). Clearly the role of heat shock proteins in the induction of exflagellation is worthy of careful examination.

However, it should also be pointed out that the role of temperature in gametogenesis of the malaria parasites of poikilothermic hosts (e.g., lizards) has not been studied and may present some interesting dilemmas.

The obvious complexity of the intracellular regulatory pathways of microgametogenesis in *Plasmodium*, combined with the simplicity with which it can be manipulated both *in vivo* and *in vitro*, suggest that it may be an area well worth studying both as a model system for cell signalling and to understand the natural infectivity of gametocyte carriers to the mosquito vector. Such studies could lead to the possible development of directed chemotherapeutic intervention measures. Ogwan'g *et al.* (1993a) have already suggested roles for quinine and chloroquine in the inhibition of the conversion of phosphatidylinositol 4,5-bis-phosphate to inositol 1,4,5-tris-phosphate (however, these compounds do not block infection of the vector *in vivo*). More recently, Jones *et al.* (1994) have shown that azadirachtin, a natural limnoid in extracts of the plant *neem* (*Azadirachta indica*), is capable of blocking some events in the exflagellation process *in vitro*, yet the molecule is not toxic to either vertebrate or mosquito hosts. Perhaps some of the unique events of exflagellation could be targets for new and selectively toxic molecules that will block malarial transmission.

5. FERTILIZATION

5.1 The Mosquito Blood-meal

The size of a full blood-meal (and therefore the number of gametocytes ingested) is correlated with the size of the mosquito (Kitthawee *et al.*, 1990). The relationship between mosquito body size and the intensity of an oocyst infection is, however, less clear; some authors (Kitthawee *et al.*, 1990; Lyimo and Koella, 1992) described a positive correlation between these characters, others (Wing *et al.*, 1985; Kittayapong *et al.*, 1992) failed to observe any correlation. Anopheline mosquitoes take blood-meals that range in volume from one to 3 µl. Using *P. falciparum* gametocyte cultures, Vaughan *et al.* (1992) showed that blood-meals in fully engorged *A. gambiae* contained between 812 and 26 115 macrogametocytes. Gametocyte rates for *P. falciparum in vivo* vary enormously but infectious individuals have been described with gametocyte densities as low as $10 \ \mu l^{-1}$, i.e., a fully fed mosquito would ingest 10–30 gametocytes of which 8–24 would probably be female. If it is assumed that such a feed produced the minimal infection of one oocyst, the effiency of gametocyte to oocyst conversion would be 1/8–1/24 — i.e., $K = 0.9$–1.38. Much lower efficiencies are, however, normally recorded (see Vaughan *et al.*, 1992).

Irrespective of the interspecific differences in size and concentration of the full blood-meal (Table 2), the absolute range in blood-meal size within any group of mosquitoes is very large, due in part to the time available for each mosquito to engorge. What is abundantly clear to all malaria workers is that the probability of infection is very low in mosquitoes that fail to take a full blood-meal. Therefore it must be concluded that either a fully blood-fed mosquito provides some stimulus to the parasite that enhances the probability of infection (e.g., a stimulatory factor such as gamete activating factor is secreted, thereby promoting efficient exflagellation), or partially fed mosquitoes provide a more hostile environment, decreasing the already slight probability of parasite survival. Other immediate effects during feeding include the mechanical haemolysis of a significant proportion of the erythrocytes by the cibarial armature in the foregut (2.3% in *A. albimanus*, 12.8% in *A. gambiae* and 21.9% in *A. stephensi*). Certainly all malarial parasites, not just gametocytes, rapidly become extracellular in the blood-meal. This lysis of infected blood cells will result in a massive release of haemoglobin and parasite products including haemazoin, and the impact of these changes on the activity of the peripheral blood mononuclear cells in the blood-meal remains largely unexplored.

5.2. Efficiency of Fertilization

Microgametes of *Plasmodium* spp. are motile *in vitro* for about 40 minutes. Fertilization is usually completed within the first hour of development in the mosquito midgut (Sinden and Croll, 1975; Sinden, 1983b; Carter and Graves, 1988). Microgametes fail to display any tactile response to the

Table 2 Abilities of *Anopheles* spp. to concentrate the blood-meal and the corresponding production of *P. berghei* ookinetes and oocysts (from Vaughan *et al.*, 1991).

	Concentration factor for human blood	Density of	
		Ookinetes	Oocysts
A. gambiae	1.8		
A. stephensi	1.7	low	low
A. arabiensis	1.4		
A. dinus	1.2		
A. freeborni	1.0	higher	higher
A. albimanus	< 1.0[a]	highest	none

[a] Due to haemolysis.

macrogamete (Sinden and Croll, 1975). In marked contrast, blood phagocytes rapidly "home in" on the extracellular gametocyte, demonstrating
the clear presence of chemotactic gradients (Sinden and Smalley, 1976).
Irrespective of whether microgametes are attracted to macrogametes or
not, one might predict that the efficiency of fertilization is density dependent. Janse *et al.* (1985) and Sinden *et al.* (1985) somewhat surprisingly
observed increasing ookinete numbers in cultures with increasing dilution
of the blood down to a haematocrit of *c.* 2%. Below 1% haematocrit the
ookinete number fell again. This, we suggest, was due to an interaction of
two factors — an inhibitory factor (see above) which is diluted out (from
40% down to 2% haematocrit), and a density dependent fall in fertilization
below 2% haematocrit. Absolute levels of fertilization (macrogamete to
ookinete conversion *in vitro*) were variously recorded as 11–44% in
optimally diluted cultures. Efficiencies *in vivo* are much more difficult to
assess, best estimates from Vaughan *et al.* (1992, 1994) being a 40- to 316-
fold reduction in parasite number between macrogamete and ookinete, i.e.
0.3–2.5% efficiency. These values are, however, not directly comparable
with those obtained *in vitro* because of the lack of controls for the presence
of natural inhibitors of exflagellation (see Sections 3.3.1 and 3.3.2).

Fertilization clearly requires a cell–cell recognition event; however, the
presence of male and female specific ligands has not yet been demonstrated. Indeed, microscopical evidence shows that large masses of mixed
activated extracellular microgametocytes and macrogametes will bind
tightly together (Yoeli and Upmanis, 1968; Sinden *et al.*, 1985). Binding
of apparently equal affinity therefore occurs both within and between
sexes.

At fertilization, membranes of the macro- and microgametocytes become
very closely apposed and ultimately fuse. Although the microgamete is
structurally polarized, there is at present no evidence of any acrosomal-like
reaction in malaria. Fusion of homogametic pairs has not been described;
however, in *P. berghei* very large "macrogametes" are found in culture:
could these be fused females? Following binding of the microgamete to the
macrogamete, the male axoneme swims into the cytoplasm of the female
taking the nucleus with it. Following fertilization, rapid depolarization of
the macrogamete plasmalemma, or slower changes in membrane composition, could be expected as mechanisms to block polyspermy, although
neither mechanism has yet been reported in *Plasmodium*.

6. POST-FERTILIZATION DEVELOPMENT

6.1. Development of the Zygote

Following fertilization the macrogamete transforms into a zygote. The initial event is the fusion of the male and female nuclei which is often achieved within 3 hours of blood-meal ingestion. Characteristic intracellular changes occur with the formation of perinuclear microtubules from the centriole-like microtubule organizing centre, which additionally may co-ordinate the assembly of the subpellicular microtubules and subpellicular membranes (as it does in the formation of the merozoite and sporozoite). Assembly of the apical complex, which occurs 5–9 hours after gamete induction, is the key event in formation of the retort-form ookinete (Canning and Sinden, 1973; Davies, 1974; Mehlhorn *et al.*, 1980; Kumar *et al.*, 1985; Sinden *et al.*, 1985) and coincides with the first and second meiotic divisions of the diploid genome (Sinden and Hartley, 1985; Sinden *et al.*, 1985).

For largely technical reasons, few rapid changes have been detected in the composition of the fertilized cell. For the first 3–5 hours the macrogamete and zygote are sensitive to complement from insusceptible hosts, but resistant to that from susceptible hosts. This resistance is lost if the parasites are subject to protease treatment, suggesting a critical role for surface proteins (Grotendorst *et al.*, 1986; Grotendorst and Carter, 1987). Susceptibility to the alternative pathway of complement (lysis) and inhibition of infectivity by heterologous complement reaches a peak at 7–12 hours but then declines with increasing maturation of the ookinete. Complement activity in the mosquito gut coincidentally declines significantly after 3–5 hours (M. Fallon, unpublished observations).

6.2. Fertilization as a Developmental Trigger

In the vast majority of eukaryotic cells fertilization induces a burst in protein synthesis, which is dramatically expanded later when the zygote genome is activated. Maturation of the malarial zygote is accompanied by numerous changes in the composition of the cell surface (see Table 3), the greatest change — the loss of all proteins of $M_r > 55$ — occurs some 10 hours after induction and appears to be fertilization dependent. This, it will be recalled, is the time at which zygote infectivity falls to its lowest value, and complement resistance is lost. It is perhaps surprising that none of the proteins identified on the ensuing ookinete surface has yet been shown to be synthesized as a direct response to fertilization (Kaushal *et al.*, 1983;

Carter and Kaushal, 1984; Kaushal and Carter, 1984; Kumar and Carter, 1985; Sinden *et al.*, 1985; Paton *et al.*, 1993); indeed the one ookinete surface molecule studied in detail (Pbs21) has been shown to be induced by gametogenesis, not fertilization (Paton *et al.* 1993; Sinden, 1994).

6.3. Ookinete Formation and Biology

Observations on the biology and ultrastructural organization of ookinete formation have been reviewed on numerous occasions (Canning and Sinden, 1973; Davies, 1974; Mehlhorn *et al.*, 1980; Kumar *et al.*, 1985; Sinden *et al.*, 1985) and will not be detailed here. The critical physiological role of the ookinete is the escape from the increasingly hostile environment of the blood-meal by invasion of the mosquito tissues. Invasion is largely mediated by the apical complex and the plasmalemma. The plasmalemma is responsible for the interaction of the parasite with mosquito tissues; thereafter the parasite cytoskeleton is the motor for ookinete motility (King, 1981; Sinden, 1982; Russell, 1983), and the rhoptry/microneme complex, which is very extensive, may be expected to induce the formation of a parasitophorous vacuole in the midgut epithelial cell. An organelle unique to the ookinete is the crystalloid, which, without good supporting evidence, has been termed an energy reserve.

Another critical and unprecedented change occurs in the ribosome population: the ribosomal RNA molecules switch from the "asexual" A type to the "mosquito" O/C/D type within the first 48 hours of sporogonic development. The biological advantage of this significant change is as yet unclear, and its impact on the regulation of protein synthesis (and hence gene expression) remains at present completely unknown (Gunderson *et al.*, 1987; Waters *et al.*, 1989; Li *et al.*, 1994; McCutchan *et al.*, 1995).

Survival of the ookinete (and other parasite stages) in the mosquito blood-meal is reportedly not constant throughout the meal, parasites at the periphery being more rapidly degraded. This has been attributed to a combination of two factors: first the parasite is susceptible to the proteolytic enzymes in the mosquito midgut (Gass, 1977; Gass and Yeates, 1979); and second, these enzymes are found at much higher concentrations in the periphery of the blood-meal (Billingsley, 1990; Feldmann *et al.*, 1990). Thus the site of development of the non-motile retort-form parasite within the blood-meal could be a significant factor in its survival.

Whilst proteolytic enzymes are without question potentially lethal to the parasite, recent studies have shown that they also play an obligatory role in successful infection of *Ae. aegypti* by *P. gallinaceum* (see Shahabuddin *et al.*, 1993, 1995). As part of their defence mechanisms against pathogens, mosquitoes secrete a chitinous structure within the gut, the peritrophic

Table 3 Proteins found on gametes and ookinetes of *P. gallinaceum* (from Howard *et al.*, 1982; Kumar and Carter, 1983; Carter and Kaushal, 1984; Kaushal and Carter, 1984) and *P. berghei* (from Winger *et al.*, 1988; Sinden *et al.*, 1987; Dearsley, 1990).

M_r	Name[a]	Location
P. gallinaceum		
260/250	Pgs1[b]	Male and female gametocytes and gametes
245	Pgs2[b]	–
225/215	Pgs3[b]	Male and female gametocytes and gametes
205	Pgs4	Female (cross-reactive antibodies bind to male)
200	Pgs5[b]	–
175	Pgs6[b]	–
155	Pgs7[b]	–
125	Pgs8[b]	–
116	Pgs9[b]	–
105	Pgs10	Ookinete
83	Pgs11	Female (cross reactive antibodies bind to male)
70	Pgs12[b]	–
59/56	Pgs13a	Female gamete–ookinete
54	Pgs13b[b]	–
52	PgO3	Late zygote and ookinete
48	Pgs14	Male and female gametocytes–ookinete
42	Pgs15	Late zygote
28	PgO1	
26	PgO2 (=Pgs16)	Late zygote and ookinete
19	Pgs17a	Zygote and ookinete
17	Pgs17b	Female gamete–ookinete
16	Pgs18	Female gamete–ookinete
P. berghei		
–	Pbs105	Internal
–	Pbs89	Surface, gamete–ookinete
–	Pbs87	Surface, gamete–ookinete
–	Pbs75	Surface, gamete–ookinete
–	Pbs71	Internal, gamete–ookinete
–	Pbs52	Surface, gamete–ookinete
–	Pbs43	Surface, gamete–ookinete
–	Pbs25	Surface, gamete–ookinete
–	Pbs21	Surface, gamete–ookinete
–	Pbs16	Internal, gamete–ookinete

[a] Pgs = *Plasmodium gallinaceum* sexual stage protein; PgO = *Plasmodium gallinaceum* ookinete stage protein; Pbs = *Plasmodium berghei* sexual stage protein.
[b] Shed by macrogametes in response to fertilization.

layer. This layer is more compact in *Aedes* than *Anopheles* (see Billingsley and Rudin, 1992; Rudin *et al.*, 1991). *Plasmodium* secretes an inactive prochitinase enzyme, which is then cleaved by the mosquito proteases to produce an active chitinase which digests the peritrophic layer. Proposals have been made to produce antibodies that inactivate these enzymes, which might act as transmission blocking antibodies (Shahabuddin and Kaslow, 1994a, 1994b; Shahabuddin, 1995). An observation of great practical use is the considerable enhancement of the success of oocyst infections when mosquitoes are fed with zygotes/ookinetes rather than gametocytes in membrane feeders (Rosenberg *et al.*, 1982; R.E. Sinden *et al.*, unpublished observations). However, it is not known whether this success is more influenced by the relative efficiencies of fertilization *in vitro* and *in vivo* than by the very rapid invasion of the midgut wall when zygotes/ookinetes are fed, before peritrophic layer formation and protease secretion.

Having successfully crossed the peritrophic layer the ookinete next meets the midgut epithelium plasma membrane. Despite much logical speculation that transmission blocking antibodies directed against the ookinete surface proteins Pfs25 or Pbs21 block the interaction of the molecules with the epithelium, there is no direct evidence that these molecules are involved in a receptor–ligand reaction. Nevertheless, specific recognition events are likely to be required to initiate the invasion process and may include protein–protein or lectin–carbohydrate interactions (Billingsley, 1994).

There is considerable debate about the route by which the ookinete crosses the midgut epithelium. Evidence for both intracellular and intercellular routes has been presented (Garnham *et al.*, 1962; Canning and Sinden, 1973; Mehlhorn *et al.*, 1980; Meis *et al.*, 1989). The most compelling evidence is for an intracellular route.

Within the midgut epithelium the ookinete is susceptible to attack by the mosquito (Weathersby, 1952). Recent studies have identified two different mechanisms by which a refractory mosquito phenotype may be produced. Collins *et al.* (1986) described ookinete melanization by the phenol oxidase/dopa/dopamine pathway. Interestingly, both refractory and susceptible lines displayed phenol oxidase in the same tissue locations, namely the basal membrane labyrinth and the basal lamina, in uninfected midguts. However, after an infective blood-meal the susceptible phenotype (only) showed reduced phenol oxidase activity (Paskewitz *et al.*, 1988, 1989). In the refractory strain, melanotic "capsules" formed around the ookinetes which were usually found in the extracellular space between the basal membrane labyrinth and the basal lamina. It is possible that intracellular parasites (perhaps in parasitophorous vacuoles) are also recognized by this process. More recently, Vernick *et al.* (1995) have described a "recessive" refractory gene that regulates a lytic mechanism which destroys ookinetes within the midgut cell. Clearly, an interesting extension of both these observations would be to select genetically refractory lines that could be used to replace susceptible vector populations in endemic areas. Laboratory data suggest that infection with *Plasmodium* significantly reduces both the fecundity and flight performance of a mosquito (Schiefer *et al.*, 1977; Hurd, 1994). However, both the melanizing and lytic refractory pheno-types are naturally occurring "mutants", and both are in a minority in nature. This therefore suggests that no significant advantage accrues to the mosquito from the refractory phenotype *per se*. A major dilemma concern-ing this interesting strategy for malaria control will therefore be to link a biological advantage to the loci for the refractory gene so that it will be driven into a susceptible population.

Having traversed the mosquito midgut epithelium the ookinete meets the basal lamina, which it does not cross (Garnham *et al.*, 1962; Freyvogel, 1966; Sinden, 1978). The insect basal lamina, which has been largely characterized (Fessler and Fessler, 1989), therefore offers either an impenetrable barrier to the ookinete or a specific trigger to stop both its migratory and invasive behaviour and the initiation of its transformation into a vegetative and sessile oocyst. Evidence that the extracellular matrix may act as a facultative stimulus (inducer?) of ookinete-to-oocyst trans-formation has been provided by the studies of Warburg and Miller (1991, 1992) and Warburg and Schneider (1993) on cultures of the sporogonic stages of *P. gallinaceum* and *P. falciparum in vitro*. Despite being sur-rounded by the basal lamina and the basement membrane, the parasite is not yet safe, and the mosquito can still react to the oocyst by melanization, and numerous oocysts may die due to as yet unknown events (Sinden and Garnham, 1973; Sinden, 1987) — but that is another story.

ACKNOWLEDGEMENTS

We thank our colleagues Drs P.F. Billingsley, R. Alejo Blanco, and R. Carter for their permission to quote unpublished data. We are grateful to the European Union, World Health Organization, Medical Research Council, Leverhulme Trust, and Stanley Thomas Johnson Foundation for financial support.

REFERENCES

Aikawa, M., Huff, C.G. and Sprinz, H. (1969). Comparative fine structure study of the gametocytes of avian, reptilian and mammalian malaria parasites. *Journal of Ultrastructure Research* **26**, 316–331.
Aikawa, M., Carter, R., Ito, Y. and Nijhout, M. (1984). New observations on

gametogenesis, fertilization, and zygote transformation in *Plasmodium gallina-ceum. Journal of Protozoology* **31**, 403–413.

Alano, P. (1991). *Plasmodium* sexual stage antigens. *Parasitology Today* **7**, 199–203.

Alano, P. and Carter, R. (1990). Sexual differentiation in malaria parasites. *Annual Review of Microbiology* **44**, 429–449.

Archer, S.L. and Hampl, V. (1992). N^G-monomethyl-L-arginine causes nitric oxide synthesis in isolated arterial rings: trouble in paradise. *Biochemical and Biophysical Research Communications* **188**, 590–596.

Arrada-Meyer, M. De, Cochrane, A.H. and Nussenzweig, R.S. (1979). Enhancement of a simian malarial infection (*Plasmodium cynomolgi*) in mosquitoes fed on rhesus (*Macaca mulatta*) previously infected with an unrelated malaria (*Plasmodium knowlesi*). *American Journal of Tropical Medicine and Hygiene* **28**, 627–633.

Ayala, S.C. (1973). The phlebotomine sandfly–protozoan parasite community of central California grasslands. *American Midland Naturalist* **89**, 266–280.

Bastien, P., Landau, I. and Baccam, D. (1987). Inhibition of infectivity of *Plasmodium* gametocytes by serum of the infected host: setting up an experimental model. *Annales de Parasitologie Humaine et Comparée* **62**, 195–208.

Bate, C.A.W. and Kwiatkowski, D. (1994). Inhibitory immunoglobulin M antibodies to tumour necrosis factor-inducing toxins in patients with malaria. *Infection and Immunity* **62**, 3086–3091.

Bate, C.A.W., Taverne, J. and Playfair, J.H.L. (1989). Soluble malarial antigens are toxic and induce the production of tumour necrosis factor *in vivo. Immunology* **66**, 600–605.

Bate, C.A.W., Taverne, J., Davé, A. and Playfair, J.H.L. (1990). Malaria exoantigens induce T-independent antibody that blocks their ability to induce tumour necrosis factor. *Immunology* **70**, 315–320.

Bianco, A.E., Crewther, P.E., Coppel, R.L., Stahl, H.D., Kemp, D.J., Anders, R.F. and Brown, G.V. (1988). Patterns of antigen expression in asexual stages and gametocytes of *Plasmodium falciparum. American Journal of Tropical Medicine and Hygiene* **38**, 258–267.

Billingsley, P.F. (1990). Blood digestion in the mosquito, *Anopheles stephensi* Liston (Diptera, Culicidae) — partial characterization and post-feeding activity of midgut aminopeptidases. *Archives of Insect Biochemistry and Physiology* **15**, 149–163.

Billingsley, P.F. (1994) Vector–parasite interactions for vaccine development. *International Journal for Parasitology* **24**, 53–58.

Billingsley, P.F. and Rudin, W. (1992). The role of the mosquito peritrophic membrane in bloodmeal digestion and infectivity of *Plasmodium* species. *Journal of Parasitology* **78**, 430–440.

Birago, C., Bucci, A., Dore, E., Frontali, C. and Zenobi, P. (1982). Mosquito infectivity is directly related to the proportion of repetitive DNA in *Plasmodium berghei. Molecular and Biochemical Parasitology* **6**, 1–12.

Bishop, A. (1955). Problems concerned with gametogenesis in Haemosporidiidae, with particular reference to the genus *Plasmodium. Parasitology* **45**, 163–185.

Bishop, A. and McConnachie, E.W. (1956). A study of the factors affecting the emergence of the gametocytes of *Plasmodium gallinaceum* from the erythrocytes and the exflagellation of the male gametocytes. *Parasitology* **46**, 192–215.

Bishop, A. and McConnachie, E.W. (1960). Further observations on the *in vitro* development of the gametocytes of *Plasmodium gallinaceum*. *Parasitology* **50**, 431–448.

Boyd, M.F. (1949). Epidemiology of malaria: factors related to the intermediate host. In: *Malariology* (M.F. Boyd, ed.). Vol. 1, pp. 551–607. Philadelphia: W.B. Saunders.

Braun-Breton, C., Blisnick, T., Barbot, P., Bulow, R., Pereira Da Silva, L. and Langsley, G. (1992a). *Plasmodium falciparum* and *Plasmodium chabaudi*: characterization of glycosylphosphatidylinositol-degrading activities. *Experimental Parasitology* **74**, 452–462.

Braun-Breton, C., Blisnick, T., Jouin, H., Barale, J.C., Rabilloud, T., Langsley, G. and Pereira Da Silva, H.L. (1992b). *Plasmodium chabaudi* p68 serine protease activity required for merozoite entry into mouse erythrocytes. *Proceedings of the National Academy of Sciences of the USA* **89**, 9647–9651.

Bray, R.S., McCrae, A.W.R. and Smalley, M.E. (1976). Lack of a circadian rhythm in the ability of the gametocytes of *Plasmodium falciparum* to infect *Anopheles gambiae*. *International Journal for Parasitology* **6**, 399–401.

Briegel, H. and Lea, A.O. (1975). Relationship between protein and proteolytic activity in the midgut of mosquitoes. *Journal of Insect Physiology* **21**, 1597–1604.

Bruce, M.C., Alano, P., Duthie, S. and Carter, R. (1990). Commitment of the malaria parasite *Plasmodium falciparum* to sexual and asexual development. *Parasitology* **100**, 191–200.

Buffinton, G.D., Hunt, N.H., Cowden, W.B., Butcher, G.A. and Clark, I.A. (1991). Lipid peroxidation processes in the immunopathology of malaria. In: *Membrane Lipid Oxidation* (C. Vigo-Pelfrey, ed.). Vol. 3, pp 45–68. Boca Raton: CRC Press.

Bullough, B.A., Hughson, F.M., Skehel, J.J. and Wiley, D.C. (1994). Structure of influenza haemagglutinin at the pH of membrane fusion. *Nature* **371**, 37–43.

Butcher, G.A. (1989). Mechanisms of immunity to malaria and the possibilities of a blood stage vaccine: a critical appraisal. *Parasitology* **98**, 315–327.

Butcher, G.A., Mitchell, G.H. and Cohen, S. (1978). Antibody-mediated mechanisms of immunity to malaria induced by vaccination with *Plasmodium knowlesi* merozoites. *Immunology* **34**, 77–86.

Butcher, G.A., Garland, T., Ajdukiewicz, A.B. and Clark I.A. (1990). Serum tumour necrosis factor associated with malaria in patients in the Solomon Islands. *Transactions of the Royal Society of Tropical Medicine and Hygiene* **84**, 658–661.

Butcher, G.A., Carr, R.E. and Fleck, S.L. (1991). The anti-malarial activity of mouse tumour necrosis serum is blocked by purines. *Annals of Tropical Medicine and Parasitology* **85**, 271–273.

Canning, E.U. and Sinden, R.E. (1973). The organization of the ookinete and observations on nuclear division in oocysts of *Plasmodium berghei*. *Parasitology* **67**, 29–40.

Cantrell, W. and Jordan, H.B. (1946). Changes in the infectiousness of gametocytes during the course of *Plasmodium gallinaceum* infections. *Journal of Infectious Diseases* **78**, 53–159.

Carter, E.H. (1983). *Studies on the gametocyes of* Plasmodium falciparum in vitro. Thesis for the Diploma of Imperial College, London.

Carter, R. and Graves, P.M. (1988). Gametocytes. In: Malaria: *Principles and Practice of Malariology* (W.H. Wernsdorfer and I.A. McGregor, eds). Vol. 1, pp. 253–306. Edinburgh: Churchill Livingstone.

Carter, R. and Kaushal, D.C. (1984). Characterization of antigens on mosquito midgut stages of *Plasmodium gallinaceum*. III. Changes in zygote surface proteins during transformation into mature ookinete. *Molecular and Biochemical Parasitology* **13**, 235–241.

Carter, R. and Miller, L.H. (1979). Evidence for environmental modulation of gametocytogenesis in *Plasmodium falciparum* in continuous culture. *Bulletin of the World Health Organization* **57**, 37–52.

Carter, R. and Nijhout, M.M. (1977). Control of gamete formation (exflagellation) in malaria parasites. *Science* **195**, 407–409.

Carter, R., Kumar, N., Quakyi, I., Good, M., Mendis, K., Graves, P. and Miller, L. (1988). Immunity to sexual stages of malaria parasites. *Progress in Allergy* **41**, 193–214.

Carter, R., Graves, P.M., Creasey, A., Byrne, K., Read, D., Alano, P. and Fenton, B. (1989a). *Plasmodium falciparum*: an abundant stage-specific protein expressed during early gametocyte development. *Experimental Parasitology* **69**, 140–149.

Carter, R., Graves, P.M., Quakyi, I. and Good, M.F. (1989b). Restricted or absent immune responses in human populations to *Plasmodium falciparum* gamete antigens that are targets of malaria transmission-blocking antibodies. *Journal of Experimental Medicine* **169**, 135–147.

Chaiyaroj, S.C., Thompson, J.K., Coppel, R.L. and Brown, G.V. (1994). Gametocytogenesis occurs in *Plasmodium falciparum* isolates carrying a chromosome 9 deletion. *Molecular and Biochemical Parasitology* **63**, 163–165.

Chorine, V. (1933). Conditions qui régissent la fécondation de *Plasmodium praecox*. *Archives de l'Institut Pasteur d'Algérie* **11**, 1–8.

Clark, I.A. (1978). Does endotoxin cause both the disease and parasite death in acute malaria and babesiosis? *Lancet* **ii**, 75–77.

Clark, I.A. (1987). Cell mediated immunity in protection and pathology of malaria. *Parasitology Today* **3**, 300–305.

Clark, I.A., Butcher, G.A., Buffinton, G.D., Hunt, N.H. and Cowden, W.B. (1987a). Toxicity of certain products of lipid peroxidation to the human malaria parasite *Plasmodium falciparum*. *Biochemical Pharmacology* **36**, 543–546.

Clark, I.A., Hunt, N.H., Butcher, G.A. and Cowden, W.B. (1987b). Inhibition of murine malaria (*Plasmodium chabaudi*) *in vivo* by recombinant interferon or tumour necrosis factor, and its enhancement by butylated hydroxyanisole. *Journal of Immunology* **139**, 3493–3496.

Collins, F.H., Sakai, R.K., Vernick, K.D., Paskewitz, S., Seeley, D.C., Miller, L.H., Collins, W.E., Campbell, C.C. and Gwadz, R.W. (1986). Genetic selection of a *Plasmodium* refractory strain of the malaria vector *Anopheles gambiae*. *Science* **234**, 607–610.

Collins, W.E., Skinner, J.C., Richardson, B.B. and Stanfill, P.S. (1975). Studies on the transmission of simian malaria VI. Mosquito infection and sporozoite transmission of *Plasmodium fragile*. *Journal of Parasitology* **61**, 718–721.

Crawford, R.M., Leilby, D.A., Green, S.J., Nacy, C.A., Forter, A.H. and Meltzer, M.S. (1994). Macrophage activation: a riddle of immunological resistance. In *Macrophage–Pathogen Interactions* (B.S. Zwilling and T.K. Eisenstein, eds), pp. 29–46. New York: Marcel Dekker.

Danilewsky, B. (1889). *La Parasitologie Comparée du Sang. I. Nouvelles Recherches sur les Parasites du Sang des Oiseaux*. Kharkoff.

Davies, E.E. (1974). Ultrastructural studies on the early ookinete stage of *Plasmo-*

dium berghei nigeriensis and its transformation into an oocyst. *Annals of Tropical Medicine and Parasitology* **68**, 283–290.

Day, K.P., Karamalis, F., Thompson, J., Barnes, D.A., Peterson, C., Brown, H., Brown, G.V. and Kemp, D.J. (1993). Genes necessary for expression of a virulence determinant and for transmission of *Plasmodium falciparum* are located on a 0.3-megabase region of chromosome-9. *Proceedings of the National Academy of Sciences of the USA* **90**, 8292–8296.

Dearsly, A.L. (1990). *Sexual and sporogonic development of* Plasmodium berghei. Ph.D. thesis, University of London.

Dearsly, A.L., Sinden, R.E. and Self, I.A. (1990). Sexual development in malarial parasites — gametocyte production, fertility and infectivity to the mosquito vector. *Parasitology* **100**, 359–368.

De Maria, R., Cifone, M.G., Trotta, R., Rippo, M.R., Festuccia, C., Santoni, A. and Test, R. (1994). Triggering of human monocyte activation through CD69, a member of the natural killer cell gene complex family of signal transducing receptors. *Journal of Experimental Medicine* **180**, 1999–2004.

Dontfraid, F.F. and Chakrabarti, D. (1994). Cloning and expression of a cDNA encoding the homologue of Ran/TC4 GTP-binding protein from *Plasmodium falciparum*. *Biochemical and Biophysical Research Communications* **201**, 423–429.

Eaton, J.W., Eckman, J.R., Berger, E. and Jacob, H.S. (1976). Suppression of malaria infection by oxidant-sensitive erythrocytes. *Nature* **264**, 758–760.

Elabbadi, N., Ancelin, M.L. and Vial, H.J. (1994). Characterization of phosphatidylinositol synthase and evidence of a polyphosphoinositide cycle in *Plasmodium*-infected erythrocytes. *Molecular and Biochemical Parasitology* **63**, 179–192.

Eyles, D.E. (1952). Studies on *Plasmodium gallinaceum*. III. Factors associated with the malaria infection of the vertebrate host which influence the degree of infection in the mosquito. *American Journal of Tropical Medicine and Hygiene* **55**, 386–391.

Eyles, D.E., Young, M.D. and Burgess, R.W. (1948). Studies on imported malarias 8. Infectivity to *Anopheles quadrimaculatus* of asymptomatic *Plasmodium vivax* parasitemias. *Journal of the National Malaria Society* **7**, 125–133.

Feldmann, A.M., Billingsley, P.F. and Savelkoul, E. (1990). Bloodmeal digestion by strains of *Anopheles stephensi* Liston (Diptera, Culicidae) of differing susceptibility to *Plasmodium falciparum*. *Parasitology* **101**, 193–200.

Fell, A.H., Currier, J. and Good, M.F. (1994). Inhibition of *Plasmodium falciparum* growth *in vitro* by CD4[+] and CD8[+] T cells from non-exposed donors. *Parasite Immunology* **16**, 579–586.

Fessler, J.H. and Fessler, L.I. (1989). *Drosophila* extracellular matrix. *Annual Review of Cell Biology* **5**, 309–339.

Field, J.W. and Shute, P.G. (1955). *The Microscopic Diagnosis of Human Malaria*. Kuala Lumpur: The Institute for Medical Research. Study no. 24.

Fleck, S.L. (1993). *Factors affecting the transmission of malaria*. Ph. D. thesis, University of London.

Fleck, S.L., Butcher, G.A. and Sinden, R.E. (1994). *Plasmodium berghei*: serum-mediated inhibition of infectivity of infected mice to *Anopheles stephensi* mosquitoes. *Experimental Parasitology* **78**, 20–27.

Foo, A., Carter, R., Lambros, C., Graves, P., Quakyi, I., Targett, G.A.T., Ponnudurai, T. and Lewis, G.E. (1991). Conserved and variant epitopes of target antigens of transmission blocking antibodies among isolates of *Plasmodium falciparum* from Malaysia. *American Journal of Tropical Medicine and Hygiene* **44**, 623–631.

Freyvogel, T.A. (1966). Shape, movement *in situ* and locomotion of plasmodial ookinetes. *Acta Tropica* **23**, 201–221.

Frontali, C. (1994). Genome plasticity in *Plasmodium*. *Genetica* **94**, 91–100.

Galun, R., Oren, N. and Zecharia, M. (1984). Effect of plasma components on the feeding response of the mosquito *Aedes aegypti* L. to adenine nucleotides. *Physiological Entomology* **9**, 403–408.

Gamage-Mendis, A.C., Rajakaruna, J., Carter, R. and Mendis, K.N. (1992). Transmission blocking immunity to human *Plasmodium vivax* malaria in an endemic population in Kataragama, Sri Lanka. *Parasite Immunology* **14**, 385–396.

Gambrell, W.E. (1937). Variations in gametocyte production in avian malaria. *American Journal of Tropical Medicine* **17**, 689–726.

Garnham, P.C.C. (1966). *Malaria Parasites and other Haemosporidia*. Oxford: Blackwell Scientific Publications.

Garnham, P.C.C. and Powers, K.G. (1974). Periodicity of infectivity of plasmodial gametocytes; the "Hawking phenomenon". *International Journal for Parasitology* **4**, 103–106.

Garnham, P.C.C., Bird, R.G. and Baker, J.R. (1962). Electron microscopic studies of motile stages of malarial parasites III. The ookinetes of *Haemamoeba* and *Plasmodium*. *Transactions of the Royal Society of Tropical Medicine and Hygiene* **56**, 116–120.

Gass, R.F. (1977). Influences of blood digestion on the development of *Plasmodium gallinaceum* (Brumpt) in the midgut of *Aedes aegypti* (L.). *Acta Tropica* **34**, 127–140.

Gass, R.F. and Yeates, R.A. (1979). *In vitro* damage of cultured ookinetes of *Plasmodium gallinaceum* by digestive proteinases from susceptible *Aedes aegypti*. *Acta Tropica* **36**, 243–252.

Georgopoulos, C. and Welch, W.J. (1993). Role of the major heat shock proteins as molecular chaperones. *Annual Review of Cell Biology* **9**, 601–634.

Gero, A.M. and Kirk, K. (1994). Nutrient transport pathways in *Plasmodium*-infected erythrocytes: what and where are they? *Parasitology Today* **10**, 395–399.

Githeko, A.K., Brandling-Bennett, A.D., Beier, M., Mbogo, C.M., Atieli, F.K., Owaga, M.L., Juma, F. and Collins, F.H. (1993). Confirmation that *Plasmodium falciparum* has aperiodic infectivity to *Anopheles gambiae*. *Medical and Veterinary Entomology* **7**, 373–376.

Good, M.F., Quakyi, I., Saul, A., Berzofsky, J.A., Carter, R. and Miller, L.H. (1987). Human T cell clones are reactive to the sexual stages of *Plasmodium falciparum* malaria: high frequency of gamete-reactive T cells in peripheral blood from non-exposed donors. *Journal of Immunology* **138**, 306–311.

Goonewardena, R., Carter, R., Gamage, C.P., Delgiudice, G., David, P.H., Howie, S. and Mendis, K.N. (1990). Human T cell proliferative responses to *Plasmodium vivax* antigens: evidence of immunosuppression following prolonged exposure to endemic malaria. *European Journal of Immunology* **20**, 1387–1391.

Gore, T.C., Pittman Noblet, G. and Noblet, R. (1982). Effects of pinealectomy and ocular enucleation on diurnal periodicity of *Leucocytozoon smithi* (Haemosporina) gametocytes in the peripheral blood of domestic turkeys. *Journal of Protozoology* **29**, 415–420.

Goto, M., Oshima, I., Tomita, T. and Ebihara, S. (1989). Melatonin content of the pineal gland in different mouse strains. *Journal of Pineal Research* **7**, 195–204.

Graves, P.M., Carter, R. and McNeill, K.M. (1984). Gametocyte production in cloned lines of *Plasmodium falciparum. American Journal of Tropical Medicine and Hygiene* **33**, 1045–1050.

Graves, P.M., Carter, R., Burkot, T. Kaushal, D.C. and Williams, J.L. (1985). Effects of different transmission-blocking antibodies on different isolates of *Plasmodium falciparum. Infection and Immunity* **48**, 611–616.

Graves, P.M., Carter, R., Burkot, T., Quakyi, I.A. and Kumar, N. (1988a). Antibodies to *Plasmodium falciparum* gamete surface antigens in Papua New Guinea sera. *Parasite Immunology* **10**, 209–218.

Graves, P.M., Burkot, T.R., Carter, R., Cattini, J.A., Parker, J., Brabin, B.J., Gibson, F.D., Bradley, D.J. and Alpers, M. (1988b). Measurements of malarial infectivity of human populations to mosquitoes in the Madang area, Papua New Guinea. *Parasitology* **96**, 251–263.

Graves, P.M., Wirtz, R.A., Carter, R., Burkot, T.R., Looker, M. and Targett, G.A.T. (1988c). Naturally occurring antibodies to an epitope on *Plasmodium falciparum* gametes detected by a monoclonal antibody-based competitive enzyme-linked immunosorbent assay. *Infection and Immunity* **56**, 2818–2821.

Graves, P.M., Bhatia, K.T., Burkot, R., Prasad, M., Wirtz, R.A. and Beckers, P. (1989). Association between HLA type and antibody response to malaria sporozoite and gametocyte epitopes is not evident in immune Papua New Guineans. *Clinical and Experimental Immunology* **78**, 418–423.

Graves, P.M., Doubrovsky, A., Carter, R., Eida, S. and Becker, P. (1990). High frequency of antibody response to *Plasmodium falciparum* gametocyte antigens during acute malaria infection in Papua New Guinea highlanders. *American Journal of Tropical Medicine and Hygiene* **42**, 515–520.

Graves, P.M., Doubrovsky, A. and Beckhurst, P. (1991). Antibody responses to *Plasmodium falciparum* gametocyte antigens during and after malaria attacks in children from Madang, Papua New Guinea. *Parasite Immunology* **13**, 291–299.

Graves, P.M., Doubrousky, A., Sattabongkot, J. and Battistutta, D. (1992). Human antibody responses to epitopes on the *Plasmodium falciparum* gametocyte antigen Pfs48/45 and their relationship to infectivity of gametocyte carriers. *American Journal of Tropical Medicine and Hygiene* **46**, 711–719.

Grotendorst, C.A. and Carter, R. (1987). Complement effects on the infectivity of *Plasmodium gallinaceum* to *Aedes aegypti* mosquitoes II. Changes in sensitivity to complement-like factors during zygote development. *Journal of Parasitology* **73**, 980–984.

Grotendorst, C.A., Carter, R., Rosenberg, R. and Koontz, L. (1986). Complement effects on the infectivity of *Plasmodium gallinaceum* to *Aedes aegypti* mosquitoes I. Resistance of zygotes to the alternative pathway of complement. *Journal of Immunology* **136**, 4270–4274.

Gunderson, J.H., Sogin, M.L., Wollett, G., Hollingdale, M., De La Cruz, V.F., Waters, A.P. and McCutchan, T.F. (1987). Structurally distinct, stage specific ribosomes occur in *Plasmodium. Science* **238**, 933–937.

Gyan, B., Troye-Blomberg, M., Perlmann, P. and Björkman, A. (1994). Human monocytes cultured with and without interferon gamma inhibit *Plasmodium falciparum* parasite growth *in vitro* via secretion of reactive nitrogen intermediates. *Parasite Immunology* **16**, 371–375.

Harte, P.G., Rogers, N.C. and Targett, G.A.T. (1985). Role of T cells in preventing transmission of rodent malaria. *Immunology* **56**, 1–7.

Hawking, F., Worms, M.J. and Gammage, K. (1968a). 24 and 48-hour cycles of malaria parasites in the blood; their purpose, production and control. *Transactions of the Royal Society of Tropical Medicine and Hygiene* **62**, 731–760.

Hawking, F., Worms, M.J. and Gammage, K. (1968b). Host temperature and control of 24-hour and 48-hour cycles in malaria parasites. *Lancet* **i**, 506–509.

Hawking, F., Worms, M.J. and Gammage, K. (1971). Evidence for cyclic development and short-lived maturity in the gametocytes of *Plasmodium falciparum. Transactions of the Royal Society of Tropical Medicine and Hygiene* **65**, 549–559.

Høgh, B., Thompson, R., Hetzel, C., Fleck, S.L., Kruse, N.A.A., Jones, I., Dgedge, M., Barreto, J. and Sinden, R.E. (1995). Specific and nonspecific responses to *Plasmodium falciparum* blood-stage parasites and observations on the gametocytemia in schoolchildren living in a malaria-endemic area of Mozambique. *American Journal of Tropical Medicine and Hygiene* **52**, 50–59.

Holloway, S.P., Min, W. and Inselburg, J.W. (1994). Isolation and characterization of a chaperonin-60 gene of the human malaria parasite *Plasmodium falciparum. Molecular and Biochemical Parasitology* **64**, 25–32.

Howard, R.J., Kaushal, D.C. and Carter, R. (1982). Radioiodination of parasite antigens with 1,3,4,6-tetrachloro-3α,6α-diphenyl-glycuril (IODOGEN); studies with zygotes of *Plasmodium gallinaceum. Journal of Protozoology* **29**, 114–123.

Huff, C.G. (1957). Studies on the infectivity of plasmodia of birds for mosquitoes, with special reference to the problem of immunity in the mosquito. *American Journal of Hygiene* **7**, 706–734.

Huff, C.G., Marchbank, D.F. and Shiroishi, T. (1958). Changes in infectiousness of malarial gametes II. Analysis of possible causative factors. *Experimental Parasitology* **7**, 399–417.

Hummert, B.A. (1994) *Plasmodium vivax* ookinetes in human peripheral blood. *Journal of Clinical Microbiology* **32**, 2578–2580.

Hunt, N.H., Kopp, M. and Stocker, R. (1992). Free radicals and antioxidants in malaria. In: *Lipid-Soluble Antioxidants: Biochemistry and Clinical Applications* (A.S.H. Ong and L. Packer, eds), pp. 337–353. Basel: Birkhauser Verlag.

Hurd, H. (1994) Interactions between parasites and insect vectors. *Memorias do Instituto Oswaldo Cruz* **89**, 27–30.

Inselberg, J. (1983). Gametocyte formation by the progeny of single *Plasmodium falciparum* schizonts. *Journal of Parasitology* **69**, 584–591.

Janse, C.J. (1993). Chromosome size polymorphism and DNA rearrangements in *Plasmodium. Parasitology Today* **9**, 19–22.

Janse, C.J. and Mons, B. (1987). DNA synthesis and genome structure of *Plasmodium*: a review. *Acta Leidensia* **56**, 1–13.

Janse, C.J., Mons, B., Croon, J.J.A.B. and Van der Kaay, H.J. (1984). Long term *in vitro* cultures of *Plasmodium berghei* and preliminary observations on gametocytogenesis. *International Journal for Parasitology* **14**, 317–320.

Janse, C.J., Mons, B., Rouwenhorst, R.J., Van der Klooster, P.F.J., Overdulve, J.P. and Van der Kaay, H.J. (1985). *In vitro* formation of ookinetes and functional maturity of *Plasmodium berghei* gametocytes. *Parasitology* **91**, 19–29.

Janse, C.J., Van der Klooster, P.F., Van der Kaay, H.J., Van der Ploeg, M. and Overdulve, J.P. (1986a). DNA synthesis in *Plasmodium berghei* during asexual and sexual development. *Molecular and Biochemical Parasitology* **20**, 173–182.

Janse, C.J., Van der Klooster, P.F.J., Van der Kaay, H.J., Van der Ploeg, M. and Overdulve, J.P. (1986b). Rapid repeated DNA replication during microgametogenesis and DNA synthesis in young zygotes of *Plasmodium berghei. Transactions of the Royal Society of Tropical Medicine and Hygiene* **80**, 154–157.

Janse, C.J., Boorsma, E.G., Ramesar, J., Van Vianen, Ph., Van Der Meer, R., Zenobi, P., Casaglia, O., Mons, B. and Van Der Berg, F.M. (1989). *Plasmodium berghei*: gametocyte production, DNA content, and chromosome-size polymorphisms during asexual multiplication *in vivo*. *Experimental Parasitology* **68**, 274–282.

Janse, C.J., Ponzi, M., Pace, T., Dore, E. and Mons, B. (1991). Variation in karyotype and gametocyte production during asexual multiplication of *Plasmodium berghei*. *Acta Leidensia* **60**, 43–48.

Jones, I.W., Denholm, A.A., Ley, S.V., Lovell, H., Wood, A. and Sinden, R.E. (1994). Sexual development of malaria parasites is inhibited *in vitro* by the Neem extract azadirachtin, and its semi-synthetic analogues. *FEMS Microbiology Letters* **120**, 267–274.

Karunaweera, N.D., Carter, R., Grau, G.E., Kwiatkowski, D., Rajakaruna, J., Delguidice, G. and Mendis, K.N. (1992a). Clinical immunity to human malaria is associated with reduced induction of cytokines and complementary parasite killing factors. *Clinical and Experimental Immunology* **88**, 499.

Karunaweera, N.D., Carter, R., Grau, G.E., Kwiatkowski, D., Delguidice, G. and Mendis, K.N. (1992b). Tumour necrosis factor dependent parasite killing effects during paroxsyms in non-immune *Plasmodium vivax* malaria patients. *Clinical and Experimental Immunology* **88**, 499–505.

Karunaweera, N.D., Grau, G.E., Gamage, P., Carter, R. and Mendis, K.N. (1992c). Dynamics of fever and serum levels of tumour necrosis factor are closely associated during clinical paroxysms in *Plasmodium vivax* malaria. *Proceedings of the National Academy of Sciences of the USA* **89**, 3200–3203.

Kaslow, D.C. (1993). Transmission blocking immunity against malaria and other vector borne diseases. *Current Opinion in Immunology* **5**, 557–565.

Kaslow, D.C., Quakyi, I.A., Syin, C., Raum, M.G., Keister, D.B., Coligan, J.E., McCutchan, T.F. and Miller, L.H. (1988). A vaccine candidate from the sexual stage of human malaria that contains EGF-like domains. *Nature* **333**, 74–76.

Kaslow, D.C., Quakyi, I.A. and Keister, D.B. (1989a). Minimal variation in a candidate vaccine from the sexual stage of *Plasmodium falciparum. Molecular and Biochemical Parasitology* **32**, 101–104.

Kaslow, D.C., Syin, C., McCutchan, T.F. and Miller, L.H. (1989b). Comparison of the primary structure of the 25kDa ookinete surface antigens of *Plasmodium falciparum* and *Plasmodium gallinaceum* reveal six conserved regions. *Molecular and Biochemical Parasitology* **33**, 283–288.

Kaushal, D.C. and Carter, R. (1984). Characterization of antigens on mosquito to midgut stages of *Plasmodium gallinaceum*. II. Comparison of surface antigens of male and female gametes and zygotes. *Molecular and Biochemical Parasitology* **11**, 145–156.

Kaushal, D.C., Carter, R., Howard, R.J. and McAuliffe, F.M. (1983). Characterization of antigens on mosquito midgut stages of *Plasmodium gallinaceum*. I. Zygote surface antigens. *Molecular and Biochemical Parasitology* **8**, 53–69.

Kawamoto, F. (1993). Ionic regulation and signal transduction system involved in the induction of gametogenesis in malaria parasites. Molecular basis of ion channels. *Annals of the New York Academy of Sciences* **707**, 431–434.

Kawamoto, F., Alejo Blanco, R., Fleck, S.L., Kawamoto, Y. and Sinden, R.E. (1990). Possible roles of Ca^{2+} and cGMP as mediators of the exflagellation of *Plasmodium berghei* and *Plasmodium falciparum. Molecular and Biochemical Parasitology* **42**, 101–108.

Kawamoto, F., Alejo Blanco, R., Fleck, S.L. and Sinden, R.E. (1991). *Plasmodium berghei* — ionic regulation and the induction of gametogenesis. *Experimental Parasitology* **72**, 33–42.

Kawamoto, F., Kido, N., Hanaichi, T., Djamgoz, M.B.A. and Sinden, R.E. (1992). Gamete development in *Plasmodium berghei* regulated by ionic exchange mechanisms. *Parasitology Research* **78**, 277–284.

Kawamoto, F., Fujioka, H., Murakami, R.I., Syafruddin, Hagiwara, M., Ishikawa, T. and Hidaka, H. (1993). The roles of Ca^{2+}/calmodulin-dependent and cGMP-dependent pathways in gametogenesis of a rodent malaria parasite, *Plasmodium berghei*. *European Journal of Cell Biology* **60**, 101–107.

Killick-Kendrick, R. and Warren, McW. (1968). Primary exoerythrocytic schizonts of a mammalian *Plasmodium* as a source of gametocytes. *Nature* **220**, 191–192.

King, C.A. (1981). Cell surface interaction of the protozoan gregarine with con-canavalin A beads — implications for models of gregarine gliding. *Cell Biology International Reports* **5**, 297–305.

Kittayapong, P., Edman, J.D., Harrison, B.A. and Delorme, D.R. (1992). Female body size, parity and malaria infection of *Anopheles maculatus* (Diptera: Culicidae) in peninsular Malaysia. *Journal of Medical Entomology* **29**, 379–383.

Kitthawee, S., Edman, J.D. and Sattabongkot, J. (1990). Evaluation of survival potential and malaria susceptibility among different size classes of laboratory-reared *Anopheles dirus*. *American Journal of Tropical Medicine and Hygiene* **43**, 328–332.

Kliger, I.J. and Mer, G. (1937). Studies on the effect of various factors on the infection rate of *Anopheles elutus* with different species of *Plasmodium*. *Annals of Tropical Medicine and Parasitology* **31**, 71–83.

Klimes, B., Rootes, D.G. and Tanielian, Z. (1972). Sexual differentiation of merozoites of *Eimeria tenella*. *Parasitology* **65**, 131–136.

Kocken, C.H.M., Milek, R.L.B., Lensen, T.H.W., Kaslow, D.C., Schoenmakers, J.G.G. and Konings, R.N.H. (1995). Minimal variation in the transmission blocking vaccine candidate Pfs 48/45 of the human malaria parasite *Plasmodium falciparum*. *Molecular and Biochemical Parasitology* **69**, 115–118.

Kumar, N. and Carter, R. (1985). Biosynthesis of two stage-specific membrane proteins during transformation of *Plasmodium gallinaceum* zygotes into ookinetes. *Molecular and Biochemical Parasitology* **14**, 127–139.

Kumar, N., Aikawa, M. and Grotendorst, C.A. (1985). *Plasmodium gallinaceum*: critical role for microtubules in the transformation of zygotes into ookinetes. *Experimental Parasitology* **59**, 239–247.

Kumar, N., Koski, G., Harada, M., Aikawa, M. and Hong, Z. (1991). Induction and localization of *Plasmodium falciparum* stress proteins related to the heat shock protein 70 family. *Molecular and Biochemical Parasitology* **48**, 47–58.

Kwiatkowski, D., Cannon, J.G., Manogue, K.R., Cerami, A., Dinarello, C.A. and Greenwood, B.M. (1989). Tumour necrosis factor production in falciparum malaria and its association with schizont rupture. *Clinical and Experimental Immunology* **77**, 361–366.

Landau, I., Miltgen, F., Boulard, Y., Chabaud, A.-G. and Baccam, D. (1979). Etudes sur les gametocytes des *Plasmodium* du groupe *"vivax"*. *Annales de Parasitologie Humaine et Comparée* **54**, 145–161.

Langhorne, J. (1989). The role of CD4 T cells in the immune response to *Plasmodium chabaudi*. *Parasitology Today* **5**, 362–384.

Laveran, M.A. (1881). De la nature parasitaire des accidents de l'impaludisme. *Comptes Rendues de la Société de Biologie* **93**, 627–630.

Lensen, A.H.W., Bolmer, M., Sauerwein, R., Eling, W. and Van Gemert, G.J.

(1996). Activated human neutrophils in a mosquito bloodmeal can reduce transmission. *Parasite Immunology* (in press).

Li, J., McConkey, G.A., Rogers, M.J., Waters, A.P. and McCutchan, T.R. (1994). *Plasmodium*: the developmentally regulated ribosome. *Experimental Parasitology* **78**, 437–441.

Lyimo, E.O. and Koella, J.C. (1992). Relationship between body size of adult *Anopheles gambiae s.l.* and infection with the malaria parasite *Plasmodium falciparum*. *Parasitology* **104**, 233–237.

Madshus, I.H. (1988). Regulation of intracellular pH in eukaryotic cells. *Biochemical Journal* **250**, 1–8.

Marchoux, E. and Chorine, V. (1932). La fécondation des gametes d'hématozoaires. *Annales de l'Institut Pasteur* **49**, 75–102.

Martin, S.K., Miller, L.H., Nijhout, M.M. and Carter, R. (1978). *Plasmodium gallinaceum*: induction of male gametocyte exflagellation by phosphodiesterase inhibitors. *Experimental Parasitology* **44**, 239–242.

Martin, S.K., Jett, M. and Schneider, I. (1994). Correlation of phosphoinositide hydrolysis with exflagellation in the malaria microgametocyte. *Journal of Parasitology* **80**, 371–378.

McCallum, W.G. (1897). On the flagellated form of the malarial parasite. *Lancet* **ii**, 1240–1241.

McGregor, I.A. (1988). Malaria and nutrition. In: *Malaria: Principles and Practice of Malariology* (W.H. Wernsdorfer and I. McGregor, eds), pp. 753–768. Edinburgh: Churchill Livingstone.

McCutchan, T.F., Li, J., McConkey, G.A., Rogers, M.J. and Waters, A.P. (1995). The cytoplasmic ribosomal RNAs of *Plasmodium* spp. *Parasitology Today* **11**, 134–138.

Mehlhorn, H., Peters, W. and Haberkorn, A. (1980). The formation of ookinetes and oocysts in *Plasmodium gallinaceum* (Haemosporidia) and considerations on phylogenetic relationships between Haemosporidia, Piroplasmida and other Coccidia. *Protistologica* **16**, 135–154.

Meis, J.F.G.M., Pool, G., Van Gemert, G.J., Lensen, A.H.W., Ponnudurai, T. and Meuwissen, J.H.E.T. (1989). *Plasmodium falciparum* ookinetes migrate intercellularly through *Anopheles stephensi* midgut epithelium. *Parasitology Research* **76**, 13–19.

Mendis, K.N. and Carter, R. (1991). Transmission blocking immunity may provide clues that antimalarial immunity is largely T cell independent. *Research in Immunology* **142**, 687–690.

Mendis, K.N. and Carter, R. (1992). The role of cytokines in *Plasmodium vivax* malaria. Memorias do Instituto Oswaldo Cruz **87**, 51–55.

Mendis, K.N., Munesinghe, Y.D., Desilva, Y.N.Y., Keragalla, I. and Carter, R. (1987a). Malaria transmission-blocking immunity induced by natural infections of *Plasmodium vivax* in humans. *Infection and Immunity* **55**, 369–372.

Mendis, K.N., Peiris, J.S.M., Premawansa, S., Udagama, P.V., Munesinghe, Y., Ranawaka, M., Carter, R. and David, P.H. (1987b). Immune modulation of parasite transmission in *Plasmodium vivax* malaria. Anti-gamete antibodies can both block and enhance transmission. In: *Molecular Strategies in Parasite Invasion* (N. Agabian, H. Goodman and N. Nogueira, eds), pp. 417–426. New York: Alan R. Liss.

Mendis, K.N., David, P.H. and Carter, R. (1990). Human immune responses against sexual stages of malaria parasites: considerations for malaria vaccines. *International Journal for Parasitology* **20**, 497–502.

Metselaar, D. (1960). Relative increase in the prevalence of *Plasmodium falciparum* some years after the beginning of a house spraying campaign in Netherlands New Guinea. *Transactions of the Royal Society of Tropical Medicine and Hygiene* **54**, 523–528.

Micks, D.W., de Caires, P.F. and Franco, L.B. (1948). The relationship of exflagellation in avian plasmodia to pH and immunity in the mosquito. *American Journal of Hygiene* **48**, 182–190.

Mons, B. (1986). Intraerythrocytic differentiation of *Plasmodium berghei. Acta Leidensia* **54**, 1–83.

Mossalayi, M.D., Paul-Eugene, N., Ouaaz, F., Arock, M., Kolb, J.P., Kilchherr, E., Debre, P. and Dugas, B. (1994). Involvement of Fce RII/CD23 and L-arginine-dependent pathway in IgE-mediated stimulation of human monocyte functions. *International Immunology* **6**, 931–934.

Motard, A., Baccam, D. and Landau, I. (1990). Temporary loss of *Plasmodium* gametocytes infectivity during schizogony. *Annales de Parasitologie Humaine et Comparée* **65**, 218–220.

Motard, A., Landau, I., Nussler, A., Grau, G.E., Baccam, D., Mazier, D. and Targett, G.A.T. (1993). The role of reactive nitrogen intermediates in modulation of gametocyte infectivity of rodent malaria parasites. *Parasite Immunology* **15**, 21–26.

Motard, A., Marussig, M., Renia, L., Baccam, D., Landau, I., Mattei, D., Targett, G. and Mazier, D. (1995). Immunization with the malaria heat shock-like protein hsp70–1 enhances transmission to the mosquito. *International Immunology* **7**, 147–150.

Mulder, B., Tchuinkam, T., Dechring, K., Verhave, J.P., Carnevale, P., Meuwissen, J.H.Th. and Roberts, V. (1994). Malaria transmission blocking activity in experimental infections of *Anopheles gambiae* from naturally infected *Plasmodium falciparum* gametocyte carriers. *Transactions of the Royal Society of Tropical Medicine and Hygiene* **88**, 121–125.

Murakami, H. (1995). A kinase from fission yeast responsible for blocking mitosis in S phase. *Nature* **374**, 817–819.

Mutabingwa, T.K. (1994). Malaria and pregnancy: epidemiology, pathophysiology and control options. *Acta Tropica* **57**, 239–254.

Naotunne, T.D., Karunaweera, N.D., Delgiudice, G., Kularante, M.U., Grau, G.E., Carter, R. and Mendis, K.N. (1991). Cytokines kill malaria parasites during infection crisis: extracellular complementary factors are essential. *Journal of Experimental Medicine* **173**, 523–529.

Naotunne, T.D., Rathnayake, K.D.L., Jayasinghe, A., Carter, R. and Mendis, K.N. (1990). *Plasmodium cynomolgi*: serum-mediated blocking and enhancement of infectivity to mosquitoes during infections in the natural host, *Macaca sinica. Experimental Parasitology* **71**, 305–313.

Naotunne, T.D., Karunaweera, N.D., Mendis, K.N. and Carter, R. (1993). Cytokine-mediated inactivation of malarial gametocytes is dependent on the presence of white blood cells and involves reactive nitrogen intermediates. *Immunology* **78**, 555–562.

Nijhout, M.M. (1979). *Plasmodium gallinaceum*: exflagellation stimulated by a mosquito factor. *Experimental Parasitology* **48**, 75–80.

Nijhout, M.M. and Carter, R. (1978). Gamete development in malarial parasites: bicarbonate-dependent stimulation by pH *in vitro. Parasitology* **76**, 39–53.

Noden, B.H., Beadle, P.S., Vaughan, J.A., Pumpuni, C.B., Kent, M.D. and Beier,

J.C. (1994). *Plasmodium falciparum*: the population structure of mature gametocyte cultures has little effect on their innate fertility. *Acta Tropica* **58**, 13–19.

O'Brien, E., Kurdi-Haidar, B., Wanachiwanawn, W., Carvajal, J.-L., Vulliamy, T.J., Cappadoro, M., Mason, P.J. and Luzzatto, L. (1994). Cloning of the glucose 6-phosphate dehydrogenase gene from *Plasmodium falciparum*. *Molecular and Biochemical Parasitology* **64**, 313–326.

Ogwan'g, R., Mwangi, J., Gachihi, G., Nwachukwu, A., Roberts, C.R. and Martin, S.K. (1993a). Use of pharmacological agents to implicate a role for phosphoinositide hydrolysis products in malaria gamete formation. *Biochemical Pharmacology* **46**, 1601–1606.

Ogwan'g, R.A., Mwangi, J.K., Githure, J., Were, J.B.O., Roberts, C.R. and Martin, S.K. (1993b). Factors affecting exflagellation of *in vitro* cultivated *Plasmodium falciparum* gametocytes. *American Journal of Tropical Medicine and Hygiene* **49**, 25–29.

Omar, M.S. (1968). Vergleichende Beobachtungen uber die Entwicklung von *Plasmodium cynomolgi bastianellii* in *Anopheles stephensi* und *Anopheles albimanus*. *Tropenmedezin und Parasitologie* **19**, 370–389.

Ong, C.S.L., Zhang, K.Y., Eida, S.J., Graves, P.M., Dow, C., Looker, M., Rogers, N.C., Chiodini, P.L. and Targett, G.A.T. (1990). The primary antibody response of malaria patients to *Plasmodium falciparum* sexual stage antigens which are potential transmission blocking vaccine candidates. *Parasite Immunology* **12**, 447–456.

Ono, T. and Nakabayashi, T. (1990). Gametocytogenesis induction by ammonium compounds in cultured *Plasmodium falciparum*. *International Journal for Parasitology* **20**, 615–618.

Ono, T., Nakai, T. and Nakabayashi, T. (1986). Induction of gametocytogenesis in *Plasmodium falciparum* by the culture supernatant of hybridoma cells producing anti-*Plasmodium falciparum* antibody. *Biken Journal* **29**, 77–82.

Ono, T., Ohnishi, Y., Nagamune, K. and Kano, M. (1993). Gametocytogenesis induction by Berenil in cultured *Plasmodium falciparum*. *Experimental Parasitology* **77**, 74–78.

Paskewitz, S.M., Brown, M.R., Lea, A.O. and Collins, F.H. (1988). Ultrastructure of the encapsulation of *Plasmodium cynomolgi* (B strain) on the midgut of a refractory strain of *Anopheles gambiae*. *Journal of Parasitology* **74**, 432–439.

Paskewitz, S.M., Brown, M.R., Collins, F.H. and Lea, A.O. (1989). Ultrastructural localization of phenoloxidase in the midgut of refractory *Anopheles gambiae* and association of the enzyme with encapsulated *Plasmodium cynomolgi*. *Journal of Parasitology* **75**, 594–600.

Paton, M.G., Barker, G.C., Matsuoka, H., Ramesar, J., Janse, C.J., Waters, A.P. and Sinden, R.E. (1993). Structure and expression of a post-transcriptionally regulated malaria gene encoding a surface protein from the sexual stages of *Plasmodium berghei*. *Molecular and Biochemical Parasitology* **59**, 263–275.

Peiris, J.S.M., Premawansa, S., Ranawaka, M.B.R., Udagama, P.V., Munesinghe, Y.D., Nanayakkara, M.V., Gamage, C.P., Carter, R., David, P.H. and Mendis, K.N. (1988). Monoclonal and polyclonal antibodies both block and enhance transmission of human *Plasmodium vivax* malaria. *American Journal of Tropical Medicine and Hygiene* **3**, 26–32.

Petit, G., Camus, D., Dei-Cas, E. and Landau, I. (1982). Inhibition immédiate de l'infectivité des gametocytes de *Plasmodium yoelii nigeriensis* par le serum de rongeurs infectés depuis 5 jours. *Annales de Parasitologie* **57**, 507–508.

Pichyangkul, S., Saengkrai, P. and Webster, H.K. (1994). *Plasmodium falciparum* pigment induces monocytes to release high levels of tumour necrosis factor-α and interleukin 1-β. *American Journal of Tropical Medicine and Hygiene* **51**, 430–435.

Playfair, J.H.L., Jones, K.R. and Taverne, J. (1989). Cell mediated immunity and its role in protection. In: *Malaria: Host Responses to Infection* (M.M. Stevenson, ed.), pp. 66–86. Boca Raton: CRC Press.

Ponnudurai, T. (1987). Plasmodiidae — erythrocytic stages. In: In vitro *Methods for Parasite Cultivation* (A.E.R. Taylor and J.R. Baker, eds), pp. 153–179. London and New York: Academic Press.

Ponnudurai, T., Meuwissen, J.H.E.T., Leeuwenberg, A.D.E.M., Verhave, J.P. and Lensen, A.H.W. (1982a). The production of mature gametocytes of *Plasmodium falciparum* in continuous cultures of different isolates infective to mosquitoes. *Transactions of the Royal Society of Tropical Medicine and Hygiene* **76**, 242–250.

Ponnudurai, T., Lensen, A.H.W., Leeuwenberg, A.D.E.M. and Meuwissen, J.H.E.T. (1982b). Cultivation of fertile *Plasmodium falciparum* gametocytes in semi-automated systems. *Transactions of the Royal Society of Tropical Medicine and Hygiene* **76**, 812–818.

Pouvelle, B., Spiegel, R., Hsiao, L., Howard, R.J., Morris, R.L., Thomas, A.P. and Taraschi, T.F. (1991). Direct access to serum macromolecules by intraerythrocytic malaria parasites. *Nature* **353**, 73–75.

Pouysségur, J., Franchi, A., l'Allemain, G. and Paris, S. (1985). Cytoplasmic pH, a key determinant of growth factor-induced DNA synthesis in quiescent fibroblasts. *FEBS Letters* **190**, 115–119.

Premawansa, S., Peiris, J.S.M., Perera, K.L.R.L., Ariyaratne, G., Carter, R. and Mendis, K.N. (1990). Target antigens of transmission blocking immunity of *Plasmodium vivax* malaria — characterization and polymorphism in natural parasite isolates. *Journal of Immunology* **144**, 4376–4383.

Premawansa, S., Gamage-Mendis, A., Perera, L., Begarnies, S., Medis, K. and Carter, R. (1994). *Plasmodium falciparum* malaria transmission-blocking immunity under conditions of low endemicity in Sri Lanka. *Parasite Immunology* **16**, 35–42.

Quakyi, I.A., Carter, R., Rener, J., Kumar, N., Good, M.F. and Miller, L.H. (1987). The 230-kDa gamete surface protein of *Plasmodium falciparum* is also a target for transmission-blocking antibodies. *Journal of Immunology* **139**, 4213–4217.

Quakyi, I.A., Matsumoto, Y., Carter, R., Udongsampetch, R., Sjolander, A., Berzins, K., Perlmann, P., Aikawa, M. and Miller, L.H. (1989). Movement of a falciparum malaria protein through the erythrocyte cytoplasm to the erythrocyte membrane is associated with lysis of the erythrocyte and release of gametes. *Infection and Immunity* **57**, 833–839.

Quarmby, L.M., Yeuh, Y.G., Cheshire, J.L., Keller, L.R., Snell, W.J. and Crain, R.C. (1992). Inositol phosphate metabolism may trigger flagellar excision in *Chlamydamonas reinhardtii*. *Journal of Cell Biology* **116**, 737–744.

Ranawaka, M.B., Munesinghe, Y.D., De Silva, D.M.R., Carter, R. and Mendis, K.N. (1988). Boosting of transmission-blocking immunity during natural *Plasmodium vivax* infections in humans depends upon frequent reinfection. *Infection and Immunity* **56**, 1820–1824.

Ranawaka, G., Alejo-Blanco, R. and Sinden, R.E. (1993). The effect of transmission-blocking antibody, ingested in primary and secondary bloodfeeds, upon the development of *Plasmodium berghei* in the mosquito vector. *Parasitology* **107**, 225–231.

Ranawaka, G.R.R., Alejo-Blanco, A.R. and Sinden, R.E. (1994a). Characterization of the effector mechanisms of a transmission-blocking antibody upon differentiation of *Plasmodium berghei* gametocytes into ookinetes *in vitro. Parasitology* **109**, 11–17.

Ranawaka, G.R.R., Fleck, S.L., Alejo-Blanco, A.R. and Sinden, R.E. (1994b). Characterization of the modes of action of anti-Pbs21 malaria transmission-blocking immunity: ookinete to oocyst differentiation *in vivo. Parasitology* **109**, 403–411.

Ranford-Cartwright, L.C., Balfe, P., Carter, R. and Walliker, D. (1993). Frequency of cross-fertilisation in the human malaria parasite *Plasmodium falciparum. Parasitology* **107**, 11–18.

Rawlings, D.J., Fujioka, H., Fried, M., Keister, D.B., Aikawa, M. and Kaslow, D.C. (1992). Alpha-tubulin-II is a male-specific protein in *Plasmodium falciparum. Molecular and Biochemical Parasitology* **56**, 239–250.

Read, D., Lensen, A.H.W., Begarnie, S., Haley, S., Raza, A. and Carter, R. (1994). Transmission blocking antibodies against multiple non-variant target epitopes of the *Plasmodium falciparum* gamete surface antigen Pfs230 are all complement fixing. *Parasite Immunology* **16**, 511–519.

Read, F.W., Nara, A., Nee, S., Keymer, A.E. and Day, K.P. (1992). Gametocyte sex ratios as indirect measures of outcrossing rates in malaria. *Parasitology* **104**, 387–395.

Read, F.W., Anwar, M., Shutler, D. and Nee, S. (1996). Sex allocation and population structure in malaria and related parasitic protozoa. *Proceedings of the Royal Society of London, B.* (in press).

Ribeiro, J.M.C., Rossignol, P.A. and Spielman, A. (1985). *Aedes aegypti*: model for blood finding strategy and prediction of parasite manipulation. *Experimental Parasitology* **60**, 118–132.

Riley, E.M., Ong, C.S.L., Olerup, O., Eida, S., Allen, S.J., Bennett, S., Andersson, G. and Targett, G.A.T. (1990). Cellular and humoral immune responses to *Plasmodium falciparum* gametocyte antigens in malaria-immune individuals — limited response to the 48/45-kilodalton surface antigen does not appear to be due to MHC restriction. *Journal of Immunology* **144**, 4810–4816.

Riley, E.M., Bennett, S., Jepson, A., Hassan-King, M., Whittle, H., Olerup, O. and Carter, R. (1994). Human antibody responses to Pfs230, a sexual stage specific antigen of *Plasmodium falciparum*: non-responsiveness is a stable phenotype but does not appear to be genetically related. *Parasite Immunology* **16**, 55–62.

Rockett, K.A., Targett, G.A.T. and Playfair, J.H.L. (1988). Killing of bloodstages of *Plasmodium falciparum* by lipid peroxides from tumour necrosis serum. *Infection and Immunity* **56**, 3180–3183.

Rockett, K.A., Awburn, M.M., Cowden, W.B. and Clark, I.A. (1991). Killing of *Plasmodium falciparum in vitro* by nitric oxide derivatives. *Infection and Immunity* **59**, 3280–3283.

Roeffen, W., Lensen, T., Mulder, B., Teelen, K., Sauerwein, R., Vandruten, J., Eling, W., Meuwissen, J.H.E.T. and Beckers, P.J.A. (1995a). A comparison of transmission-blocking activity with reactivity in a *Plasmodium falciparum* 48/45-kD molecule-specific competition enzyme-linked immunosorbent assay. *American Journal of Tropical Medicine and Hygiene* **52**, 60–65.

Roeffen, W., Beckers, P.J.A., Teelen, K., Lensen, T., Sauerwein, R.W., Meuwissen, J.H.E.Th. and Eling, W. (1995b). *Plasmodium falciparum*: a comparison of the activity of Pfs230-specific antibodies in an assay of transmission blocking immunity and specific competition ELISAs. *Experimental Parasitology* **80**, 15–26.

Roller, N.R. and Desser, S.S. (1973). The effect of temperature, age and density of gametocytes, and changes in gas composition on exflagellation of *Leucocytozoon simondi*. *Canadian Journal of Zoology* **51**, 577–587.

Rook, G.A.W. (1994). Macrophages and *Mycobacterium tuberculosis*; the key to pathogenesis. In: *Macrophage–Pathogen Interactions* (B.S. Zwilling and T.K. Eisenstein, eds), pp. 249–262. New York: Marcel Dekker.

Rosenberg, R. and Koontz, L.C. (1984). *Plasmodium gallinaceum*: density dependent limits on infectivity to *Aedes aegypti*. *Experimental Parasitology* **57**, 234–238.

Rosenberg, R., Koontz, L.C. and Carter, R. (1982). Infection of *Aedes aegypti* with zygotes of *Plasmodium gallinaceum* fertilized *in vitro*. *Journal of Parasitology* **68**, 653–656.

Rosenberg, R., Koontz, L.C., Alston, K. and Friedman, F.K. (1984). *Plasmodium gallinaceum*: erythrocyte factor essential for zygote infection of *Aedes aegypti*. *Experimental Parasitology* **57**, 158–164.

Ross, R. (1897). Observations on a condition necessary to the transformation of the malaria crescent. *British Medical Journal* **i**, 251–255.

Rossignol, P.A., Ribeiro, J.M.C. and Spielman, A. (1984). Increased intradermal probing time in sporozoite-infected mosquitoes. *American Journal of Tropical Medicine and Hygiene* **33**, 17–20.

Rossignol, P.A., Ribeiro, J.M.C., Jungery, M., Turell, M.J., Spielman, A. and Bailey, C.L. (1985). Enhanced mosquito blood-finding success on parasitemic hosts: evidence for vector–parasite mutualism. *Proceedings of the National Academy of Sciences of the USA* **82**, 7725–7727.

Rudin, W., Billingsley, P.F. and Saladin, S. (1991). The fate of *Plasmodium gallinaceum* in *Anopheles stephensi* Liston and possible barriers to transmission. *Annales de la Société Belge de Médecine Tropicale* **71**, 167–177.

Russell, D.G. (1983). Host cell invasion by Apicomplexa: an expression of the parasite's contractile system. *Parasitology* **87**, 199–209.

Rutledge, L.C., Gould, D.J. and Tantichareon, B. (1969). Factors affecting the infection of anophelines with human malaria in Thailand. *Transactions of the Royal Society of Tropical Medicine and Hygiene* **63**, 613–619.

Saul, A., Graves, P. and Edser, L. (1990). Refractoriness of erythrocytes infected with *Plasmodium falciparum* gametocytes to lysis by sorbitol. *International Journal for Parasitology* **20**, 1095–1097.

Schiefer, B.A., Ward, R.A. and Eldridge, B.F. (1977). *Plasmodium cynomolgi*: effects of malaria infection on laboratory flight performance of *Anopheles stephensi* mosquitoes. *Experimental Parasitology* **41**, 397–404.

Schneweis, S., Maier, W.A. and Seitz, H.M. (1991). Haemolysis of infected erythrocytes — a trigger for formation of *Plasmodium falciparum* gametocytes. *Parasitology Research* **77**, 458–460.

Shahabuddin, M. (1995). Chitinase as a vaccine. *Parasitology Today* **11**, 46–47.

Shahabuddin, M. and Kaslow, D.C. (1994a). *Plasmodium*: parasite chitinase and its role in malaria transmission. *Experimental Parasitology* **79**, 85–88.

Shahabuddin, M. and Kaslow, D.C. (1994b). Biology of the development of *Plasmodium* in the mosquito midgut: a molecular and cellular view. *Bulletin de l'Institut Pasteur* **92**, 119–132.

Shahabuddin, M., Toyoshima, T., Aikawa, M. and Kaslow, D.C. (1993). Transmission-blocking activity of a chitinase inhibitor and activation of malarial parasite chitinase by mosquito protease. *Proceedings of the National Academy of Sciences of the USA* **90**, 4266–4270.

Shahabuddin, M., Criscio, M. and Kaslow, D.C. (1995). Unique specificity of *in vitro* inhibition of mosquito midgut trypsin-like activity correlates with *in vivo* inhibition of malaria parasite infectivity. *Experimental Parasitology* **80**, 212–219.

Shen, S.S. and Steinhardt, R.A. (1978). Direct meaurement of intracellular pH during metabolic depression of the sea urchin egg. *Nature* **272**, 253–254.

Sinden, R.E. (1978). Cell biology. In: *Rodent Malaria* (R. Killick-Kendrick and W. Peters, eds), pp. 85–186. New York: Academic Press.

Sinden, R.E. (1982). The role of cell motility in host cell invasion by the Apicomplexa. In: *Parasitological Topics* (E.U. Canning, ed.), pp. 242–247. Kansas: Allen Press (Society of Protozoologists, Special Publication no. 1).

Sinden, R.E. (1983a). The cell biology of sexual development in *Plasmodium. Parasitology* **86**, 7–28.

Sinden, R.E. (1983b). Sexual development of malarial parasites. *Advances in Parasitology* **22**, 153–216.

Sinden, R.E. (1987). The cellular and molecular interactions of malaria species with their mosquito vectors. In: *Host–Parasite Cellular and Molecular Interactions in Protozoal Infections* (K.P. Chang and D. Snary, eds), pp. 407–415. Berlin and Heidelberg: Springer Verlag.

Sinden, R.E. (1991). Asexual blood stages of malaria modulate gametocyte infectivity to the mosquito vector — possible implications for control strategies. *Parasitology* **103**, 191–196.

Sinden, R.E. (1994). Antimalarial transmission-blocking vaccines. *Science Progress* **77**, 1–14.

Sinden, R.E. and Croll, N.A. (1975). Cytology and kinetics of microgametogenesis and fertilisation of *Plasmodium yoelii nigeriensis*. *Parasitology* **70**, 53–65.

Sinden, R.E. and Garnham, P.C.C. (1973). A comparative study on the ultrastructure of *Plasmodium* sporozoites within the oocyst and salivary glands, with particular reference to the incidence of the micropore. *Transactions of the Royal Society of Tropical Medicine and Hygiene* **67**, 631–637.

Sinden, R.E. and Hartley, R.H. (1985). Identification of the meiotic division of malarial parasites. *Journal of Protozoology* **32**, 742–744.

Sinden, R.E. and Smalley, M.E. (1976). Gametocytes of *Plasmodium falciparum*: phagocytosis by leucocytes *in vivo* and *in vitro*. *Transactions of the Royal Society of Tropical Medicine and Hygiene* **70**, 344–345.

Sinden, R.E. and Smalley, M.E. (1979). Gametocytogenesis of *Plasmodium falciparum in vitro*: the cell cycle. *Parasitology* **79**, 277–296.

Sinden, R.E., Canning, E.U. and Spain, B. (1976). Gametogenesis and fertilisation in *Plasmodium yoelii nigeriensis*: a transmission electron microscope study. *Proceedings of the Royal Society of London, B* **193**, 55–76.

Sinden, R.E., Canning, E.U., Bray, R.S. and Smalley, M.E. (1978). Gametocyte and gamete development in *Plasmodium falciparum*. *Proceedings of the Royal Society of London, B* **201**, 375–399.

Sinden, R.E., Hartley, R.H. and Winger, L. (1985). The development of *Plasmodium* ookinetes *in vitro*: an ultrastructural study including a description of meiotic division. *Parasitology* **91**, 227–244.

Sinden, R.E., Winger, L.A., Hartley, R.H., Carter, H.E., Tirawanchai, N., Davies, C.S. and Sluiters, J.G. (1987). Ookinete antigens of *Plasmodium berghei*: a light and electron microscope immunogold study of the 21kD determinant recognised by transmission blocking antibodies. *Proceedings of the Royal Society of London, B* **230**, 443–458.

Smalley, M.E. and Brown, J. (1981). *Plasmodium falciparum* gametocytogenesis stimulated by lymphocytes and serum from infected Gambian children. *Transactions of the Royal Society of Tropical Medicine and Hygiene* **75**, 316–317.

Smalley, M.E. and Sinden, R.E. (1977). *Plasmodium falciparum* gametocytes: their longevity and infectivity. *Parasitology* **74**, 1–8.

Smalley, M.E., Abadalla, S. and Brown, J. (1981). The distribution of *Plasmodium falciparum* in the peripheral blood and bone marrow of Gambian children. *Transactions of the Royal Society of Tropical Medicine and Hygiene* **75**, 103–105.

Suhrbier, A., Janse, C.J., Mons, B., Fleck, S.L., Nicholas, J., Davies, C.S. and Sinden, R.E. (1987). The complete development *in vitro* of the vertebrate stages of the rodent malaria parasite *Plasmodium berghei*. *Transactions of the Royal Society of Tropical Medicine and Hygiene* **81**, 907–910.

Taliaferro, W.H. and Taliaferro, L.G. (1944). The effect of immunity on the asexual reproduction of *Plasmodium brasilianum*. *Journal of Infectious Diseases* **75**, 1–32.

Taverne, J., Depledge, P. and Playfair, J.H.L. (1982). Differential sensitivity *in vivo* of lethal and nonlethal malarial parasites to endotoxin-induced serum factor. *Infection and Immunity* **37**, 927–934.

Taylor-Robinson, A.W., Phillips, R.S., Severn, A., Moncada, S. and Liew, F.Y. (1993). The role of TH_1 and TH_2 cells in a rodent malaria infection. *Science* **260**, 1931–1934.

Theander, T.G., Hviid, L., Abu-Zeid, Y.A., Abdulhadi, N.H., Saeed, B.D., Jakobsen, P.H., Reihert, C.M., Jepsen, S. and Bayoumi, R.A.L. (1990). Reduced cellular immune reactivity in healthy individuals during the malaria transmission season. *Immunology Letters* **25**, 237–242.

Thelu, J., Bracchi, V., Burnod, J. and Ambroise-Thomas, P. (1994). Evidence for expression of a Ras-like and a stage specific GTP binding homologous protein by *Plasmodium falciparum*. *Cellular Signalling* **6**, 777–782.

Thompson, J. and Sinden, R.E. (1994). *In situ* detection of Pbs21 mRNA during sexual development of *Plasmodium berghei*. *Molecular and Biochemical Parasitology* **68**, 189–196.

Toyé, P.J., Sinden, R.E. and Canning, E.U. (1977). The action of metabolic inhibitors on microgametogenesis in *Plasmodium yoelii nigeriensis*. *Zeitschrift für Parasitenkunde* **53**, 133–141.

Trager, W. and Gill, G.S. (1989). *Plasmodium falciparum* gametocyte formation *in vitro* — its stimulation by phorbol diesters and by 8-bromo cyclic adenosine monophosphate. *Journal of Protozoology* **36**, 451–454.

Tsuji, M., Mattei, D., Nussenzweig, R.S. Eichinger, D. and Zavala, F. (1994). Demonstration of heat-shock protein-70 in the sporozoite stage of malaria parasites. *Parasitology Research* **80**, 16–21.

Vaidya, A.B., Muratova, O., Guinet, F., Keister, D., Wellems, T.E. and Kaslow, D.C. (1995). A genetic locus on *Plasmodium falciparum* chromosome 12 linked to a defect in mosquito-infectivity and male gametogenesis. *Molecular and Biochemical Parasitology* **69**, 65–71.

Vaughan, J.A., Noden, B.H. and Beier, J.C. (1991). Concentration of human erythrocytes by anopheline mosquitoes (Diptera: Culicidae) during feeding. *Journal of Medical Entomology* **28**, 780–786.

Vaughan, J.A., Noden, B.H. and Beier, J.C. (1992). Population dynamics of

Plasmodium falciparum sporogony in laboratory-infected *Anopheles gambiae*. *Parasitology* **78**, 716–724.

Vaughan, J.A., Hensley, L. and Beier, J.C. (1994). Sporogonic development of *Plasmodium yoelii* in five anopheline species. *Journal of Parasitology* **80**, 674–681.

Vernick, K.D., Fujioka, H., Seeley, D.C., Tandler, B., Aikawa, M. and Miller, L.H. (1995). *Plasmodium gallinaceum*: a refractory mechanism of ookinete killing in the mosquito, *Anopheles gambiae*. *Experimental Parasitology* **80**, 583–595.

Warburg, A. and Miller, L.H. (1991). Critical stages in the development of *Plasmodium* in mosquitoes. *Parasitology Today* **7**, 179–181.

Warburg, A. and Miller, L.H. (1992). Sporogonic development of a malaria parasite *in vitro*. *Science* **255**, 448–450.

Warburg, A. and Schneider, I. (1993). *In vitro* culture of the mosquito stages of *Plasmodium falciparum*. *Experimental Parasitology* **76**, 121–126.

Waters, A.P., Syin, C. and McCutchan, T.F. (1989). Developmental regulation of stage-specific ribosome populations in *Plasmodium*. *Nature* **342**, 438–440.

Weathersby, A.B. (1952). The role of the stomach wall in the exogenous development of *Plasmodium gallinaceum* as studied by means of haemocoel injections of susceptible and refractory mosquitoes. *Journal of Infectious Diseases* **91**, 198–205.

Weiss, L. (1990). The spleen in malaria — the role of barrier cells. *Immunology Letters* **25**, 165–172.

Williamson, K.C. and Kaslow, D.C. (1993). Strain polymorphism of *Plasmodium falciparum* transmission-blocking target antigen Pfs230. *Molecular and Biochemical Parasitology* **62**, 125–127.

Wing, S.R., Young, M.D., Mitchell, S.E. and Seawright, J.A. (1985). Comparative susceptibilities of *Anopheles quadrimaculatus* mutants to *Plasmodium yoelii*. *Journal of the American Mosquito Control Association* **1**, 511–513.

Winger, L.A., Tirawanchai, N., Nicholas, J., Carter, E.H., Smith, J.E. and Sinden, R.E. (1988). Ookinete antigens of *Plasmodium berghei*: appearance on the zygote surface of an M_r 21kD determinant recognised by transmission-blocking monoclonal antibodies. *Parasite Immunology* **10**, 193–207.

Yoeli, M. and Upmanis, R.S. (1968). *Plasmodium berghei* ookinete formation *in vitro*. *Experimental Parasitology* **22**, 122–128.

Mouse–Parasite Interactions: from Gene to Population

Catherine Moulia, Nathalie Le Brun and François Renaud

Laboratoire de Parasitologie Comparée, URA CNRS 698, Université Montpellier II, Place Eugène Bataillon, 34095 Montpellier, Cédex 05, France

1. Introduction 119
2. One Kind of Immune Response, One Set of T Cells. 122
3. One Gene, One Cell, One Phenotype 127
4. Involvement of the Major Histocompatibility Complex (MHC) 133
   4.1. The foreign antigen's recognition is determined by its presentation 133
   4.2. MHC polymorphism: the host strategy 138
5. The Mouse Hybrid Zone: a Natural Research Laboratory 140
   5.1. Description and genetic analysis of the zone 140
   5.2. Parasitism in the hybrid zone: evidence of "wormy" mice 143
   5.3. Laboratory strains: are they representative samples of the mouse genomes? 147
   5.4. Parasite susceptibility and hybrid fitness 150
6. Conclusion and Perspectives 153
Acknowledgements 155
References 155

1. INTRODUCTION

According to the tenets of genetic and evolutionary biology, the host–parasite system consists of the direct confrontation between two interacting genomes, each exerting a selective pressure on the other. A parasite is an organism living at the expense of another, hence, the term "parasite" refers

ADVANCES IN PARASITOLOGY VOL 38
ISBN 0–12–031738–9

to all the existing biotic aggressors from viruses and bacteria (Procaryotes) to protozoa and metazoa (Eucaryotes).

From the host standpoint, being exploited by parasites may lead to significant decrease of fitness (fecundity and survival) (Renaud and De Meeus, 1991). Thus, hosts try to defend themselves against biotic aggressors in order to limit the parasite impact on their fitness. But, within a host species susceptible to a specific parasite, are many populations and individuals, which differ genetically from each other. This intraspecific variability implies greater or lesser degrees of susceptibility to infection (Wakelin, 1978), i.e. a polymorphism of compatibility of the hosts (Renaud et al., 1992). This difference of proneness is shown at two levels in the host organism (Wakelin, 1978, 1992): physiological and immunological. First, the physiological and biochemical parameters provided by the host for the parasites may be compared with those of the environment of free-living species. The parasite infecting a new host will find more or less appropriate conditions or materials to establish, invade and/or mature in this new environment. One example may be found in the human species and its relative resistance to *Plasmodium falciparum* infection due to haemoglobinopathy genes, especially Hbs (Wakelin, 1978, 1992; Markell et al., 1986). A further example is *Trypanosoma musculi* infection in laboratory mice: different strains present different levels of parasitaemia. Albright and Albright (1989) suggest that this phenomenon may be partly explained by physiological parameters more or less favourable to the parasite growth in each strain. These parameters are fixed characteristics of the host which can not control them (i.e. to change, modulate, activate or deactivate them). In other words, this "physiological resistance/susceptibility" is not an immune phenomenon.

However, genetic polymorphism leads to variability in the host ability to control actively the infection by the immune system. The genetically determined differences between individuals in a population affect the efficiency of the immune response and thus, the host phenotype of susceptibility. Once again, the human species provides a good example of such variability in the immune response against parasites: Hill et al. (1991) showed that two alleles, respectively on HLA class I and class II loci, are independently associated with protection from severe malaria in West African populations. Under natural constraints exercised by *P. falciparum*, these alleles are more frequent than in other racial groups. This suggests that parasites play a role in maintaining the MHC polymorphism, a phenomenon still under debate and which will be developed later.

The variance of host response against parasites is a main parameter of the host–parasite interactions, which greatly interest immunologists and parasitologists. However, this variability must be first restrained to understand the fundamental mechanisms involved in the response: as the immune system requires many cellular and molecular components in complex interactions, only laboratory studies, with both controlled environmental and genetic parameters, give us the means to observe each stage leading to a variability of the immune response. The second, more difficult, step tries to apply the artificial results of the laboratory to natural populations, according to the immune variability. To realize such a complete study from the gene to the population, it is fundamental to find a host which can be studied both in the laboratory and in the field. The mouse appears to be the most relevant model: it is actually the ideal experimental mammal model being easily bred and having a short generation time. Many inbred strains with known genetic features have been selected and allow precise genetic studies. The most interesting works attempt to link cellular and/or genetic components with the phenotype of the immune response (i.e. susceptibility/resistance) of these strains. As Blackwell (1985) explained, "In effect, what we are doing is using the mice as the 'test-tube' in which we vary both the composition of genes and the efficiency of various immunoregulatory cell populations, add the parasite and determine the outcome of the infection. By altering just one gene at a time, we can then determine the relative contribution of each gene to the overall response of the host." In addition the mouse is a wild animal and the link between the laboratory and the natural population is easily made. Not only do many different taxa of mice of the genus *Mus* exist, but their mixing also provides a wide range of new genetic combinations of different murine genomes, as for example the hybrid zone between the two European commensal mice *Mus musculus musculus* and *M. m. domesticus*. Such a hybrid zone represents an ideal situation to determine the ecological and genetic components acting respectively in the parasite–mice interactions.

In the first part of this review, we evoke the main events of the immune response in mammals. Then, we discuss experimental works on the genetics of the resistance of laboratory mice. We focus especially on the mouse genetic control of the early response against *Leishmania/Salmonella/Mycobacterium* infections, which provides one of the best examples of successful linkage between the phenotype of susceptibility/resistance of inbred mice and their genetic background. In the second part, we have studied the influence of the genes in the major histocompatibility complex on the development of a specific immune response. These two examples of laboratory studies clearly illustrate the difficulty in applying laboratory results to natural populations.

In the last part of this review, we have developed the reverse approach of the genetic interaction of the mouse and its parasites: from a natural situation to the laboratory studies, illustrating the need to develop this kind of process.

2. ONE KIND OF IMMUNE RESPONSE, ONE SET OF T CELLS

Immune response is based on the discrimination between "self" and "non-self" components at the molecular level. Two different, but tightly linked, processes are then involved.

1. Innate (or natural) immunity. This mechanism is not specific to the particular parasite and takes place immediately.
2. Acquired immunity occurs one to two weeks after infection. It is characterized by its specificity (it acts against a definite parasite) and memory (the response is quicker and more effective if a second infection occurs).

Many immunological and cellular studies attempted to determine a framework for the mammalian specific immune response against parasites. From these works, it now appears that T cell subsets and their specific cytokines play a critical role in the infection outcome.

Three main steps in the mounting of an immunological response may be identified (Figure 1).

1. Foreign antigens are recognized as such by B and T cells;
2. Macrophages and T helper lymphocytes (T_H) play the main role of selecting one set of effector mechanisms (humoral or cell-mediated);
3. Immune response occurs in order to eliminate the "foreigner".

Thus, the recognition of an antigen by T cells is the key to the induction of the host response against a definite biotic aggressor. However, unlike B cells, T cells are unable to recognize foreign antigens. To trigger the response, these antigens have to be presented to T lymphocytes in association with histocompatibility molecules on the surface of other particular host cells.

There are two kinds of T cells:

1. The cytotoxic T lymphocytes ($CD8^+$), involved in the elimination of intracellular parasites (virus) or tumours, intervene directly against the infected/tumoral cells. They recognize foreign antigen when presented by Class I histocompatibility molecules on the surface of the infected or tumoral cells (Male et al., 1987; Roitt et al., 1993).
2. The T helper cells (T_H) ($CD4^+$ lymphocytes) control the immune response. Antigen-Presenting Cells (APCs), such as macrophages, dendritic or B cells etc. (Steinman and Nussenzweig, 1980; Chesnut and Grey, 1981; Unanue, 1984; Grey and Chesnut, 1985) catch and ingest the foreign antigens, process them and finally present the peptidic fragments to T_H cell receptors (TCR) in association with their surface Class II histocompatibility antigens (Figure 2) (Ziegler and Unanue,

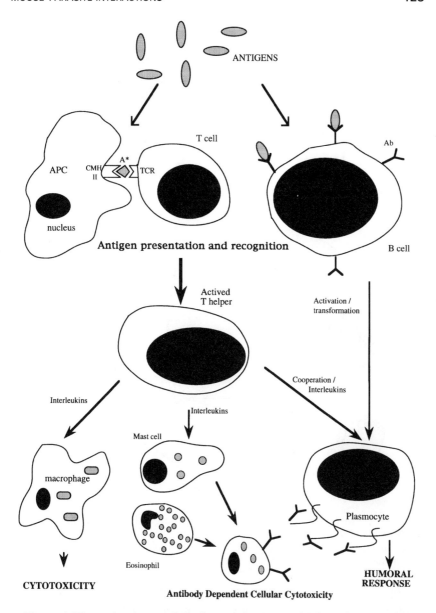

Figure 1 The main phases of the immune response. APC: antigen presenting cell; Ab: antibody; CMH II: Class II histocompatibility molecules; A*: fragment of digested antigen; TCR: T cell receptor.

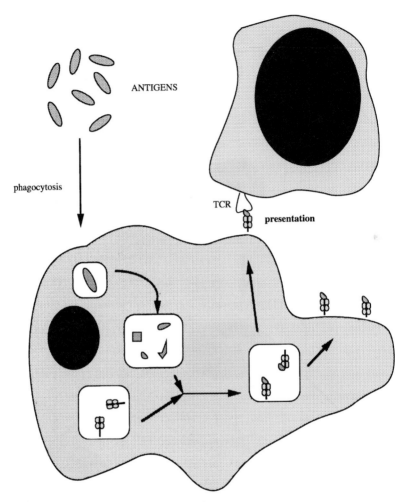

Figure 2 Processing and presentation of the antigen by Class II histocompatibility molecules on antigen presenting Cells (APC). TCR: T cell receptor. Fragments of digested antigens are linked to Class II molecules in the APC vacuoles
and presented to T cell on the APC surface.

1981; Allen *et al.*, 1985; Babbit *et al.*, 1985; Shwartz, 1985). At this
stage, complex cell interactions occur, involving co-stimulatory ligands
(B7 family and CD28/CTLA-4, respectively) on APCs and T cells (see
Allison, 1994 for reviews) and leading to activation and proliferation of
the T lymphocytes.

It is now well established that these T_H are divided into two functional
populations: T_H1 and T_H2 (Figure 3). Each subset is characterized by its

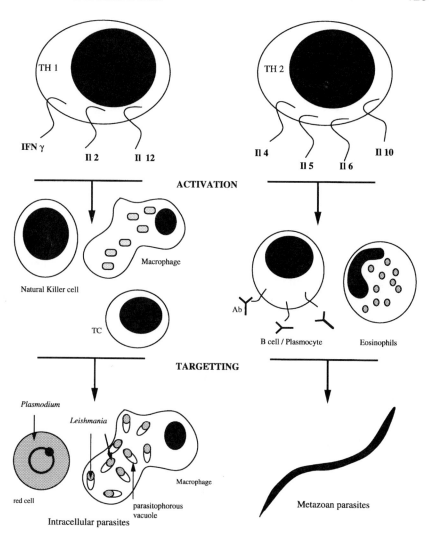

Figure 3 Profiles of cytokine secretion of the T_H subsets and induced immune response. TH: T helper lymphocyte; TC: T cytotoxic lymphocyte; Il interleukins; IFNγ: interferon gamma; Ab: antibody. The presentation of the antigen to T_H cells induces the activation of one of the T subsets. Then they secreted their specific cytokines, and one kind of effector cells is activated, targeting one kind of parasite.

pattern of cytokine secretion. Being key molecules of the immune system, the cytokines are involved in the cell-to-cell communication, activation, maturation and inhibition. Indeed, gamma interferon (IFN-γ) dominates in T_H1 subset and interleukin 4 (Il-4) in T_H2 (Mosmann *et al.*, 1986;

Mosmann and Coffman, 1989; Cox and Liew, 1992; Sher and Coffman, 1992). A similar dichotomy may be observed in the cytotoxic T cells thanks to recent works (Bloom *et al.*, 1992; Erard *et al.*, 1993; Le Gros and Erard, 1994), explaining why Cox and Liew (1992) now refer to T1 and T2 cells.

Each T_H subset determines the activation of one kind of effector mechanism to control the parasites' growth or to eliminate them: T_H1 lymphocytes are responsible for the cell-mediated immune response whereas T_H2 cells control the antibody-mediated immunity (Figure 3) (Mosmann *et al.*, 1986; Mosmann and Coffman, 1989; Cox and Liew, 1992; Sher and Coffman, 1992).

The activation of the appropriate T_H subset and immune effectors against a specific parasite is an important feature of the host defences, some of them being more affected by a cellular response (i.e. T_H1) than by a humoral one (i.e. T_H2) and vice versa (Garra and Murphy, 1994). Several studies have provided precise results leading to the following general concept: T_H1 and their cytokines would be important to protect mice against intracellular infections like cutaneous leishmaniasis caused by *Leishmania major* (Liew, 1987; Scott *et al.*, 1988; Heinzel *et al.*, 1989; Locksley and Scott, 1991; Sher *et al.*, 1992; Garra and Murphy, 1994) and toxoplasmosis due to *Toxoplasma gondii* (Gazzinelli *et al.*, 1991; Subauste and Remington, 1991). Conversely, the activation of the T_H2 subset would lead to resistance to multicellular parasites such as *Trichuris muris* (Else and Grencis, 1991) and *Heligmosomoides polygyrus* (Monroy and Enriquez, 1992) (Figure 3).

However, this does not seem to be the general rule. Indeed, T_H2 activation may block the T_H1 protective immune response in schistosomiasis (*Schistosoma mansoni*) (Reynolds *et al.*, 1990; Sher *et al.*, 1990; Finkelmann *et al.*, 1991), and, at this stage, it is still unclear which of the two subsets is involved in the protection against *Trichinella spiralis* infection (Pond *et al.*, 1989; Grencis *et al.*, 1991). Moreover, in the case of the intracellular protozoan *Eimeria vermiformis*, the T_H1 response dominates in high- and low-responder inbred mice and no evidence was found for a T_H2-mediated interference in these immune phenotypes. Here, the response phenotype of inbred mice would reflect a kinetic, rather than a qualitative, difference in the mounting of protective T_H subsets (Wakelin *et al.*, 1992).

The different role played by each T_H subset to regulate different parasitoses such as *Leishmania* and *Eimeria* infections in experimentally infected inbred mice (i.e. standardized conditions), indicates that the genetic controls are not the same. However, whether we succeed or not in linking some cellular components (as T_H subsets) with high- or low-responder phenotypes, the variation of resistance between inbred strains of mice points out the genetic basis of this response. Given the difficulty in correlating genetic background, cellular events and response phenotype to a definite parasite, several authors used inbred strains of mice to carry out direct genetic studies. Some of the most exhaustive works have connected the linkage of one gene on chromosome 1 with resistance in the early phase of infection by three different parasites.

3. ONE GENE, ONE CELL, ONE PHENOTYPE

Leishmania donovani (visceral leishmaniasis), *Salmonella typhimurium* (murine typhoid) and *Mycobacterium bovis* (Bacillus of Calmette-Guérin BCG, the most common vaccine used against *Mycobacterium tuberculosis* (tuberculosis) and *M. leprae* (leprosy)), are three phylogenetically distinct biotic aggressors sharing the same habitat in their host: they live inside cells of the reticuloendothelial system (RES) and inside macrophages. In the mouse, the courses of the infection of these intracellular parasites are comparable and can be divided into two phases: the early stage when parasites invade the host cells and grow (establishment of the infection), and the later stage when they disseminate and cause systemic infection.

During the early phase of an infection with any of these three parasites, inbred strains of mice segregate into two distinct groups according to their phenotypes of susceptibility. Therefore, some of them will not develop the disease whereas others will. This phenomenon is usually described as natural/innate resistance (Blackwell, 1985; Buschman and Skamene, 1988; Wassom and Kelly, 1990; Vidal *et al.*, 1993) or as early-stage response (O'Brien, 1986), the latter being a more appropriate term.

Studies of the genetic control of these inbred strains against the three intracellular pathogens have used the protocol described by Skamene and Pietrangeli (1991). First, the pattern of disease susceptibility of F_2 and backcross hybrids from resistant and susceptible mice permitted the evaluation of the number of genes involved in the control of the response against the parasite. In the three models, the study demonstrated that the early-stage resistance to infection is controlled by one gene named *Bcg* or *Ity* or *Lsh* (after the model used to detect its effects), and localized on mouse chromosome 1 (Plant and Glynn, 1976, 1979; Bradley, 1977; Bradley *et al.*, 1979; O'Brien *et al.*, 1980; Gros *et al.*, 1981; Plant *et al.*, 1982; Taylor and O'Brien, 1982; Skamene *et al.*, 1982, 1984). This gene presents two allelic forms designated as *Bcg/Ity/Lsh^r* and *Bcg/Ity/Lsh^s* according to the phenotype they determine.

Secondly, recombinant inbred strains (RIS) were used to perform a more precise gene mapping. These mice are inbred descendants of F_2 of a susceptible and resistant strain, and their genomes present homozygous

CHROMOSOME 1   MACROPHAGE   MICE

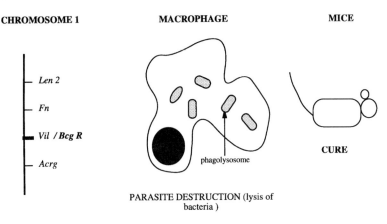

PARASITE DESTRUCTION (lysis of
bacteria)

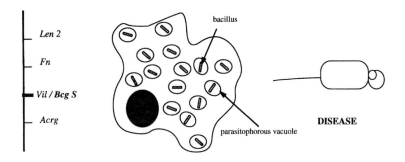

PARASITE MULTIPLICATION

Figure 4 Resistance/susceptibility in the early-phase of infection with BCG:
from allele to cell phenotype and disease expression. Loci close to *Bcg*: *Len 2*: lens
protein 2; *Fn*: fibronectin; *Vil*: villin; *Acrg*: acetylcholine receptor gamma subunit.

F_2-like recombinations between several marker genes located on chromosome 1 and the putative gene of resistance. Linkage analysis of the
phenotypic feature of resistance with the different genetic markers showed
that a group of loci encoding for cytoskeleton proteins, cosegregate with
Bcg/Ity/Lsh alleles (Figure 4) (Schurr *et al.*, 1989; Malo *et al.*, 1991). This
last result is specially interesting and while genetic studies were performed,
some workers have tried to find the host component or cell where this
resistance could be expressed. The same gene controls the host suscept-

ibility to the early growth of three unrelated parasites which share the same intracellular location. Consequently, it has been suggested that macrophages were the cells expressing the *Bcg/Ity/Lsh* gene. Studies *in vivo* and *in vitro* confirmed this hypothesis, and demonstrated that the events leading to the restriction of the parasite growth occurred after their phagocytosis (O'Brien *et al.*, 1979; Gros *et al.*, 1983; Crocker *et al.*, 1984; Stach *et al.*, 1984). The fine mechanisms by which macrophages are or are not able to stop the infection were not elucidated, but the knowledge that the *Bcg/Ity/Lsh* gene is closely linked to cytoskeleton genes allowed Schurr *et al.* (1989) to speculate about the nature of the *Bcg* product: it would have to be another cytoskeleton protein, involved in the cell to cell and cell–environment interactions, essential for macrophage functions. This hypothesis was not confirmed by succeeding studies, but it was a first indication of how *Bcg* could work.

In the past few years, the model that really helps our understanding of the *Bcg/Ity/Lsh* gene is BCG. Subsequent linkage analyses precisely located the *Bcg* locus (Malo *et al.*, 1993a, b), and Vidal *et al.* (1993) succeeded in cloning a candidate gene for *Bcg* called *Nramp*. The product thus encoded is exclusively expressed in macrophages of the RE organs and is a membrane protein identical to those of the transport systems. The *Nramp* products of Bcg^r and Bcg^s strains differ in one amino acid in what is believed to be the second transmembrane domain (TM2) of the protein. This presents a glycine (Gly) (in Bcg^r strains) to an aspartic acid (Asp) (in Bcg^s mice) substitution. The work of Malo *et al.* (1994) showed an absolute association of this allelic variation with the *Bcg* phenotype observed in the 27 inbred strains tested for their resistance/susceptibility to BCG infection, the Gly to Asp mutation in *Nramp* as the only polymorphism that displays absolute allelic association with the resistance/susceptibility phenotype and the *Nramp* Gly allele as the probable wild type form of the gene.

To summarize, the *Bcg* gene encodes a transport system protein (possibly a nitrogen compound needed for the lytic activities of macrophages, Vidal *et al.*, 1993). It displays altered physical properties in susceptible macrophages (Bcg^s individuals) due to an amino acid substitution (Gly to Asp). This phenomenon explains some deficiencies of the cell killing activity in mice of Bcg^s strains (Figure 5).

These results on the genetics of resistance of the mouse to intracellular parasites may have direct consequences in both human genetics and medicine and on mouse population genetics.

In humans, evidence of a genetic component in resistance to leprosy was provided by observations of disease occurrence among certain ethnic groups and families (Hopkins and Denny, 1929; Aycock and McKinley, 1938; Aycock, 1940; Belknap and Haynes, 1961; Spickett, 1962; Chakravartti and Vogel, 1973; Serjeantson *et al.*, 1979; Barreyro *et al.*, 1982;

Resistant mice : *Bcg R*

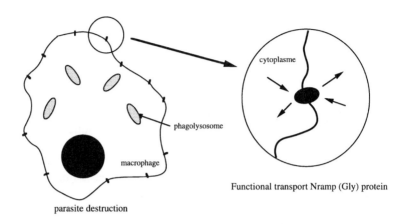

Susceptible mice : *Bcg S*

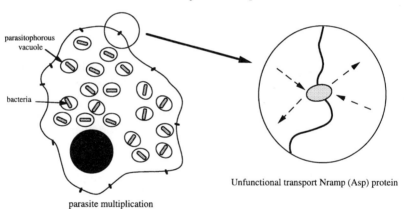

Figure 5 Hypothetic role of Nramp protein in the Bcg-resistant (*Bcg R*) and susceptible (*Bcg S*) mice during the early phase of infection (from the suggestion of Vidal *et al.* (1993)).

Shields *et al.*, 1987; Stead, 1992). Models have thus been developed to determine the mode and influence of genetic factors on the outcome of the infection, and more precisely, the number of genes involved (Morton and McLean, 1974; Smith, 1979; Haile *et al.*, 1985; Abel and Demenais, 1988; Wagener *et al.*, 1988). It appears that the genes which control leprosy in humans may be distinguished in two groups of loci. One of them determines the susceptibility to disease *per se*, i.e. at the early-stage of infection

(Schurr *et al.*, 1991). The same conclusions were also reached after studies of human susceptibility to tuberculosis (Comstock, 1978; Fine, 1981; Skamene, 1986, 1991). These observations become important when we compare the gene maps of mouse and human species. Indeed, the cluster of five genes tightly linked to *Bcg* on mouse chromosome 1 is precisely conserved on the telomeric end of human chromosome 2q (Nadeau and Taylor, 1984; Nadeau, 1989; Schurr *et al.*, 1989, 1990). The resistance to leprosy and tuberculosis *per se* observed in some human individuals could thus be linked to a human homologue of the *Bcg* gene (Schurr *et al.*, 1989; Skamene, 1991; Vidal *et al.*, 1993). Cellier *et al.* (1994) have succeeded in cloning such a human *Nramp* homologue on chromosome 2q. This gene is expressed in the reticuloendothelial organs, suggesting that the macrophage could be one of the main sites of expression of the encoded product. Moreover, the human protein could present transmembrane domains in the same way as its murine homologue does. The comparison of the protein sequences of these *Nramp* products reveals that the predicted TM 2 domains are not only perfectly conserved in humans and mice, but also in other species such as rats and chickens. This emphasizes how this domain could be critical for protein function (Cellier *et al.*, 1994; Malo *et al.*, 1994). Thus, the cellular expression of mouse susceptibility, probably concerning the *Nramp* product, seems to be the same in humans, and better knowledge of mouse susceptibility mechanisms would be a first step in a more efficient control of human disease.

However, we must be cautious and bear in mind that these clear and promising laboratory results were obtained under specific conditions. Only the extreme simplicity of the determinism of the early-stage resistance (one gene, two alleles) permits such a complete work. In fact, most of the genetic systems controlling the resistance against parasites are probably more complex, especially when considered in natural populations. Actually, few studies comparing the mechanisms observed in the laboratory with those existing in nature are available, and the only data to determine the real involvement of the *Bcg* gene to regulate infection in wild populations come from O'Brien *et al.* (1986). They inoculated *S. typhimurium* into wild *Mus musculus musculus* from Czechoslovakia (Czech I), into *M. m. Molossinus* from Japan, and into *M. m. domesticus* from Maryland, USA. Even if the size of each sample was small, this study showed that Czech I mice were susceptible whereas the others were resistant. Crosses were then realized between Czech I and susceptible inbred (Ity^s) mice on the one hand, and Czech I and resistant (Ity^r) mice on the other. Analysis of the F_1 phenotype indicated that the wild Czech I mice were not Ity^s. This preliminary study suggests that wild mice would express resistance genes which might be different from those selected in inbred mice genomes. Immediately, two questions arise. Do these genes exist in other wild populations at a significant level? Do the range of inbred genomes used in our genetics studies represent only a small sample of all the wild mouse genes? The latter is highly probable knowing that inbred strains do not have a simple and clear origin. Indeed, recent genetic studies have shown that laboratory mice result from interspecific recombinant inbreeding of the wild subspecies of the *Mus musculus* complex (Bishop *et al.*, 1985; Bonhomme *et al.*, 1987; Morgado *et al.*, 1993).

Work on the *Bcg/Lsh/Ity* resistance gene is essential in describing the linkage between genotype and cellular and individual phenotypes, even if it is proved that other varieties of genes influence the early-stage or natural resistance against many infectious agents. In many cases, the exact location of the gene is not determined, and the mechanisms involved still have to be investigated. However, the following genes have been identified.

1. The *Rmp* (1, 2, 3 and 4) genes control the resistance to the ectromelia virus. The activity of each of them in the host response depends on the entire genotype of the mouse strain. One (*Rmp 3*) is linked to the H-2 complex (Brownstein *et al.*, 1992).
2. The *Lsr1* gene located on chromosome 2 controls the early-stage resistance to the bacterium *Listeria monocytogenes* (Skamene *et al.*, 1979; Gervais *et al.*, 1984).
3. The *Pchr* gene influences response to infection with the rodent protozoan parasite *Plasmodium chabaudi*. Its exact chromosome location is still unknown (Stevenson, 1989).

Most of the genes mentioned here were involved in the genetic control of the early phase of the infection, and the obvious question now is: could there be another mechanism that would act in the later stages (i.e. specific responses) and finally that would defeat the parasite? Once again, the *Bcg/Lsh/Ity* gene model provides most of the information.

The laboratory strains characterized as susceptible to the early phase of the infection with *L. donovani*, *S. typhimurium* and BCG can be further subdivided according to their ability to control the parasite burden in the late stages of infection.

It is not surprising that the *Bcg* gene expressed at the macrophage level, i.e. in an APC, plays a regulatory role on the generation of lymphocyte responses to BCG and influences the acquired immune response (Buschman and Skamene, 1988). Indeed, an *in vitro* study proved that the Bcg^r macrophages present the antigens more efficiently to T cells during BCG infection than do the Bcg^s ones (Denis *et al.*, 1988). Zwilling *et al.* (1987) showed that peritoneal macrophages of Bcg^r animals contained a higher proportion of Class II-positive cells than Bcg^s mice. Moreover, the accumulation of a high degree of bacterial antigens during the innate phase of host response in the Bcg^s mice results in the development of an immunosuppression not observed at the same level in Bcg^r animals (Nakamura *et al.*, 1989; Skamene, 1991).

These results indicate that the first step in the immune events (the antigen presentation and T-cell activation) determining the subsequent immune response, could be more or less effective in some mice. Hence, the Major Histocompatibility Complex and its encoded products play the central role and many studies have been carried out to determine their exact influence on the outcome of different infections.

4. INVOLVEMENT OF THE MAJOR HISTOCOMPATIBILITY COMPLEX (MHC)

The genes of the MHC, also named H-2 (mouse) or HLA (human) complex, are located on chromosome 17 of the mouse genome. Among them, three kinds of loci may be distinguished according to the molecule they encode: the products of class I genes are involved in the cytotoxic T cell activation; the molecules of class II genes present foreign Ag to helper T cells; the class III genes encode different components of complement. We will focus especially on the class II loci itself divided into four functional genes $A\alpha$, $A\beta$, $E\alpha$ and $E\beta$. Those encode for the α (heavy) and β (light) chains necessary to the formation of the complete heterodimeric histocompatibility molecule. According to the encoding loci, these histocompatibility antigens are named I-A ($A\alpha$ and $A\beta$ chains) or I-E ($E\alpha E\beta$ dimer) molecules (Figure 6) (Hood *et al.*, 1983; Klein *et al.*, 1983; Male *et al.*, 1987).

The inbred strains of mice present different H-2 haplotypes, and each strain has a specific collection of alleles on all the MHC loci (Figure 7). Some of them share the same H-2 haplotype but express different genetic backgrounds, and vice versa (Klein *et al*, 1982; Male *et al.*, 1987; Wassom and Kelly, 1990). This is essential to determine the respective influence of H-2 and background genes in the outcome of infection by comparing the response of the different strains to a definite parasite (Wassom and Kelly, 1990).

4.1. The Foreign Antigen's Recognition is Determined by its Presentation

One of the best demonstrations of the influence of H-2 molecules on the immune response was shown by the infection of mice with the helminth parasite *Trichinella spiralis*. This nematode has a large range of hosts

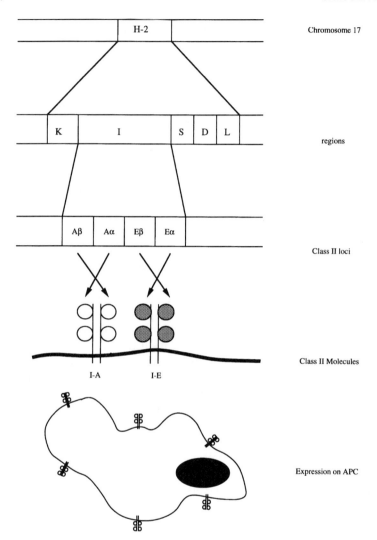

Figure 6 The major histocompatibility complex (H-2), Class II genes and molecules. The I region presents four loci: the Aα and Aβ encode the I-A molecules, the Eα and Eβ the I-E ones. These two kinds of Class II molecules are expressed on the surface of the antigen presenting cells (APC).

including humans, rats, pigs and horses. Inside their host, the adult worms live and reproduce in the columnar epithelial cells of the small intestine. The newborn larvae migrate and penetrate the cells of the striated muscles. There they mature and encyst, and may remain viable inside the muscles for many years. Resistant and susceptible mice were designated according

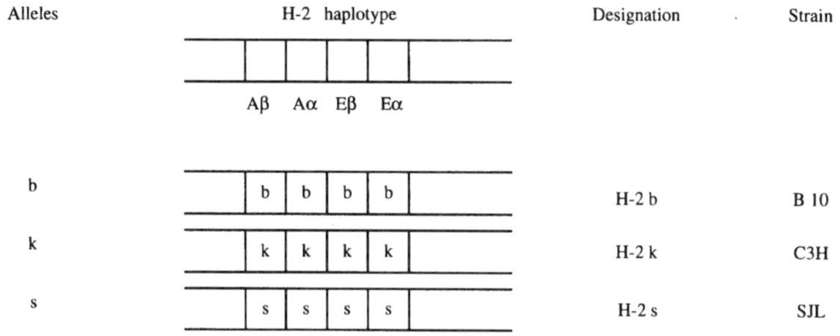

Alleles	H-2 haplotype				Designation	Strain
	Aβ	Aα	Eβ	Eα		
b	b	b	b	b	H-2 b	B 10
k	k	k	k	k	H-2 k	C3H
s	s	s	s	s	H-2 s	SJL

Figure 7 Examples of haplotypes of the H-2 complex (I region) of three inbred strains of mice.

to their ability to expel their intestinal worm burden quickly and/or to limit the production of newborn larvae (Wakelin, 1980; Bell *et al.*, 1982, 1984). Thanks to the congenic strains, two genes, influencing the susceptibility to *T. spiralis* were mapped within the MHC: the Ts-1 gene was located in the I-A region (Wassom *et al.*, 1979, 1980, 1984) and the Ts-2 between the S (class III) and D (class I) loci (Wassom *et al.*, 1983, 1984). According to the H-2 haplotype, i.e. the alleles presented at the Ts-1 and Ts-2 loci, mice expel worms more or less slowly and exhibit weak or high muscle larval burdens. It is worth noting that some of these haplotypes do not lead to the expression of functional I-E molecules on the surface of APCs and that the mice lacking these class II molecules are more resistant to the parasite than the I-E positive ones (Wassom *et al.*, 1987; Wassom and Kelly, 1990). Thus, it seems that the presentation of the Ag in the I-E context leads to susceptibility (Figure 8) (Wassom and Kelly, 1990).

There is an interesting similarity between the influence of H-2 on the nematode *T. spiralis* and the protozoan *L. donovani* infections. Here as well, the late phase of the infection in *Bcg^s* mice is under dose-dependent H-2 control, and important genes were mapped in the I-A and I-E regions (Blackwell *et al.*, 1980; Blackwell, 1983; Ulczack and Blackwell, 1983). Without generalizing, several studies tend to demonstrate that the I-A gene could determine the resistance/susceptibility phenotype, and that the I-E gene could modify the expression of this response according to the earlier dose of infection (low dose induced no I-E effect on the phenotype and vice versa) (Figure 9) (Blackwell and Roberts, 1987; Wassom and Kelly, 1990).

The influence of the MHC on the outcome of numerous parasitoses is now well documented. In addition to *L. donovani* and *T. spiralis* infections previously mentioned, we can quote many examples: the ectromelia virus (Brownstein *et al.*, 1991); the cytomegalovirus (Scalzo *et al.*, 1990); the virus of the murine acquired immunodeficiency syndrome (MAIDS) (Hamelin-Bourassa *et al.*, 1989); the bacteria and protozoans *Mycobacterium*

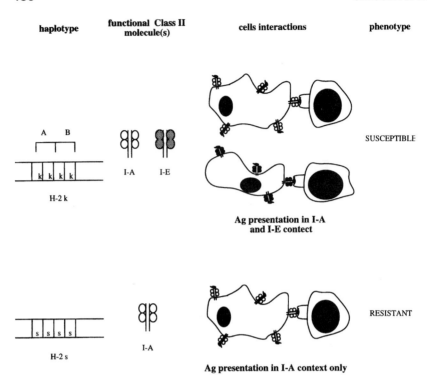

Figure 8 Involvement of the MHC in resistance to *Trichinella spiralis*. The resistance (susceptibility) seems to be linked to the absence (presence) of I-E molecules on the APC surface to present the antigen to T cells. (From the works and suggestions of Wassom *et al.* (1987) and Wassom and Kelly (1990).)

lepraemurium (Adu *et al.*, 1983), *S. typhimurium* (O'Brien *et al.*, 1984; Nauciel *et al.*, 1990); *L. major* (Blackwell, 1985) and *Toxoplasma gondii* (Williams *et al.*, 1978; McLeod *et al.*, 1989). Moreover, H-2 control was demonstrated during infections using the following helminths; *Heligmosomoides polygyrus* (Wassom *et al.*, 1987; Enriquez *et al.*, 1988; Keymer *et al.*, 1990; Behnke and Wahid, 1991), *Trichuris muris* (Else and Wakelin, 1988; Else *et al.*, 1990a, b) and *Nippostrongylus brasiliensis* (Kennedy *et al.*, 1990, 1991).

Therefore, in most cases, it appears that both H-2 and non-H-2 genes influence the entire host response against the parasites. For example, besides the Ts-1 and Ts-2 H-2 loci, two genes designated Ts-3 and Ts-4 and located outside of the MHC, could influence the outcome of the *Trichinella* infection (Wassom *et al.*, 1987; Bell, 1988). This H-2 and non-H-2 mixed influences appear normal as they control two different

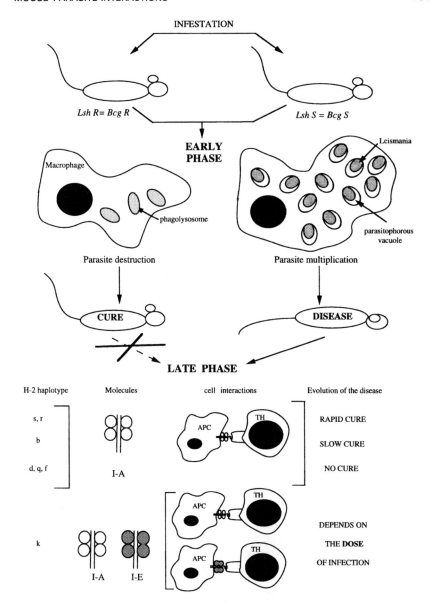

Figure 9 The evolution of *Leishmania donovani* infection in different inbred strains of mice. *Lsh R (S)*: allele of resistance (susceptibility) at the *Lsh/Bcg* locus; I-A, I-E: the two kinds of Ia molecules encoded by the I region of the MHC; expressed on the antigen presenting cells (APC) surface to present the antigen to T helper (TH) cells. The *Lsh* gene controls the early-stage resistance and the MHC products influence the late phase of the disease. (From the review and suggestions of Wassom and Kelly (1990).)

steps of the immune response: the T-cell activation and the effector cells and molecules mediation (Wassom and Kelly, 1990).

4.2. MHC Polymorphism: the Host Strategy

At this stage, it is noteworthy that not only the laboratory strains of inbred mice present different alleles on these loci, but that the wild mammals present as well the most polymorphic genes of their entire genome on the class I and II loci. Moreover, this high variability is well documented in natural populations of mice, rats and humans (Duncan and Klein, 1980; Kaufman *et al.*, 1984; Klein, 1986; Figueroa *et al.*, 1988; Lawlor *et al.*, 1988; McConnel *et al.*, 1988; She *et al.*, 1990; Wakeland *et al.*, 1990). The question remains why this polymorphism exists and is maintained and what could be its impact on the parasitism in host populations. In 1980, Klein suggested that most of the MHC alleles represent ancient polymorphism that preceded the species divergence. This hypothesis was named the "trans-species evolution" and was later supported by studies by McConnel *et al.* (1988), Figueroa *et al.* (1988), She *et al.* (1990) and Wakeland *et al.* (1990) who analysed MHC polymorphism at the $A\beta$ gene in several mouse and rat species, and by Lawlor *et al.* (1988) who worked on primates. Moreover, Wakeland *et al.* (1990) and She *et al.* (1991) demonstrated that intra-exonic recombination amplifies the diversity by shuffling five poly- morphic motifs located in the exon 2 of this $A\beta$ gene encoding the A molecule's antigen binding site, into countless different combinations (Figure 10). Stable lineages of the entire exon are found within a species or closely related taxa and are lost between distant species, suggesting that this phenomenon is spread over long evolutionary periods.

The very fact that this exon encodes for the antigen binding site allows us to understand the impact of the polymorphism on the antigen presenta- tion and recognition. The genetic variability of Exon 2 leads to class II molecules which differ in the tertiary structure of their antigen binding site. As they do not share the same structural properties, the class II molecules bind efficiently with one set of Ag but associate poorly with another. The subsequent immune response of the host will undoubtedly depend on the quality of the antigen binding and presentation. Thus, the host benefit to present such a polymorphism at immune loci is quite obvious; keeping numerous alleles at MHC loci in mouse populations appears necessary to face the large spectrum of parasites encountered.

With this view of the MHC polymorphism in natural populations of mice, we leave the laboratory for the wild and turn our attention to the two European taxa of commensal mice, *Mus musculus musculus* and *Mus musculus domesticus*, that naturally hybridize where they meet. The result-

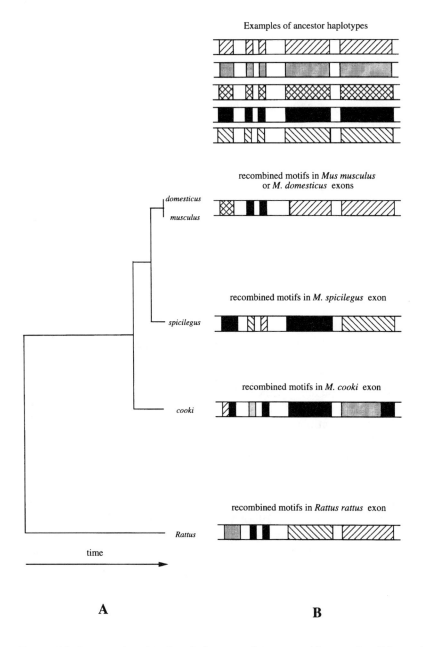

Figure 10 Compared molecular phylogeny of some muridae species (*Mus* and *Rattus*) (A) and intra-exonic recombinations of the Aβ exon (B). (Modified from Boursot *et al.* (1993) (A) and Wakeland *et al.* (1990) (B).)

ing hybrid individuals present mosaic genomes, mixed products of the two parental ones. Wakeland *et al.* (1990) and She *et al.* (1991) showed that closely related species share the same allele lineages at the $A\beta$ gene, and the same result could be applied to these two subspecies of mice. However, they differ in some obvious characters such as morphology and behaviour. Therefore, could they not also differ in other components of the immune system? The hybrid zone gives the opportunity to realize a genetic study in a natural laboratory and to explore the interactions between the host mouse and its parasites. First, let us present the European mouse hybrid zone.

5. THE MOUSE HYBRID ZONE: A NATURAL RESEARCH LABORATORY

5.1. Description and Genetic Analysis of the Zone

The main subspecies of the *Mus musculus* (house mouse) species complex probably originated in the Indian subcontinent where they started their radiation and their individualization about 0.5 million years ago (Boursot *et al.*, 1993; F. Bonhomme, personal communication). Today, the most commensal subspecies, *Mus musculus domesticus*, spreads all over the world, including Western Europe, whereas *M. m. musculus* is found only in Eastern Europe and Northern Asia. After a secondary contact, these taxa mix together within a 40 km wide zone of hybridization stretching across the European continent from Denmark to Bulgaria (Figure 11) (Hunt and Selander, 1973; Ferris *et al.*, 1983; Boursot *et al.*, 1984; Sage *et al.*, 1986a; Vanlerberghe *et al.*, 1988a, b). Within this zone, hybrids of the first generation (F_1) do not exist, but the hybrid genotypes found are mosaics stemming from recombinations of the two parental genomes. Genetic studies in Denmark (Hunt and Selander, 1973; Ferris *et al.*, 1983; Boursot *et al.*, 1984; Nancé *et al.*, 1990; Vanlerberghe *et al.*, 1986, 1988a; Dod *et al.*, 1993), West Germany and Austria (Sage *et al.*, 1986a; Tucker *et al.*, 1992) and Bulgaria (Vanlerberghe *et al.*, 1986, 1988b) have observed restricted genetic exchange between the two subspecies (Figure 12). Thus, ecological or genetic parameters could be at the origin of selective pressures against hybrid genomes. These parameters would oppose the spreading of the zone because of the migration of the mice and the constant crosses it implies at the edges of the zone. It may be that the two subspecies of mice are adapted to different environments (e.g. different climates). Some workers have attempted to correlate the geographical pattern of environmental variables with the pattern of genomic introgression (i.e.

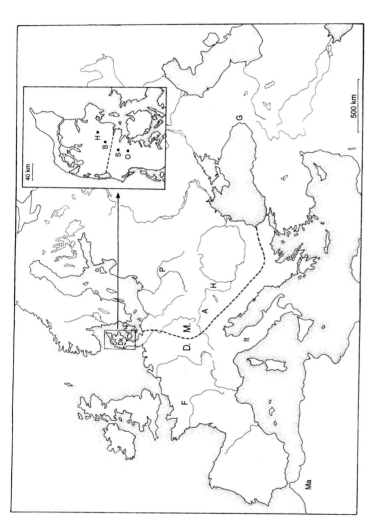

Figure 11 The hybrid zone between *Mus musculus musculus* and *M. m. domesticus* and the trapping sites of wild-derived mice. – –: hybrid zone; D.: *domesticus* territory; M.: *musculus* territory. The *domesticus* mice were from France (F), the *musculus* ones from Georgia (G) and/or Austria (A), and the hybrids from Denmark (Dk). Trapping sites in the Danish hybrid zone: H: Hov; B: Bastrup; S: Store Lihme; O: Ödis.

frequency index (%)

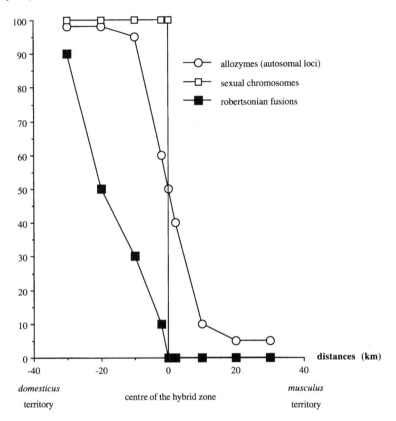

Figure 12 Schematic representation of the distribution of three genetic markers in the hybrid zone. Note that hybrid zone *per se* is not more than 40 km wide. For each marker, the frequency index varies from 100% to 0 of the *domesticus* type. (Modified from Nancé *et al.* (1990).)

the mixing of the two taxa genomes) (Boursot *et al.*, 1984; Klein *et al.*, 1987), but with little success. Indeed, it seems unlikely that the climatic gradient is sharp enough to maintain the hybrid zone. Concurrently, other workers (Sage *et al.*, 1986a, b; Vanlerberghe *et al.*, 1986, 1988a, b; Tucker *et al.*, 1992; Dod *et al.*, 1993) have suggested a genetic "evolutionary hypothesis" concluding that different co-adapted gene systems could have evolved between the two taxa during the adaptive radiation of the *M. musculus* species. Hence, hybridization of the subspecies occurred; the two genomes mixed and the co-adapted systems disrupted because of the recombination events. In the recombinant genomes of hybrids, the new gene reassortments could be less functional, leading to hybrid dysgenesis and unfitness. The hybrid zone would then result from the equilibrium between a low migration rate and the hybrid counterselection. Consequently, the hybrid zone could not be extended and the two subspecies would not mix. This low dispersal rate explains why F_1 hybrids no longer exist in the zone and why hybrid mice get more opportunities to meet and reproduce with another hybrid than with a parental individual.

In 1986, in order to analyse the genetic barriers between the species, Sage and collaborators had the simple but inspired idea of considering the burden of helminth parasites of the wild mice within a transect along the hybrid zone in Germany.

5.2. Parasitism in the Hybrid Zone: Evidence of "Wormy" Mice

Sage *et al.* (1986a) mostly found the very common intestinal pinworms *Aspiculuris tetraptera* and *Syphacia obvelata*, and some other helminths such as the cestodes *Hymenolepis* and *Taenia*. For each trapping locality along the transect, they calculated first the mean hybrid index based on allelic distribution of diagnostic enzymatic loci between the two subspecies. This index relates to the entire sample of mice from a trapping site, and not to one individual. Therefore, it represents the mean rate of introgression of the hybrid genomes from a definite locality of the zone. Then the authors tried to link it with parasite burdens of trapped individuals from these populations. A clear excessive worminess was found among the hybrid mice of the centre of the zone, whereas the mice at the edges with mainly *musculus* or *domesticus* genomes were normally infected (Figure 13). As the authors suggested, these "wormy" recombinant animals might be more susceptible than the parental ones because they no longer have the functional parental genetic combinations leading to resistance. This result provides indirect evidence in favour of the evolutionary hypothesis suggesting that hybrid genomes could be counterselected (Sage *et al.*, 1986b). However, this phenomenon could have been restricted to the German part of the zone due to an unknown ecological or genetic parameter. It was then necessary to investigate the hybrid parasitism in other locations of the zone. For this reason, Moulia *et al.* (1991) performed a similar study in Jutland (Denmark). In this work, in order to exclude ecological parameters involved in the dynamic of heteroxenous cycles of cestodes, where several hosts are needed for the completion of the life cycle of the parasite, they only examined the intestinal pinworms (*A. tetraptera* and *S. obvelata*). These parasites with a direct cycle are only dependent on the presence of their host, the mouse. Then, they verified in Denmark the same overinfection of hybrid mice when compared with the individuals of the two parental taxa as Sage *et al.* (1986b) did in Germany (Figure 14). This excessive

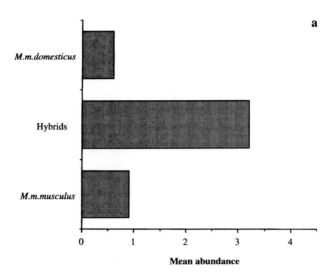

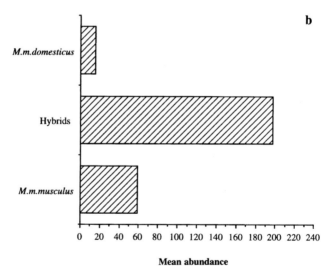

Figure 13 Abundance of cestodes (a) and intestinal nematodes (b) in *M. m. musculus, M. m. domesticus* and hybrid mice of the German part of the hybrid zone. (Data from Sage *et al.* (1986b).) The abundance is the ratio of the total number of parasites collected on the total number of examined hosts from a definite sample. It represents the mean number of parasites that any potential host of the sample may shelter (Margolis *et al.*, 1982).

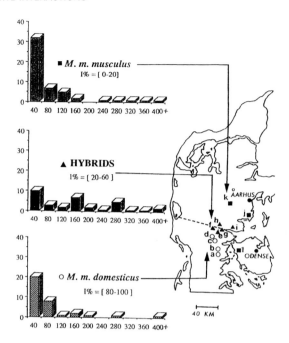

Figure 14 Danish hybrid zone and frequency distributions of parasite loads (pinworms) in mice from three groups of collecting localities: *domesticus* (a, b, c, d), hybrid (e, f, g, h, i), and *musculus* (j, k, l) territories respectively. – – –: centre of the zone. (From Moulia *et al.* (1991).)

worminess with intestinal parasites was thus a general feature of the hybrid mice resulting from *M. m. musculus* and *M. m. domesticus* (Moulia *et al.*, 1991). Moreover, these two parasitological studies showed that a variability of response exists in the two parental mouse taxa.

However, the evolutionary hypothesis of the dysfunction of hybrid recombinant genomes was still under debate. According to Klein (1988), it is probably a climatic parameter that explains the restricted gene flow between the two taxa. The study of the phenotype of resistance/susceptibility to parasites appeared to be the best way to test the genetic hypothesis. Indeed, if the *in situ* studies of hybrid parasitism could not demonstrate the genetic origin of the high susceptibility of hybrids to pinworms, an experimental approach in the laboratory, with controlled infections of mice from different origins, would determine the respective part of the ecological and genetic parameters in this phenomenon. Moulia *et al.* (1993) infected parental and hybrid mice with the pinworm *A. tetraptera*. These mice were wild-derived and out-bred animals maintained for a few generations in the laboratory. The *musculus* mice originated from Georgia, the *domesticus* ones from France and the hybrids from

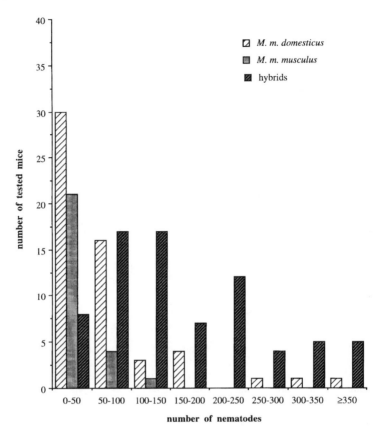

Figure 15 Frequency distributions of *Aspiculuris tetraptera* in experimentally infected *musculus*, *domesticus* and hybrid mice. (Modified from Moulia *et al.* (1993).)

four different localities within the Danish zone (Figure 11). This standardized experimental study confirmed the high susceptibility of hybrid mice compared with the *musculus* or *domesticus* ones, and the response variability among the parental taxa (Figure 15). This study demonstrated indisputably, on one hand, the genetic basis of the control of the intestinal parasite burden and, on the other, the hybrid overinfection (Moulia *et al.*, 1993). Thus, the genetic dysfunction of the hybrid genome remains the most probable basis of the high hybrid susceptibility observed *in situ*. But, this study could not clearly determine when this hybrid dysgenesis occurs. Does this susceptibility appear as soon as the two taxa genomes are confronted without being recombined, or is the break-up of functional gene interactions the result of the genetic recombination, as previously proposed? In the first alternative, the F_1 hybrids, the direct progeny from *musculus* and *domesticus* crosses, are more susceptible to pinworms than their parents. To test this on the F_1 phenotype, Moulia *et al.* (1995) realized an experimental study of the resistance/susceptibility to *A. tetraptera* of F_1 from crosses between mice of the two taxa which originated from different localities of their respective distribution area (*musculus* mice from Austria and Georgia; *domesticus* mice from France (Figure 11)). All these F_1 hybrids appear to be significantly more resistant than their parents, and reveal a phenomenon of hybrid vigour when confronted with the parasites (Figure 16).

The results on the genetic basis of hybrid susceptibility and F_1 heterosis are easily integrated into the evolutionary hypothesis proposed above. Indeed, Jacobson and Reed (1974), Behnke (1975) and more recently Lewis *et al.* (1991) underlined the role of the immune system and especially of T cells in the control of the intestinal burden of pinworms in mice. We previously emphasized the fact that the immune system, composed of many different cells and molecules, is under the control of many interacting genes. These functional genic associations may be different in *musculus* and *domesticus* genomes. In those of F_1 hybrids, not only the two unbroken parental associations are functional but the complementation between alleles on each concerned locus could also be at the origin of hybrid vigour. In the second generation of hybrids, the genetic recombination would disrupt these functional gene systems and lead to "bad" new associations incapable of controlling the parasite burden. That probably explains why some mice of the hybrid zone with highly recombined genotypes harbour such parasite loads. Studies of the parasite phenotype of experimentally infected F_2 hybrids, in progress, will indisputably demonstrate the role of the genetic recombination in the hybrid susceptibility.

5.3. Laboratory Strains: Are They Representative Samples of the Mouse Genomes?

In order to find a reproducible, standardized model for the genetic studies of parasite susceptibility and hybrid dysgenesis, the phenotype of inbred and outbred strains of mice has to be determined. A total of 17 strains were infected with *A. tetraptera*, according to the same protocol used to infect the wild mice. The results of these studies are quite interesting. After comparison with wild mice (resistant parental/susceptible hybrid), two results must be stressed (Figure 17). First, the inter- and intra-strain variability of parasite loads for these laboratory mice is largely reduced, even if it exists. This is not surprising as the inbred and outbred mice present little or no polymorphism when compared to the highly variable

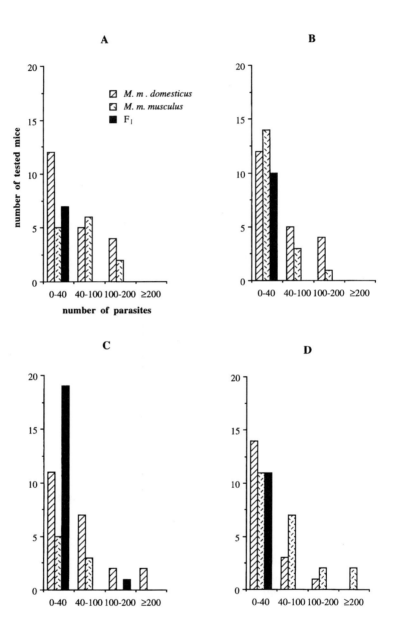

Figure 16 Frequency distributions of *Aspiculuris tetraptera* in experimentally infected *musculus*, *domesticus* and F$_1$ from four different crosses of mice of the two subspecies (see Moulia *et al.*, 1995).

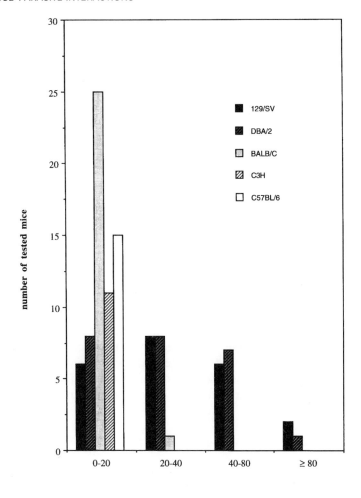

Figure 17 Frequency distributions of *Aspiculuris tetraptera* in five inbred strains of mice. Note the low parasite loads when compared to wild mice (Figures 14 and 15) (see Moulia *et al.*, 1995).

wild individuals. Second, all the tested laboratory (in- and outbred) strains must be regarded as resistant to the nematode, even if they differ from each other (Moulia *et al.*, 1995; C. Moulia, unpublished data).

To explain this last result, Moulia *et al.* (1995) suggested that the resistant allelic associations were selected because of the breeding conditions of the mice in the laboratory. Indeed, the inbreeding process leads to the accumulation of mutations at the homozygous state. The contacts between potential hosts, favouring direct cycles of oxyuroids, are increased when animals are maintained in captivity (limited area in the cage, soiled litters with faeces and parasite eggs). Then, the exposure of the mice to the parasites takes place daily. This way, inbred mice with genic systems of resistance were selected whereas others with systems conferring suscept-ibility to pinworms were counter-selected. Once more, this last result on the parasite phenotype clearly raised the question of the adequacy of these laboratory mice as immunological and genetic models of the mouse. Not only do they have inbred recombinant genotypes from mixed genomes of the *Mus musculus* subspecies (Bonhomme *et al.*, 1987; Morgado *et al.*, 1993), but their controlled breeding in the particular conditions of the laboratory has also selected several characters in all the strains. Hence, all the polymorphism of the wild populations is not represented in these different mice. In the case of oxyuroids, the parasite pressure selected the resistant genes. On the contrary, the rare exposure to *Mycobacterium*, *Salmonella* or *Leishmania*, and then the little or non-persistent survival disadvantage of susceptible Bcg^s mice leads the susceptible allele to randomly fix itself on any strains (Malo *et al.*, 1994).

The obvious selection of resistance to oxyuroids in laboratory mice genomes seems to indicate that these intestinal parasites, regarded as non-pathogenic, may in some way affect the fitness of that host. This would be especially important when the parasite burden is high, as for the mice of the hybrid zone that no longer have the functional allele associations in their immune system to control the intestinal load.

5.4. Parasite Susceptibility and Hybrid Fitness

The question remains if the resistance/susceptibility against intestinal parasites is specifically modified under the hybridization process or if the resistance against all the biotic aggressors is no longer efficient in hybrid mice. In other words, the high susceptibility of recombinant mice to intestinal pinworms might reflect either a defect in a local intestinal component of the immune response, or a more general immunodeficiency at some key-levels of the elaboration of this response. In this latter hypothesis, the selective pressure of parasites on hybrid populations may be important in understanding the occurrence of the hybrid zone. Suppos-ing that a viral or bacterial epidemic, seriously affecting host fitness (mortality) spreads in the natural populations of mice, and assuming that the immune response of the hybrid is less efficient than that of the parental mice, a higher mortality rate would reduce the hybrid population size and then limit the extension of the hybrid zone.

Intestinal immunodeficiency alone may also strongly affect hybrid fitness. Many studies have attempted to find a relationship between host infection with gut parasites and decrease in fitness. Indeed, the idea was that their location within the gastrointestinal tract could affect the digestive physiology of the host (for instance, they might divert some of the host's food or disturb its assimilation by the host's intestine). The digestive disorders would then affect some fitness parameters. Some studies focused particularly on the parasitism by gastrointestinal helminths of two kinds of host models: first, domestic animals with an important economic value, such as pigs, cattle, goats and sheep and secondly, micromammals representing good laboratory models. In the case of domestic animals, the clinical signs of pathology (weight loss, anorexia, diarrhoea or haematological changes) as well as productivity loss (milk, wool, meat) were significant (Randall and Gibbs, 1981; Symons and Hennessy, 1981; Williams *et al.*, 1983; Holmes, 1985; Symons, 1985; Abbot *et al.*, 1986; Vercruysse *et al.*, 1988; Stewart and Gasbarre, 1989; Maclean *et al.*, 1992; Hoste and Chartier, 1993; Ploeger and Kloosterman, 1993). Moreover, many cases of mortality were documented (Pandey *et al.*, 1984; Gulland, 1992; Ideguchi *et al.*, 1992). Any health professional, either of humans or animals, has the certitude that parasites affect the general health status of its host. But it is quite interesting to translate in genetic terms these veterinary notions as well as the economical concept of productivity loss usually referring to the milk production and the number of saleable young. They clearly describe some parameters of the reproductive functions of the mammal host, an important component of its fitness. Moreover, it has been demonstrated that: (i) miscarriages, stillbirth and mortality of ewes and of their offspring were clearly reduced by anthelminthic treatment when they were infected by gut and lung worms (Pandey *et al.*, 1984); and (ii) ecto- and endoparasites had adverse effects on weight gain of calves (Devaney *et al.*, 1992) and on the metabolism of heifers, increasing the age of their puberty and reducing their rate of calving, if these animals were not fed supplements (O'Kelly *et al.*, 1988).

These last studies point out the difficulties of having precise information of the consequences of intestinal parasite infection on the fitness of the domestic farming animals. Being naturally infected, they often harbour many different parasites (such as lung and intestinal helminths of ewes and their lambs, ecto- and endoparasites of heifers). Once again, experimental studies are easily controlled with micromammals, and this takes us back to the mouse and its relatives. Although Heasley (1983) clearly showed that reduced feeding affects the reproduction characteristics of laboratory mice (female weight loss, lower birth weights, lower milk production and increased cannibalism), he could not establish any link with intestinal parasite infection. Then, a study was made by Munger and Karasov (1989) on the host–parasite association of the white-footed mouse, *Peromyscus leucopus*, and the intestinal tapeworm, *Hymenolepis citedelli*. They based their arguments on their observations of non-reproductive parasitized individuals, and proposed that: "a tapeworm would have the potential (if compensations do not occur) to decrease by 20% the production of young during one pregnancy or lactation for two-thirds of the mice, and during two such events for one-third of the mice". Moreover, they noted that: "if environmental conditions preclude compensations by increased food intake, then an actual reduction in reproduction might be expected to occur"; "for the majority of mice (which are born and reproduce within one summer season, then die during the following winter), this loss would represent a substantial fraction of the lifetime reproductive output". If *H. citedelli* would have little effect because of its low prevalence in natural populations, effects might exist for gut parasites that are at higher prevalences, for example, whipworms (*Trichuris* sp.) in *P. leucopus*, or tapeworms in kangaroo rats, or tapeworms in *Microtus*. We note that this would be particularly true for mice of the hybrid zone that present high parasite burden.

All these findings insist on the fact that in natural conditions, with the difficulties of finding food and the risk of predation, any parasite could be significant as a selective factor exerted on a micromammalian population.

Intestinal protozoan parasites may have the same effects as helminths on their host's fitness. A productivity loss similar to that caused by helminthiasis has been noted in domestic animals infected by the intestinal coccidian of the genus *Eimeria* (Catchpole *et al.*, 1976; Friend and Stockdale, 1980; Stockdale *et al.*, 1981; Harleman and Meyer, 1983; Aumont *et al.*, 1984; Lindsay *et al.*, 1985; Norton, 1986; Vercruysse *et al.*, 1988). A high mortality rate may even occur in ruminants and pigs when heavily infected (Vercruysse *et al.*, 1988). Laboratory studies have also been carried out on mice infected by *E. vermiformis*. Mice subclinically infected (presenting no apparent digestive clinical symptoms) with this parasite seem to live very well with their coccidia. Actually, their behaviour and physiology such as endogenous analgesic mechanisms, are modified (Colwells and Kavaliers, 1993) and their defence reactions when exposed to a predator, are altered (Kavaliers and Colwell, 1993, 1994). Recently, Kavaliers and Colwells (1993, 1994) demonstrated that intestinal parasites, even if the digestive disorder seems weak, may strongly affect the host's fitness. Supposing that an unfortunate encounter of the infected host with a predator takes place in the field, the chances of the mouse escaping would probably be inferior to those of a healthy individual.

Perhaps pinworms recovered from mice of the hybrid zone would not cause the same physiological changes to their hosts as *E. vermiformis*, but still intestinal disorders of the infected mice (rectal prolapse, occlusions) (Harwell and Boyd, 1968; Jacobson and Reed, 1974; Taffs, 1976) and decrease in their health status (Harwell and Boyd, 1968; Eaton, 1972) are well documented. These clinical signs are similar to those of domestic animals and micromammals with other parasites presented earlier in this review. Therefore, it is highly probable that pinworms have a negative effect on the reproductive success of the hyperinfected hybrid mouse, resulting from unknown but real physiological changes. Moreover, hybrid mice have more liver cysts of taeniid cestodes than their parents (Sage *et al.*, 1986b), and this calls for two important remarks. First, Lin *et al.* (1990) showed that male and female rats experimentally infected with *Taenia taeniaformis* presented reproductive dysfunctions (irregular oestrous cycles, reduced male fertility, less mating success, reduced embryo implantation, reduced litter size). Secondly, even if hybrid mice were found not to be more susceptible to parasites (viruses, bacteria, protozoans, helminths) when compared to parental ones, it seems they are more susceptible to helminth parasites of the alimentary system (intestine and the liver). Perhaps, some of the particular immune mechanisms involved in the control of these helminthiases of the digestive system (for example, one T cell subset) is less functional in the hybrids, due to the mixing of the two subspecies genomes.

Thus, parasitism could be looked upon as an important factor able to maintain the hybrid zone.

6. CONCLUSION AND PERSPECTIVES

Our purpose was to show that if controlled laboratory studies are essential in understanding the cellular and genetic mechanisms of host–parasite interactions, the use of non-natural hosts may lead to erroneous conclusions of the immune response of the natural host(s) and the immunological status of the experimental host confronted by its natural parasites. Hence, *Meriones unguiculatus*, the Mongolian gerbil, would have been regarded too hastily as defective in local intestinal immunity when infected by the rat parasite, *Strongyloides venezuelensis* (Khan *et al.*, 1993). The parasite forced to invade an unnatural host (a new host genome), as a free organism living in a new environment, may quickly adapt in a few generations and thus differ genetically from the natural parasites. This uncontrolled selection of new parasite genomes in unnatural hosts is well described by the work of Brémond *et al.* (1993). These authors showed that modifications of the genetic polymorphism (measured by the frequency variation of one malate dehydrogenase allele) occurs in strains of *Schistosoma mansoni*, a parasite of the human blood system, when it is experimentally maintained in mice. This result confirms previous analyses of the modifications of the enzymatic polymorphism when the schistosomes are transferred to the murine experimental host (Fletcher *et al.*, 1981; LoVerde *et al.*, 1985). More generally, the work of Gregory *et al.* (1990) on the intestinal nematode parasite *Heligmosomoides polygyrus* in its natural host *Apodemus sylvaticus* showed that many aspects (parasite fecundity, male/female worm survival, duration of the infection, etc.) of the parasite–laboratory mouse association were modified when compared to the natural association. The results obtained with such host–parasite models are rarely extrapolated to natural situations, as host and parasite differ from those in natural locations and the genetic polymorphism of natural populations is disregarded. These difficulties are regrettably obvious in vaccine development against human parasites such as *Plasmodium* sp. (Lockyer *et al.*, 1989; Arnot, 1990; Hui and Chang, 1992).

Although insufficient by itself, the mouse model with its natural parasites remains useful for experimental studies of mammal/parasite interactions. This micromammal is the host of numerous parasites, among which many have medical importance and/or may be good models for the study of similar parasites of humans. But, again, we must be cautious with the results obtained from laboratory strains as we now have several indications showing that they do not represent good genetic samples of the wild mice. Thus, it is essential to develop studies with wild-derived strains. They entitle us to consider the greatest part of the genetic diversity of the murine genome justifying why the hybrid zone model (the entire panel of recombinant hybrid genomes between the two parental taxa) is of particular interest. This model transposed from *in situ* to laboratory, enabled us to show the following. First, genes within the immune system were selected in mammal host genomes to control (regulate) the infrapopulations of parasites. On the one hand, the parasite burden is not too high and does not compromise the survival and reproduction of the host (i.e. its fitness) and on the other hand, the parasite is able to realize its cycle. Each of the two partners benefits from this compromise, but the range of parasites is large (pinworms and other helminths are not the only parasites of mice) and the cost of resistance against a definite parasite is real even if hardly measurable. Hence, inside the host population, a response polymorphism allows the mouse population to adapt its response to each possible biotic aggressor and to limit its impact. Secondly, different co-adapted systems were selected in the *musculus* and *domesticus* genomes to regulate the parasite load. They express a coevolutionary phenomenon between hosts and their parasites. These gene associations are sufficiently different in closely related taxa to generate hybrids unable to control the parasite infections. Thus, the genetic disharmony between the two mice subspecies, which probably prevent them from mixing, would be partly expressed by their modalities of interaction with parasites, i.e. at the immune level.

Further studies must be carried out to understand the genetic basis of the resistance/susceptibility to parasites within the hybrid zone and in the parental populations, as well as the distribution of these immune genes in populations subjected to many different aggressors.

Studies of the interactions between a host and a parasite need a new approach which we have called Evolutionary Immunology. This will establish the bridge between laboratory and field researches. Indeed, the necessary experimental investigations are limited by their restricted point of view (the cell or the organism), and it is essential to work on natural populations. They are the only possible products of the evolving story of the association between a host and a parasite.

ACKNOWLEDGEMENTS

The authors are especially grateful to Prof. D. Wakelin and Prof. C. Combes for their always constructive comments and suggestions about the manuscript. We wish to acknowledge Prof. J. Marti and Dr J.-P. Liautard for their helpful comments about the immunological mechanisms explained in this review. Special thanks to Eva Chancel for kindly reviewing the English. During the preparation of this manuscript, C. Moulia was in receipt of a grant from the Foundation for Medical Research.

REFERENCES

Abbot, E.M., Parkins, J.J. and Holmes, P.H. (1986). Influence of dietary protein on the pathophysiology of acute ovine haemonchosis. *Veterinary Parasitology* **20,** 291–306.

Abel, L. and Demenais, F. (1988). Detection of major genes for susceptibility to leprosy and its subtypes in a Caribbean island, Desirade Island. *American Journal of Human Genetics* **42,** 256–266.

Adu, H.O., Curtis, J. and Turk, J.L. (1983). Role of the major histocompatibility complex in resistance and granuloma formation in response to *Mycobacterium leprae* infection. *Infection and Immunity* **49,** 720–725.

Albright, J.W. and Albright, J.F. (1989). Immunological and non-immunological control of severity of *Trypanosoma musculi* infections in C3H and C57BL/6 inbred mice. *Infection and Immunity* **57,** 1647–1655.

Allen, P.M., McKean, D.J., Beck, B.N., Sheffield, J. and Glincher, L.H. (1985). Direct evidence that class II molecules and a simple globular protein generate multiple determinants. *Journal of Experimental Medicine* **162,** 1264–1274.

Allison, J.P. (1994). CD28-B7 interactions in T-cell activation. *Current Opinion in Immunology* **6,** 414–419.

Arnot, D.E. (1990). Polymorphism in the circumsporozoite protein and anti-sporozoite malaria vaccines. *Parasitology Today* **6**, 64–65.

Aumont, G., Yvore, P. and Esnault, A. (1984). Experimental coccidiosis in goats. 1. Experimental model. Effects of parasitism on the feeding behaviour and the growth of animals; intestinal lesions. *Annales de Recherche Vétérinaire* **15**, 467–473.

Aycock, W.L. (1940). Familial susceptibility as a factor in the propagation of leprosy in North America. *International Journal of Leprosy* **8**, 137–150.

Aycock, W.L. and McKinley, E.B. (1938). The roles of familial susceptibility and contagion in the epidemiology of leprosy. *International Journal of Leprosy* **6**, 169–184.

Babbit, B.P., Allen, P.M., Matsueda, G., Haber, E. and Unanue, E.R. (1985). Binding of immunogenetic peptides to Ia histocompatibility molecules. *Nature* **317**, 359–361.

Barreyro, D.A., Baras, M., Squires, P., Walerstein, M., Yodfat, Y. and Levy, L. (1982). Familial clustering of leprosy patients in an Israeli village. *Leprosy Review* **53**, 277–283.

Behnke, J.M. (1975). Immune expulsion of the nematode *Aspiculuris tetraptera* from mice given primary and challenge infections. *International Journal for Parasitology* **5**, 511–515.

Behnke, J.M. and Wahid, F.N. (1991). Immunological relationships during primary infection with *Heligmosomoides polygyrus* (*Nematospiroides dubius*), H-2 linked genes determine worm survival. *Parasitology* **103**, 157–164.

Belknap, N.R. and Haynes, W.G. (1961). A genetic analysis of families in which leprosy occurs. *International Journal of Leprosy* **29**, 375–383.

Bell, R.G. (1988). Genetic analysis of expulsion of adult *Trichinella spiralis* in NFS, C3H/He, and B10-BR mice. *Experimental Parasitology* **66**, 57–65.

Bell, R.G., McGregor, D.D. and Adams, L.S. (1982). *Trichinella spiralis*: genetic basis for differential expression of phase-specific intestinal immunity in inbred mice. *Experimental Parasitology* **53**, 315–325.

Bell, R.G., Adams, L.S. and Ogden, R.W. (1984). *Trichinella spiralis*: genetics of worm expulsion in inbred and F1 mice infected with different worm doses. *Experimental Parasitology* **58**, 345–355.

Bishop, C.E., Boursot, P., Baron, B., Bonhomme, F. and Hatat, D. (1985). Most classical *Mus musculus domesticus* laboratory mouse strains carry a *Mus musculus musculus* Y chromosome. *Nature* **325**, 70–72.

Blackwell, J.M. (1983). *Leishmania donovani* infection in heterozygous and recombinant H-2 haplotype mice. *Immunogenetics* **18**, 101–109.

Blackwell, J.M. (1985). A murine model of genetically controlled host responses to leishmaniasis. In: *Ecology and Genetics of Host–Parasite Interactions*. pp. 147–155. Linnean Society of London.

Blackwell, J.M. and Roberts, M.B. (1987). Immunomodulation of murine visceral leischmaniasis by administration of monoclonal anti-I-A *vs.* anti-I-E antibodies. *European Journal of Immunology* **17**, 1669–1672.

Blackwell, J., Freeman, J.C. and Bradley, D.J. (1980). Influence of H-2 complex on acquired resistance to *Leishmania donovani* infection in mice. *Nature* **283**, 72–74.

Bloom, B.R., Salgame, P. and Diamond, B. (1992). Revisiting and revising suppressor T cells. *Immunology Today* **13**, 131–136.

Bonhomme, F., Guénet, J.L., Dod, B., Morivaki, K. and Bufield, G. (1987). The polyphyletic origin of laboratory inbred mice and their rate of evolution. *Biological Journal of the Linnean Society* **30,** 51–58.

Boursot, P., Auffray, J.-C., Britton-Davidian, J. and Bonhomme, F. (1993). The evolution of house mice. *Annual Review of Ecology and Systematics* **24,** 119–152.

Boursot, P., Bonhomme, F., Britton-Davidian, J., Catalan, J., Yonekawa, H., Orsini, P., Gerasimov, S. and Thaler, L. (1984). Introgression différentelle des génomes nucléaires et mitochondriaux chez deux semi-espèces de souris. *Comptes Rendus de l'Academie des Sciences de Paris* **299,** 365–370.

Bradley, D.J. (1977). Regulation of *Leishmania* populations within the host. II. Genetic control of acute susceptibility of mice to *Leishmania donovani* infection. *Clinical and Experimental Immunology* **30,** 130–140.

Bradley, D.J., Taylor, B.A., Blackwell, J.M., Evans, E.P. and Freeman, J. (1979). Regulation of *Leishmania* populations within the host. III. Mapping of the locus controlling susceptibility to visceral leismaniasis in the mouse. *Clinical and Experimental Immunology* **37,** 7–14.

Brémond, P., Pasteur, N., Combes, C., Renaud, F. and Théron, A. (1993). Experimental host-induced selection in *Schistosoma mansoni* strains from Guadeloupe and comparison with natural observations. *Heredity* **70,** 33–37.

Brownstein, D.G., Bhatt, P.N., Gras, L. and Jacoby, R.O. (1991). Chromosomal locations and gonadal dependence of genes that mediate resistance to Ectromelia (Mousepox) virus-induced mortality. *Journal of Virology* **65,** 1946–1951.

Brownstein, D.G., Bhatt, P.N., Gras, L. and Budries, T. (1992). Serial backcross analysis of genetic resistance to Mousepox using marker loci for *Rmp-2* and *Rmp-3*. *Journal of Virology* **66,** 7073–7079.

Buschman, E. and Skamene, E. (1988). Immunological consequences of innate resistance and susceptibility to BCG. *Immunology Letters* **19,** 199–210.

Catchpole, J., Norton, C.C. and Joyner, L.P. (1976). Experiments with defined multispecific coccidial infections in lambs. *Parasitology* **72,** 137–147.

Cellier, M., Govoni, G., Vidal, S., Kwan, T., Groulx, N., Liu, J., Sanchez, F., Skamene, E., Schurr, E. and Gros, P. (1994). Human natural resistance-associated macrophage protein: cDNA cloning, chromosomal mapping, genomic organisation, and tissue-specific expression. *Journal of Experimental Medicine* **180,** 1741–1752.

Chakravarrti, M.R. and Vogel, F. (1973). A twin study on leprosy. In: *Topics in Human Genetics* (Becker, P.E. and others, eds), pp. 1–123. Stuttgart: Georg Thieme Verlag.

Chesnut, R.W. and Grey, H.M. (1981). Studies on the capacity of B cells to serve as antigen-presenting cells. *Journal of Immunology* **126,** 1075–1079.

Colwells, D.D. and Kavalier, M. (1993). Altered nociceptive responses of mice infected with *Eimeria vermiformis*. Evidence for involvement of endogenous opioid systems. *Journal of Parasitology* **79,** 751–756.

Comstock, G.W. (1978). Tuberculosis in twins, a re-analysis of the Prophit study. *American Review of Respiratory Disease* **117,** 621–624.

Cox, F.E.G. and Liew, E.Y. (1992). T-cell subsets and cytokines in parasitic infections. *Parasitology Today* **8,** 371–374.

Crocker, P.R., Blackwell, J.M. and Bradley, D.J. (1984). Expression of the natural resistance gene *Lsh* in resident liver macrophages. *Infection and Immunity* **43,** 1033–1040.

Denis, M., Forget, A., Pelletier, M. and Skamene, E. (1988) Pleiotropic effects of the *Bcg* gene. I. Antigen presentation in genetically susceptible and resistant mouse strains. *Journal of Immunology* **140,** 2395–2400.

Devaney, J.A., Graig, T.M., Rowe, L.D., Wade, C. and Miller, D.K. (1992). Effects of low levels of lice and internal nematodes on weight gain and blood parameters in calves in central Texas. *Journal of Economic Entomology* **85,** 144–149.

Dod, B., Jermiin, L.S., Boursot, P., Chapman, V.H., Nielsen, J.T. and Bonhomme, F. (1993). Counterselection on sex chromosomes in the *Mus musculus* European hybrid zone. *Journal of Evolutionary Biology* **6,** 529–546.

Duncan, W.R. and Klein, J. (1980). Histocompatibility-2 system in wild mice. IX. Serological analysis of 13 new B10.W congenic lines. *Immunogenetics* **10,** 45–65.

Eaton, G.J. (1972). Intestinal helminths in inbred strains of mice. *Laboratory Animal Science* **22,** 850–853.

Else, K. and Grencis, R.K. (1991). Helper T-cell subsets in mouse trichuriasis. *Parasitology Today* **7,** 313–316.

Else, K. and Wakelin, D. (1988). The effects of H-2 and non-H-2 genes on the expulsion of the nematode *Trichuris muris* from inbred and congenic mice. *Parasitology* **9,** 543–550.

Else, K.J., Wakelin, D., Wassom, D.L. and Hauda, K.M. (1990a). The influence of genes mapping within the major histocompatibility complex on the resistance to *Trichuris muris* in mice. *Parasitology* **101,** 61–67.

Else, K.J., Wakelin, D., Wassom, D.L. and Hauda, K.M. (1990b). MHC-restricted antibody responses to *Trichuris muris* excretory/secretory (E/S) antigen. *Parasite Immunology* **12,** 509–527.

Enriquez, F.J., Brooks, B.O., Cypess, R.H., David, C.S. and Wassom, D.L. (1988). *Nematospiroides dubius,* Two H-2-linked genes influence levels of resistance to infection in mice. *Experimental Parasitology* **67,** 221–226.

Erard, F., Wild, M.T., Garcia-Sanz, J.A. and Le Gros, G. (1993). Switch of CD8 T cells to noncytolytic CD8-CD4 cells that make Th2 cytokines and help B cells. *Science* **260,** 1802–1805.

Ferris, S.D., Sage, R.D., Huang, C.M., Nielsen, J.T., Ritte, U. and Wilson, A.C. (1983). Flow of mitochondrial DNA across a species boundary. *Proceedings of the National Academy of Sciences, USA* **80,** 2290–2294.

Figueroa, F., Gunther, E. and Klein, J. (1988). MHC polymorphism pre-date speciation. *Nature* **335,** 265–267.

Fine, P.E.M. (1981). Immunogenetics of susceptibility to leprosy, tuberculosis, and leishmaniasis; an epidemiological perspective. *International Journal of Leprosy* **49,** 437–454.

Finkelman, F.D., Pearce, E.J., Urban, J.F. and Sher, A. (1991). Regulation and biological function of helminth-induced cytokine responses. In: *Immunoparasitology Today* (C. Ash and R.B. Gallagher, eds), pp. A62–A66. Cambridge: Elsevier Trends Journals.

Fletcher, M., LoVerde, P.T. and Woodruff, D.S. (1981). Genetic variation in *Schistosoma mansoni,* enzyme polymorphisms in populations from Africa, Southwest Asia, South America and the West Indies. *American Journal of Tropical Medicine and Hygiene* **30,** 406–421.

Friend, S.C.E. and Stockdale, P.H.G. (1980). Experimental *Eimeria bovis* infections in calves, a histopathological study. *Canadian Journal of Comparative Medicine* **44,** 129–140.

Garra, A.O. and Murphy, K. (1994). Role of cytokines in determining T-lymphocyte function. *Current Opinion in Immunology* **1994,** 458–466.

Gazzinelli, R.T., Hakim, F.T., Hieny, S., Shearer, G.M. and Sher, A. (1991). Synergistic role of CDA+ and CD8+ T lymphocytes in IFNγ production and protective immunity induced by an attenuated *Toxoplasma gondii* vaccine. *Journal of Immunology* **146**, 286–292.

Gervais, F., Stevenson, M. and Skamene, E. (1984). Genetic control of resistance to *Listeria monocytogenes*, regulation of leukocytes inflammatory responses by the Hc locus. *Journal of Immunology* **132**, 2078–2083.

Gregory, R.D., Keymer, A.E. and Clarke, J.R. (1990). Genetics, sex and exposure: the ecology of *Heligmosomoides polygyrus* (Nematoda) in the wood mouse. *Journal of Animal Ecology* **59**, 363–378.

Grencis, R.K., Hültner, L. and Else, K. (1991). Host protective immunity to *Trichinella spiralis* in mice, activation of Th subsets and lymphokine secretion in mice expressing different response phenotypes. *Immunology* **74**, 329–332.

Grey, H.M. and Chesnut, R. (1985). Antigen processing and presentation of T cells. *Immunology Today* **6**, 101–106.

Gros, P., Skamene, E. and Forget, A. (1981). Genetic control of natural resistance to *Mycobacterium bovis* (BCG) in mice. *Journal of Immunology* **127**, 2417–2421.

Gros, P., Skamene, E. and Forget, A. (1983). Cellular mechanisms of genetically controlled host resistance to *Mycobacterium bovis* (BCG). *Journal of Immunology* **131**, 1966–1972.

Gulland, F.M.D. (1992). The role of nematode parasites in Soay sheep (*Ovis aries* L.) mortality during a population crash. *Parasitology* **105**, 493–503.

Haile, R.W., Iselius, L., Fine, P.E. and Monton, N.E. (1985). Segregation and linkage analysis of 72 leprosy pedigrees. *Human Heredity* **35**, 43–52.

Hamelin-Bourassa, D., Skamene, E. and Gervais, F. (1989). Susceptibility to a mouse acquired immunodeficiency syndrome is influenced by the *H-2*. *Immunogenetics* **30**, 266–272.

Harleman, J.H. and Meyer, R.C. (1983). *Isospora suis* infection in piglets. A review. *Veterinary Questions* **5**, 178–183.

Harwell, J.F. and Boyd, D.D. (1968). Naturally occurring oxyuriasis in mice. *Journal of American Veterinary Medical Association* **153**, 950–953.

Heasley, J.H. (1983). Energy allocation in response to reduced food intake in pregnant and lactating laboratory mice. *Acta Theriologica* **28**, 55–71.

Heinzel, F.P., Sadick, M.D., Holaday, B.J., Coffman, R.L. and Locksley, R.M. (1989). Reciprocal expression of Interferon gamma or Interleukine 4 during the resolution or progression of murine leishmaniasis. *Journal of Experimental Medicine* **169**, 59–72.

Hill, A.V.S., Allsopp, C.E.M., Kwiatkowski, D., Anstey, N.M., Twumasi, P., Rowe, P.A., Bennett, S., Brewster, D., McMichael, A.J. and Greenwood, B.M. (1991). Common West African HLA antigens are associated with protection from severe malaria. *Nature* **352**, 595–600.

Holmes, P.H. (1985). Pathogenesis of trichostrongylosis. *Veterinary Parasitology* **18**, 89–101.

Hood, L., Steinmetz, M. and Malissen, B. (1983). Genes of the major histocompatibility complex of the mouse. *Annual Review of Immunology* **1**, 529–568.

Hopkins, R. and Denny, O.E. (1929). Leprosy in the United States. *Journal of the American Medical Association* **92**, 191–198.

Hoste, H. and Chartier, C. (1993). Comparison of the effects on milk production of concurrent infection with *Haemonchus contortus* and *Trichostrongylus colubriformis* in high- and low-producing dairy goats. *American Journal of Veterinary Research* **54**, 1886–1893.

Hui, S.N. and Chang, S.P. (1992). *Plasmodium falciparum*: induction of biologically active antibodies to gp195 is dependent on the choice of adjuvants. *Experimental Parasitology* **75**, 155–157.

Hunt, W.G. and Selander, R.K. (1973). Biochemical genetics of hybridisation in European house mouse. *Heredity* **31**, 11–33.

Ideguchi, H., Matsuda, M., Taira, N., Nishitateno, H. and Nishi, S. (1992). Periodical parasitological survey of calves on a farm where "sudden death" occurred. *Journal of the Japan Veterinary Medical Association* **45**, 747–751.

Jacobson, R.H. and Reed, N.D. (1974). The thymus dependency of resistance to pinworm infection in mice. *Journal of Parasitology* **60**, 976–979.

Kaufman, J.F., Auffray, C., Korman, A.J., Shackleford, D.A. and Strominger, J. (1984). The class II molecules of the human and murine major histocompatibility complex. *Cell* **36**, 1–13.

Kavaliers, M. and Colwell, D.D. (1993). Multiple opioid system involvement in the mediation of parasitic-infection induced analgesia. *Brain Research* **623**, 316–320.

Kavaliers, M. and Colwell, D.D. (1994). Parasite infection attenuates nonopioid mediated predator-induced analgesia in mice. *Physiology and Behaviour* **55**, 505–510.

Kennedy, M.W., McIntosh, A.E., Blair, A.J. and McLaughlin, D. (1990). MHC (RT1) restriction of the antibody repertoire to infection with the nematode *Nippostrongylus brasiliensis* in the rat. *Immunology* **71**, 317–322.

Kennedy, M.W., Wassom, D.L., McIntosh, A.E. and Thomas, J.C. (1991). H-2 (I-A) control of the antibody repertoire to secreted antigens of *Trichinella spiralis* in infection and its relevance to resistance and susceptibility. *Immunology* **73**, 36–43.

Keymer, A.E., Tarton, A.B., Hiorns, R.W., Lawrence, C.E. and Pritchard, D.I. (1990). Immunogenetic correlates of susceptibility to infection with *Heligmosomoides polygyrus* in outbred mice. *Parasitology* **101**, 69–73.

Khan, A.I., Horii, Y. and Nawa, Y. (1993). Defective mucosal immunity and normal systemic immunity of Mongolian gerbils, *Meriones unguiculatus*, to reinfection with *Strongyloides venezuelensis*. *Parasite Immunology* **15**, 565–571.

Klein, D., Tewarson, S., Figueroa, F. and Klein, J. (1982). The minimal length of the differential segment in H-2 congenic lines. *Immunogenetics* **16**, 319–328.

Klein, J. (1980). Generation of diversity at MHC loci: implications for T-cell receptor repertoires. In: *Immunology 80* (M. Fougereau and J. Dausset, eds), pp. 239–253. London: Academic Press.

Klein, J. (1986). *Natural History of the Major Histocompatibility Complex*. New York: Wiley.

Klein, J. (1988). Debate about three pages. *Immunogenetics* **28**, 67–68.

Klein, J., Figueroa, F. and David, C.S. (1983). H-2 haplotypes, genes and antigens, second listing. II. The H-2 complex. *Immunogenetics* **17**, 553–596.

Klein, J., Tichy, H. and Figueroa, F. (1987). On the origin of mice. *Annual of the University of Chile* **5**, 91–120.

Lawlor, D.A., Ward, F.E., Ennis, P.D., Jackson, A.P. and Parham, P.P. (1988). HLA-A and B polymorphisms predate the divergence of humans and chimpanzees. *Nature* **335**, 268–271.

Lewis, D.B., Yu, C.C., Forbush, K.A., Carpenter, J., Sato, T.A., Grossman, A., Liggit, D.H. and Perlmutter, R.M. (1991). Interleukine 4 expressed in situ selectively alters thymocyte development. *Journal of Experimental Medicine* **173**, 89–100.

Le Gros, G. and Erard, F. (1994). Non-cytotoxic, Il-4, IL-5, Il-10 producing CD8[+] T cells, their activation and effector functions. *Current Opinion in Immunology* **6**, 453–457.

Liew, F.Y. (1987). Regulation of cell-mediated immunity in cutaneous leishmaniasis. *Immunology Letters* **16**, 321–328.

Lin, Y.C., Rikihisa, Y., Kono, H. and Gu, Y. (1990). Effects of larval tapeworm (*Taenia taeniaformis*) infection on reproductive functions in male and female host rats. *Experimental Parasitology* **70**, 344–352.

Lindsay, D.S., Current, W.L. and Taylor, J.R. (1985). Effects of experimentally induced *Isospora suis* infection on morbidity, mortality and weight gains in nursing pigs. *American Journal of Veterinary Research* **46**, 1511–1512.

Locksley, R.M. and Scott, P. (1991). Helper T-cell subsets in mouse leishmaniasis, induction, expansion and effector function. In: *Immunoparasitology Today* (C. Ash, and R.B. Gallagher, eds), pp. A58–A61. Cambridge: Elsevier Trends Journal.

Lockyer, M.J., Marsh, K. and Newbold, C.I. (1989). Wild isolates of *Plasmodium falciparum* show extensive polymorphism in T cell epitopes of the circumsporozoite protein. *Molecular and Biochemical Parasitology* **37**, 275–280.

LoVerde, P.T., Dewald, J., Minchela, D.J., Bosshardt, S.C. and Damian, R.T. (1985). Evidence for host-induced selection in *Schistosoma mansoni*. *Journal of Parasitology* **71**, 297–301.

Maclean, J.M., Bairden, K., Holmes, P.H., Mulligan, W. and McWilliam, P.N. (1992). Sequential *in vivo* measurements of body composition of calves exposed to natural infection with gastro-intestinal nematodes. *Research in Veterinary Science* **53**, 381–389.

Male, D., Champion, B. and Cooke, A. (1987). *Advanced Immunology*. London: Gower Medical Publishing Ltd.

Malo, D., Schurr, E., Esptein, D.J., Vekmanns, M., Skamene, E. and Gros, P. (1991). The host resistance locus *Bcg* is tightly linked to a group of cytoskeletal-associated protein genes which include Villin and Desmin. *Genomics* **10**, 356–364.

Malo, D., Vidal, S., Hu, J., Skamene, E. and Gros, P. (1993a). High resolution linkage map in the vicinity of the host resistance locus *Bcg*. *Genomics* **16**, 655–663.

Malo, D., Vidal, S., Lieman, J.H., Ward, D. and Gros, P. (1993b). Physical delineation of the minimal chromosomal segment encompassing the murine host resistance locus *Bcg*. *Genomics* **17**, 667–675.

Malo, D., Vogan, K., Vidal, S., Hu, J., Cellier, M., Schurr, E., Fuks, A., Bumstead, N., Morgan, K. and Gros, P. (1994). Haplotype mapping and sequence analysis of the mouse *Nramp* gene predict susceptibility to infection with intracellular parasites. *Genomics* **23**, 51–661.

Margolis, L., Esch, G.W., Holmes, J.C., Kuris, A.M. and Schad, G.A. (1982). The use of ecological terms in parasitology. *Journal of Parasitology* **68**, 131–133.

Markell, E.K., Voge, M. and John, D.T. (1986). *Medical Parasitology*, 6th edition. Philadelphia: W.B. Saunders.

McConnel, T.J., Talbot, W.S., McIndoe, R.A. and Wakeland, E.K. (1988). The origin of MHC class II gene polymorphism within the genus *Mus*. *Nature* **332**, 651–654.

McLeod, R., Skamene, E., Brown, C.R., Eisenhauer, P.B. and Mack, D.G. (1989).

Genetic regulation of early survival and cyst number after peroral *Toxoplasma gondii* infection of A × B/B × A recombinant inbred and B10 congenic mice. *Journal of Immunology* **143**, 3031–3034.

Monroy, F.G. and Enriquez, F.J. (1992). *Heligmosomoides polygyrus*, a model for chronic gastrointestinal helminthiasis. *Parasitology Today* **8**, 49–54.

Morgado, M.G., Jouvin-Marche, E., Gris-Liebe, C., Bonhomme, F., Anabd, R., Talwar, G.P. and Cazenave, P.-A. (1993). Restriction fragment length polymorphism and evolution of the mouse immunoglobulin constant region gamma loci. *Immunogenetics* **38**, 184–192.

Morton, N.E. and McLean, C.J. (1974). Analysis of family resemblance. 3. Complex segregation of quantitative traits. *American Journal of Human Genetics* **26**, 489–503.

Mosmann, T.R. and Coffman, R.L. (1989). Heterogeneity of cytokine secretion patterns and functions of helper T cells. *Advances in Immunology* **46**, 111–147.

Mosmann, T.R., Cherwinski, H., Bond, M.W., Biedlin, M.A. and Coffman, R.L. (1986). Two types of murine helper T cell clone, 1. Definition according to the profiles of lymphokine activities and secreted proteins. *Journal of Immunology* **136**, 2348–2357.

Moulia, C., Aussel, J.P., Bonhomme, F., Boursot, P., Nielsen, J.T. and Renaud, F. (1991). Wormy mice in a hybrid zone, a genetic control of susceptibility to parasite infection. *Journal of Evolutionary Biology* **4**, 679–687.

Moulia, C., Le Brun, N., Dallas, J., Orth, A. and Renaud, F. (1993). Experimental evidence of genetic determinism in high susceptibility to intestinal pinworm infection in mice, a hybrid zone model. *Parasitology* **106**, 387–393.

Moulia, C., Le Brun, N., Loubes, C., Marin, R. and Renaud, F. (1995). Hybrid vigour against parasites in interspecific crosses between two mice species. *Heredity* **74**, 48–52.

Munger, J.C. and Karasov, W.H. (1989). Sublethal parasites and host energy budgets, tapeworm infection in white-footed mice. *Ecology* **70**, 904–921.

Nadeau, J.H. (1989). Maps of linkage and synteny homologies between mouse and man. *Trends in Genetics* **5**, 82–86.

Nadeau, J.H. and Taylor, B.A. (1984). Lengths of chromosomal segments conserved since divergence of man and mouse. *Proceedings of the National Academy of Sciences USA* **81**, 814–818.

Nakamura, R.M., Goto, Y. and Kitamura, K. (1989). Two types of supressor T cells that inhibit delayed-type hypersensitivity to *Mycobacterium intracellulare* in mice. *Infection and Immunity* **57**, 779–784.

Nancé, V., Vanlerberghe, F., Nielsen, J.T., Bonhomme, F. and Britton-Davidian, J. (1990). Chromosomal introgression in house mice from the hybrid zone between *M. m. domesticus* and *M. m. musculus* in Denmark. *Biological Journal of the Linnean Society* **41**, 215–227.

Nauciel, C., Ronco, E. and Pia, M. (1990). Influence of different regions of H-2 complex on the rate of clearance of *Salmonella typhimurium*. *Infection and Immunity* **58**, 573–574.

Norton, C.C. (1986). Coccidia of the domestic goat *Capra hircus*, with notes on *Eimeria ovinoidalis* and *E. bakuensis* (syn. *E. ovina*) from the sheep *Ovis aries*. *Parasitology* **92**, 279–289.

O'Brien, A.D. (1986). Influence of host genes on resistance of inbred mice to lethal infection with *Salmonella typhimurium*. *Current Topics in Microbiology and Immunology* **124**, 37–48.

O'Brien, A.D., Scher, I. and Formal, S.B. (1979). Effect of silica on the innate resistance of inbred mice to *Salmonella typhimurium* infection. *Infection and Immunity* **25**, 513–520.

O'Brien, A.D., Rosenstreich, D.L. and Taylor, B.A. (1980). Control of natural resistance to *Salmonella typhimurium* and *Leishmania donovani* in mice by closely linked but distinct genetic loci. *Nature* **287**, 440–442.

O'Brien, A.D., Taylor, B.A. and Rosenstreich, D.L. (1984). Genetic control of natural resistance to *Salmonella typhimurium* in mice during the late phase of infection. *Journal of Immunology* **133**, 3313–3318.

O'Brien, A.D., Weinstein, D.L., D'Hoostelaere, L.A. and Potter, M. (1986). Susceptibility of *Mus musculus musculus* (Czech I) mice to *Salmonella typhimurium* infection. *Current Topics in Microbiology and Immunology* **127**, 309–312.

O'Kelly, J.C., Post, T.B. and Bryan, R.P. (1988). The influence of parasitic infestations on metabolism. Puberty and first mating performance of heifers grazing in a tropical area. *Animal Reproduction Science* **16**, 177–189.

Pandey, V.S., Cabaret, J. and Fikri, A. (1984). The effect of strategic anthelmintic treatment on the breeding performance and survival of ewes naturally infected with gastro-intestinal strongyles and protostrongylids. *Annales de Recherche Vétérinaire* **15**, 491–496.

Plant, J. and Glynn, A.A. (1976). Genetics of resistance to infection with *Salmonella typhimurium* in mice. *Journal of Infectious Disease* **133**, 72–78.

Plant, J. and Glynn, A.A. (1979). Locating *Salmonella* resistance gene on mouse chromosome 1. *Clinical and Experimental Immunology* **37**, 1–6.

Plant, J.E., Blackwell, J.M., O'Brien, A.D., Bradley, D.J. and Glynn, A.A. (1982). Are *Lsh* and *Ity* at one locus on mouse chromosome 1? *Nature* **297**, 510–511.

Ploeger, H.W. and Kloosterman, A. (1993). Gastrointestinal nematode infections and weight gain in dairy replacement stock, first-year calves. *Veterinary Parasitology* **46**, 223–241.

Pond, L., Wassom, D.L. and Hayes, C.E. (1989). Evidence for differential induction of helper T cell subsets during *Trichinella spiralis* infection. *Journal of Immunology* **143**, 4232–4237.

Randall, R.W. and Gibbs, H.C. (1981). Effects of clinical and subclinical gastro-intestinal helminthiasis on digestion and energy metabolism in calves. *American Journal of Veterinary Research* **42**, 1730–1734.

Renaud, F. and De Meeus, T. (1991). A simple model of host–parasite evolutionary relationships. Parasitism, compromise or conflict? *Journal of Theoretical Biology* **152**, 319–327.

Renaud, F., Coustau, C., Le Brun, N. and Moulia, C. (1992). Parasitism in host hybrid zone. *Research and Reviews in Parasitology* **52**, 13–20.

Reynolds, S.R., Kunkel, S.L., Thomas, D.W. and Higashi, G.I. (1990). T cell clones for antigen selection and lymphokine production in murine *Schistosomiasis mansoni*. *Journal of Immunology* **144**, 2757–2762.

Roitt, I.M., Brostoff, J. and Male, D.K. (1993). *Immunology*, 3rd edition. Mosby YearBook Europe Ltd.

Sage, R.D., Whitney, J.B. and Wilson, A.C. (1986a). Genetic analysis of a hybrid zone between *domesticus* and *musculus* mice (*Mus musculus* complex), hemoglobin polymorphism. *Current Topics in Microbiology and Immunology* **127**, 78–85.

Sage, R.D., Heyneman, D., Lim, K.C. and Wilson, A.C. (1986b). Wormy mice in a hybrid zone. *Nature* **324**, 60–63.

Scalzo, A.A., Fitzgerald, N.A., Simmons, A., LaVista, A.B. and Shellam, G.R.

(1990). *Cmv-1*, a genetic locus that controls murine cytomegalovirus replication in the spleen. *Journal of Experimental Medicine* **171**, 1469–1483.

Schurr, E., Skamene, E., Forget, A. and Gros, P. (1989). Linkage analysis of the *Bcg* gene on mouse chromosome 1, identification of a tightly linked marker. *Journal of Immunology* **142**, 4507–4513.

Schurr, E., Skamene, E., Morgan, K., Chu, M.-L. and Gros, P. (1990). Mapping of *Col3al* and *Col6a3* to proximal murine chromosome 1 identifies conserved linkage of structural protein genes between murine chromosome 1 and human chromosome 2q. *Genomics* **8**, 477–486.

Schurr, E., Morgan, K., Gros, P. and Skamene, E. (1991). Genetics of leprosy. *American Journal of Tropical Medicine and Hygiene* **44**, 4–11.

Scott, P., Natovitz, P., Coffman, R.L., Pearce, E. and Sher, A. (1988). Immunoregulation of cutaneous leishmaniasis. T cell lines that transfer protective immunity or exacerbation belong to different T helper subsets and respond to distinct parasite antigens. *Journal of Experimental Medicine* **168**, 1675–1684.

Serjeantson, S., Wilson, S.R. and Keats, B.J. (1979). The genetics of leprosy. *Annal of Human Biology* **6**, 375–383.

She, J.X., Boehme, S., Wang, T.W., Bonhomme, F. and Wakeland, E.K. (1990). The generation of MHC class II gene polymorphism in the genus *Mus. Biological Journal of the Linnean Society* **41**, 141–161.

She, J.X., Boehme, S.A., Wang, T.W., Bonhomme, F. and Wakeland, E.K. (1991). Amplification of the major histocompatibility complex class II gene diversity by intraexonic recombination. *Proceedings of the National Academy of Sciences, USA* **88**, 453–457.

Sher, A. and Coffman, R.L. (1992). Regulation of immunity to parasites by T cells and T cell-derived cytokines. *Annual Review in Immunology* **10**, 385–409.

Sher, A., Coffman, R.L., Hieny, S. and Cheever, A.W. (1990). Ablation of eosinophil and IgE responses with anti-Il-5 or anti-Il-4 antibodies fails to affect immunity against *Schistosoma mansoni* in the mouse. *Journal of Immunology* **145**, 3911–3916.

Sher, A., Gazzinelli, R.T., Oswald, I.P., Clerici, M., Kullberg, M., Pearce, E.J., Berzovsky, J.A., Mosmann, T.R., James, S.L., Mores III, H.C. and Shearer, G.M. (1992). Role of T-cell derived cytokines in the downregulation of immune responses in parasitic and retroviral infections. *Immunological Review* **127**, 183–204.

Shields, E.D., Russel, D.A. and Pericak-Vance, M.A. (1987). Genetic epidemiology of the susceptibility to leprosy. *Journal of Clinical Investigation* **79**, 1139–1143.

Shwartz, R.H. (1985). T-lymphocyte recognition of antigen in association with gene products of the major histocompatibility complex. *Annual Review of Immunology* **3**, 237–261.

Skamene, E. (1986). Genetic control of resistance to mycobacterial infection. *Current Topics in Microbiology* **124**, 49–66.

Skamene, E. (1991). Population and molecular genetics of susceptibility to tuberculosis. *Clinical Investigation and Medicine* **14**, 160–166.

Skamene, E. and Pietrangeli, C.E. (1991). Genetics of the immune response to infectious pathogens. *Current Opinion in Immunology* **3**, 511–517.

Skamene, E., Kongshavn, P.A.L. and Sachs, D.H. (1979). Resistance to *Listeria monocytogenes* in mice is genetically-controlled by genes which are not linked to the H-2 complex. *Journal of Infectious Diseases* **139**, 228–231.

Skamene, E., Gros, A., Forget, P., Kongshavn, C., Charles, S. and Taylor, B.A.

(1982). Genetic regulation of resistance to intracellular pathogens. *Nature* **297**, 506–509.

Skamene, E., Gros, P., Forget, A., Patel, P.J. and Nesbitt, M.N. (1984). Regulation of resistance to leprosy by chromosome 1 locus in the mouse. *Immunogenetics* **19**, 117–124.

Smith, D.G. (1979). The genetic hypothesis for susceptibility to lepromatous leprosy. *Human Genetics* **50**, 163–177.

Spickett, S.G. (1962). Genetics and the epidemiology of leprosy. I. The incidence of leprosy. *Leprosy Review* **33**, 76–93.

Stach, J.L., Gros, P., Forget, A. and Skamene, E. (1984). Phenotypic expression of genetically-controlled natural resistance to *Mycobacterium bovis* (BCG). *Journal of Immunology* **132**, 888–892.

Stead, W.W. (1992). Genetics and resistance to tuberculosis. Could resistance be enhanced by genetic engineering? *Annals of Internal Medicine* **116**, 937–941.

Steinman, R.M. and Nussenzweig, M.C. (1980). Dendritic cells. Features and functions. *Immunological Reviews* **53**, 125–147.

Stevenson, M.M. (1989). *Malaria, Host Responses to Infection.* Boca Raton: CRC Press.

Stewart, T.B. and Gasbarre, L.C. (1989). The veterinary importance of nodular worms (*Oesophagostomum* spp.). *Parasitology Today* **5**, 209–213.

Stockdale, P.H.G., Bainborough, A.R., Bailey, C.B. and Niilo, L. (1981). Some physiopathological changes associated with infection of *Eimeria zuernii* in calves. *Canadian Journal of Comparative Medicine* **45**, 34–37.

Subauste, C.S. and Remington, J.S. (1991). Role of gamma interferon in *Toxoplasma gondii* infection. *European Journal of Clinical Microbiology and Infectious Disease* **10**, 58–67.

Symons, L.E.A. (1985). Anorexia. Occurrence, pathophysiology, and possible causes in parasitic infections. *Advances in Parasitology* **24**, 103–133.

Symons, L.E.A. and Hennessy, D.R. (1981). Cholecystokinin and anorexia in sheep infected by the intestinal nematode *Trichostrongylus colubriformis*. *International Journal for Parasitology* **11**, 55–58.

Taffs, L.F. (1976). Pinworm infections in laboratory rodents, a review. *Laboratory Animals Science* **10**, 1–13.

Taylor, B.A. and O'Brien, A.D. (1982). Position on mouse chromosome 1 of a gene that controls resistance to *Salmonella typhimurium*. *Infection and Immunology* **36**, 1257–1260.

Tucker, P.K., Sage, R.D., Warner, J., Wilson, A.C. and Eicher, E.M. (1992). Abrupt cline for sex chromosomes in a hybrid zone between two subspecies of mice. *Evolution* **46**, 1146–1163.

Ulczak, O.M. and Blackwell, J.M. (1983). Immunoregulation of genetically controlled acquired responses to *Leishmania donovani* infection in mice, the effects of parasite dose, cyclophosphamide and sublethal irradiation. *Parasite Immunology* **5**, 449–463.

Unanue, E.R. (1984). Antigen-presenting function of the macrophage. *Annual Review of Immunology* **2**, 395–428.

Vanlerberghe, F., Dod, B., Boursot, P., Bellis, M. and Bonhomme, F. (1986). Absence of Y-chromosome introgression across the hybrid zone between *Mus musculus domesticus* and *Mus musculus musculus*. *Genetical Research* **48**, 191–197.

Vanlerberghe, F., Boursot, P., Nielsen, J.T. and Bonhomme, F. (1988a). A steep cline for mitochondrial DNA in Danish mice. *Genetical Research* **52**, 185–193.

Vanlerberghe, F., Boursot, P., Catalan, J., Gerasimov, S., Bonhomme, F., Botev, A. and Thaler, L. (1988b). Analyse génétique de la zone d'hybridation entre les deux sous-espèces de souris *Mus musculus domesticus* et *Mus musculus musculus* en Bulgarie. *Genome* **30**, 427–437.

Vercruysse, J., Taraschewski, H. and Voigt, W.P. (1988). Main clinical and pathological signs of parasitic infections in domestic animals. In: *Parasitology in Focus. Facts and Trends* (H. Mehlhorn, ed.), pp. 477–537. Berlin: Springer-Verlag.

Vidal, S.M., Malo, D., Vogan, K., Skamene, E. and Gros, P. (1993). Natural resistance to infection with intracellular parasites, isolation of a candidate for *Bcg. Cell* **73**, 469–485.

Wagener, D.K., Schauf, V., Nelson, K.E., Scollard, D., Brown, A. and Smith, T. (1988). Segregation analysis of leprosy in families of Northern Thailand. *Genetics and Epidemiology* **5**, 95–105.

Wakeland, E.K., Boehme, S. and She, J.X. (1990). The generation and maintenance of MHC class II gene polymorphism in rodents. *Immunological Reviews* **113**, 207–226.

Wakelin, D. (1978). Genetic control of susceptibility and resistance to parasitic infection. *Advances in Parasitology* **16**, 219–308.

Wakelin, D. (1980). Genetic control of immunity to parasites. Infection with *Trichinella spiralis* in inbred and congenic mice showing rapid and slow responses to infection. *Parasite Immunology* **2**, 85–98.

Wakelin, D. (1992). Genetic variation in resistance to parasite infection, experimental approaches and practical applications. *Research in Veterinary Science* **53**, 139–147.

Wakelin, D., Rose, M.E., Hesketh, P., Else, K.J. and Grencis, R.K. (1992). Immunity to coccidiosis, genetic influences on lymphocyte and cytokine responses to infection with *Eimeria vermiformis* in inbred mice. *Parasite Immunology* **15**, 11–19.

Wassom, D.L. and Kelly, E.A.B. (1990). The role of the major histocompatibility complex in resistance to parasite infections. *Immunology* **10**, 31–52.

Wassom, D.L., David, C.S. and Gleich, G.J. (1979). Genes within the major histocompatibility complex influence susceptibility to *Trichinella spiralis* in the mouse. *Immunogenetics* **9**, 491–496.

Wassom, D.L., David, C.S. and Gleich, G.J. (1980). MHC-linked genetic control of the immune response to parasites, *Trichinella spiralis* in the mouse. In: *Genetic Control of Natural Resistance to Infection and Malignancy* (E. Skamene and P.A.L. Kongshavn, eds), pp. 75–82. New York: Academic Press.

Wassom, D.L., Brooks, B.O., Babish, J.G. and David, C.S. (1983). A gene mapping between the S and D regions of the H-2 complex influences resistance to *Trichinella spiralis* infections of mice. *Journal of Immunogenetics* **10**, 371–378.

Wassom, D.L., Wakelin, D., Brooks, B.O., Krco, C.J. and David, C.S. (1984). Genetic control of immunity to *Trichinella spiralis* infections in mice, hypothesis to explain the role of H-2 genes in primary and challenge infections. *Immunology* **51**, 625–631.

Wassom, D.L., Kcro, C.J. and David, C.S. (1987). I-E expression and susceptibility to parasite infection. *Immunology Today* **8**, 39–43.

Williams, D.M., Grumet, F.C. and Remington, J.S. (1978). Genetic control of murine resistance to *Toxoplasma gondii. Infection and Immunity* **19**, 416–420.

Williams, J.C., Sheehan, D.S., Fuselier, R.H. and Knox, J.W. (1983). Experimental and natural infection of calves with *Bunostonum phlebotomum. Veterinary Parasitology* **13**, 225–237.

Ziegler, K. and Unanue, E.R. (1981). Identification of a macrophage antigen-processing event required for I region-restricted antigen presentation to T lymphocytes. *Journal of Immunology* **127**, 1869–1875.

Zwilling, B.S., Vespa, L. and Massie, M. (1987). Regulation of I-A expression by murine peritoneal macrophages: differences linked to the *Bcg* gene. *Journal of Immunology* **138**, 1372–1376.

Detection, Screening and Community Epidemiology of Taeniid Cestode Zoonoses: Cystic Echinococcosis, Alveolar Echinococcosis and Neurocysticercosis

P.S. Craig, M.T. Rogan and J.C. Allan

Department of Biological Sciences, University of Salford, Salford M5 4WT, UK

1. Introduction 170
   1.1. Parasites, distribution and public health importance 170
   1.2. Pathology, diagnosis and treatment 179
   1.3. Problems in community-based studies 184
2. Life-cycle Biology and Transmission Dynamics 185
   2.1. *Echinococcus* spp. 185
   2.2. *Taenia solium* 186
   2.3. Transmission dynamics 186
3. Measurement of Human Infection with Taeniid Cestodes in endemic communities 189
   3.1. Retrospective case finding 190
   3.2. Active screening 191
4. Detection of Taeniid Infection in Animal Definitive and Intermediate Hosts 200
   4.1. Animal intermediate hosts 201
   4.2. Animal definitive hosts 202
5. Cystic Echinococcosis in Endemic Communities 203
   5.1. Cystic echinococcosis in a highly endemic nomadic community in East Africa 204
   5.2. Cystic echinococcosis among transhumant and settled communities in northwest China 208
   5.3. Cystic echinococcosis in the British Isles — low endemicity 211
   5.4. Cystic echinococcosis transmission in an urban setting — Uruguayan town and Kathmandu City 213
6. Alveolar Echinococcosis in Endemic Communities 215

ADVANCES IN PARASITOLOGY VOL 38
ISBN 0–12–031738–9

6.1. Alveolar echinococcosis on St Lawrence Island, Alaska 216
6.2. Alveolar echinococcosis in northwest China 218
6.3. Alveolar echinococcosis in Japan — a newly endemic country 221
7. *T. solium* Cysticercosis and Taeniasis in Communities 223
7.1. *T. solium* in rural Latin America 224
7.2. *T. solium* and neurocysticercosis in a non-endemic community 228
8. Conclusions 229
Acknowledgements 230
References 231

1. INTRODUCTION

1.1. Parasites, Distribution and Public Health Importance

There are approximately 30 species of Cyclophyllidean and Pseudophyllidean tapeworms which can infect humans (Miyazaki, 1991). The most important species from a public health viewpoint belong to the family *Taeniidae* (Cyclophyllidea) which includes the zoonotic species that cause larval (metacestode) infection in human tissues, i.e. *Taenia solium* (Linnaeus, 1758), *Echinococcus granulosus* (Batsch, 1786) and *E. multilocularis* (Leuckart, 1863). *T. solium* and *E. granulosus* were almost certainly known in antiquity, at least in ancient Greece (~ 400BC), but were only properly described in the second half of the nineteenth century (for historical review see Grove, 1990). Confusion between *E. granulosus* and *E. multilocularis* meant that it was another 100 years before the studies of Rausch and Schiller (1951) and Vogel (1955) confirmed that *E. multilocularis* was a separate species and was the cause of hepatic lesions in people on St Lawrence Island (Alaska) and in Germany.

Echinococcus granulosus and *E. multilocularis* are classical zoonoses in that their parasitic life cycles can only normally be maintained between vertebrate animal (mammalian) hosts. Humans are susceptible to infection, but they are not required to perpetuate the life cycle. In contrast, for *T. solium* (commonly called the human pork tapeworm), a human definitive host is obligatory in nature. *T. solium* (with *T. saginata*) is therefore a unique zoonotic infection. Table 1 summarizes some of the main features of the parasites and associated infections.

E. granulosus is the most geographically widespread of the three taeniid species, being truly cosmopolitan with transmission able to occur in all climatic zones (Figure 1). There are two main biotypes, i.e. Northern and European, the latter with a number of strains or subspecies which have now been defined by DNA analysis (Bowles and McManus, 1993). The "European" biotype is exemplified by synanthropic cycles, especially the

Table 1 Summary of the human and animal host aspects of *Taenia solium*, *Echinococcus granulosus* and *E. multilocularis*.

Scientific name	Common name	Common pathology in humans	Pathology in animals	Prevalence range in animals	Main treatment in humans	Most effective drugs	Morbidity/ mortality in humans	Est. total infections humans	Est. total no. of livestock infections	Highest prevalence in humans, incidence	Transmission in Europe	Global distribution
Taenia solium	Human pork tapeworm	Cysts (cysticerci) in muscles, subcutaneous, and brain — neurocysticercosis	Cysts in pig meat (muscle) — Porcine cysticercosis	5–30% pigs	Surgery, primarily chemo therapy for cysticercosis and taeniasis	Praziquantel, albendazole up to 70% for cysticercosis	50% of neurocysticercosis with convulsions	Adult tapeworm ~ 5–10 million cysticercosis, 5–6 million	> 50 million pigs	~ 5–10% central America	Small focus Spain and Eastern Europe	Mexico, central and S. America, central and S. Africa, India, SE Asia, China
Echinococcus granulosus	Small, dog tapeworm	Cysts (hydatids) in liver, lungs other organs. Cystic echinococcosis	Cysts in liver and lungs of sheep and other livestock — ovine hydatidosis	10–80% sheep, 5–50% dogs	Surgery, cystectomy, drainage, resection	Albendazole ~ 40% effective	2–10% fatal, average 6 months in hospital	~ 2–3 million	> 200 million sheep (other livestock)	5–9%, E. Africa, South America ~ 200/ 100 000 per year	Includes UK, France, Spain, Italy, Yugoslavia, Rumania, Bulgaria, Greece	Worldwide cosmopolitan, in tropics and temperate zones. Especially pastoral regions
Echinococcus multilocularis	Small, fox tapeworm	Cysts in liver, with secondary lesions in lungs/brain Alveolar echinococcosis	Cysts in liver of rodents e.g. arvicolids	< 1–30% rodents, < 5–90% foxes	Surgery, liver resection	Albendazole ~ 10% effective	95% fatal mortality if untreated by 10–15 years	~ 100 000– 300 000	Not applicable	Up to 5% in China, Russia 80–100/ 100 000 per year	Eastern France, S. Germany, N. Switzerland, W. Austria	Northern hemisphere only esp. N. America, Siberia, China, Japan, W. Europe

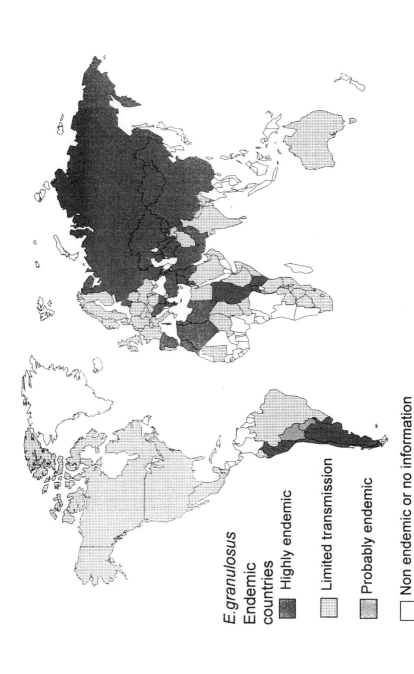

E.granulosus
Endemic
countries

■ Highly endemic

▨ Limited transmission

▨ Probably endemic

□ Non endemic or no information

Figure 1 Global distribution of *E. granulosus* by country (specific endemic regions not shown for individual countries).

predominant dog–sheep cycle. This probably arose from the Northern biotype, in which wolf–moose (*Alces alces*) or wolf–reindeer (*Rangifer tarandus*) cycles occur, soon after the domestication of dogs and ungulates approximately 8000–10000 years ago (Rausch, 1995). The Northern biotype continues to exist in tundra and taiga zones of North America and Russia as the cervid strain but appears to have only benign infection capabilities in humans and, therefore, is not considered to be of significant public health importance (Wilson *et al.*, 1968). In contrast, the dog–sheep strain of *E. granulosus* is relatively infectious in humans, often with significant pathology and is transmitted worldwide, primarily in pastoral regions. Highly endemic areas include large parts of Chile, Argentina, Peru, Brazil and Uruguay; Morocco, Algeria, Tunisia and Libya; parts of Kenya, Ethiopia, Uganda and Tanzania; southern European countries especially Spain, Greece, parts of Italy, former Yugoslavia and Turkey; Iraq, Iran, Jordan and Syria, Bulgaria, Romania, Russia and the former soviet republics Uzbekistan, Kazakstan, Tajikstan, Kirgizstan and Turkmenistan; and China (Schantz *et al.*, 1995). The European biotype of *E. granulosus* also occurs in other regions of North and Central America, Europe, Africa, India, South East Asia and Australia, both in the dog–sheep cycle as well as other domestic "strains", e.g. dog–pig, dog–horse, dog–camel and dog–cattle. Wildlife cycles in Central and East Africa (e.g. lion–wart hog) (Eckert and Thompson, 1988; Macpherson and Craig, 1991) and in China–Tibet (e.g. wolf–blue sheep) (Guo *et al.*, 1993) have probably also originated from this biotype. Epidemiological evidence suggests that the horse and camel strains of *E. granulosus* have low infectivity for humans (Nelson, 1972; McManus *et al.*, 1987; Wachira *et al.*, 1993). Recent DNA hybridization studies actually show enough nucleic acid variability in some of the "strains" of *E. granulosus* to warrant consideration of separate species status (Lymbery, 1992; Bowles *et al.*, 1995).

Echinococcus multilocularis is confined to the northern hemisphere where it is maintained in wildlife cycles principally between the arctic fox (*Alopex lagopus*), the red fox (*Vulpes vulpes*) and to a lesser extent the coyote (*Canis latrans*), as definitive hosts and arvicolid rodents (voles and lemmings) as intermediate hosts. Other canid species including the domestic dog are susceptible to infection, as are rodent species in seven other families (Rausch, 1995; Schantz *et al.*, 1995). Evidence for the existence of strains of *E. multilocularis* is not very strong and the epidemiological significance, if any, is unclear, though some genetic variability in mitochondrial DNA has been demonstrated between isolates from Europe, North America and China (Bowles and McManus, 1993). Transmission of *E. multilocularis* principally occurs over the vast tundra and taiga regions of North America (Alaska and Canada) and Russia (as far as the Pacific coast), northern Japan, central and northwest China, and the central

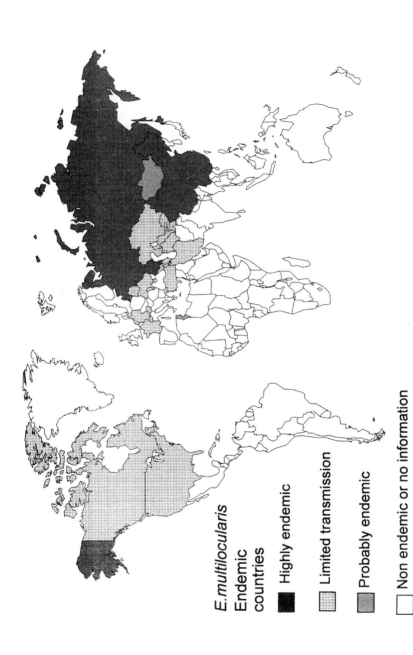

Figure 2 Global distribution of *E. multilocularis* by country (specific endemic regions not shown for individual countries).

Asian republics extending through to Iran and Turkey, with an apparently discontinuous focus in central Western Europe (primarily France, Switzerland and Germany) (Figure 2). Two human cases have also been reported in Tunisia (Robbana *et al.*, 1981; Zitouna *et al.*, 1985). A large, apparently isolated, focus also occurs in central North America which extends from Manitoba through North Dakota and Minnesota down to Missouri (reviewed by Schantz *et al.*, 1995). Concern is growing in Europe about human alveolar echinococcosis, especially with the increase in red fox populations, in part, as a result of rabies control programmes (Lucius and Bilger, 1995). There is also concern in Japan as *E. multilocularis* appears to be spreading southwards from Hokkaido to Honshu Island (Suzuki *et al.*, 1993).

Taenia solium, in contrast to *E. multilocularis* and to some extent *E. granulosus*, is now primarily a problem in the underdeveloped tropics though formerly its range included much of Europe (Grove, 1990). Most of the reported studies on *T. solium* are from a relatively small number of countries, primarily in Latin America and from Africa, India, China and Indonesia (Figure 3). This means that currently the true status of the parasite in large areas of the world, where it is thought to be endemic, is largely unknown. This is particularly true of sub-Saharan Africa where, with a few notable exceptions, the parasite has remained largely unstudied despite what may, in fact, be significant levels of infection (Geerts, 1995). For instance, in the first study of its kind in Tanzania, a country in which the disease had never before been formally recorded, over 13% of pigs at abattoirs were recently found to be infected (Boa *et al.*, 1995). Most studies from this region have taken the form of occasional case reports of human neurocysticercosis or observations on the occurrence of cysts in pigs (Harrison and Sewell, 1991). However, these studies have indicated that the parasite is present in practically all of the sub-Saharan countries of this continent. Neuroepidemiological studies have indicated that it is implicated in a very large percentage of cases of epilepsy in south, west and central Africa (Powell *et al.*, 1966; Dumas *et al.*, 1989). Like *E. multilocularis*, evidence for the existence of strains of *T. solium* is scanty. However, the rarity of human subcutaneous cysticercosis in Latin America, but not in Asia, has led to the suggestion of possible strain differences (Cruz *et al.*, 1994a; Flisser, 1994). The description of a third form of human *Taenia*, called "Asian *Taenia*" in Taiwan aboriginals and other isolated southeast Asian populations has prompted the suggestion that an unusual form of *T. saginata* or *T. solium* exists in that region (Fan, 1988). Recent nucleic acid sequence data indicate that Asian *Taenia* is more closely related to *T. saginata* than to *T. solium* (Bowles and McManus, 1994).

The geographic distribution of these three taeniid zoonoses is such that overlap occurs in a number of regions. *E. granulosus* and *E. multilocularis*

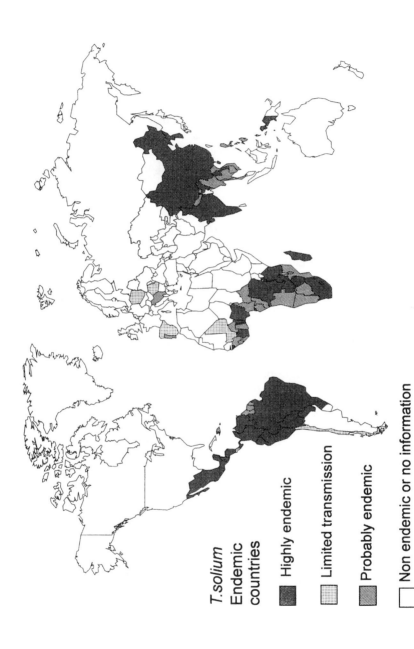

Figure 3 Global distribution of *T. solium* by country (specific endemic regions not shown for individual countries).

T.solium
Endemic
countries

■ Highly endemic

▦ Limited transmission

▨ Probably endemic

□ Non endemic or no information

occur sympatrically over much of the range of the latter species but both parasites cause significant public health problems in only relatively few countries, notably, Russia, China and the Central Asian Republics (Craig *et al.*, 1991; Schantz *et al.*, 1995). Autochthonous human cases of cysticercosis and cystic echinococcosis occur together in a few country regions including Peru (Moro *et al.*, 1994), Spain (Garcia-Albea, 1989), India and China. A few central or northwest Chinese provinces notably Gansu, Qinghai and Sichuan may be the only regions in the world where all three species, *E. granulosus, E. multilocularis* and *T. solium* occur together and result in human infections (P.S. Craig and D. Liu, unpublished observations).

From a public health viewpoint it is very difficult even to approximate the global numbers of human cases of cystic and alveolar echinococcosis and human cysticercosis. An attempt is shown in Table 1 based on extrapolations from known endemic areas and other calculations (PAHO, 1994; Roberts *et al.*, 1994; Schantz *et al.*, 1995). For all three larval taeniid infections human prevalence rates >1% may be considered high due to the chronicity and significant morbidity of infection. The highest prevalence and annual incidence rates for human cystic echinococcosis are around 2–6% and >50 per 100000 respectively, for example amongst pastoralists in northwest Kenya and Uruguay (Purriel *et al.*, 1974; French and Nelson, 1982). Human infection rates for alveolar echinococcosis are usually significantly lower, although similar rates have been reported in communities in eastern Siberia, Alaska and Central China (Schantz *et al.*, 1991; Craig *et al.*, 1992). For *T. solium* cysticercosis post-mortem prevalence rates of >6% have been reported in Mexico (Rabiela *et al.*, 1972) and community epileptic prevalence rates (with CT confirmation) of 2–3% in Guatemala (Garcia-Noval *et al.*, 1996). It appears, however, that human cysticercosis is largely under-reported in most endemic countries (Tsang and Wilson, 1995). Figure 4 shows age prevalence distribution for the three cestode zoonoses species when data were obtained from mass screening in endemic communities.

The economic importance of taeniid zoonoses is also significant as a result of medical treatment (hospitalization and lost working days) in addition to those losses due to livestock condemnation. For example, in Uruguay surgical treatment for cystic echinococcosis results in an average 34 days hospitalization at a cost of US $2200 per patient (PAHO, 1994). Radical surgery, hospitalization and medication for human alveolar echinococcosis is very costly, approximately US $250000 per patient in Alaska (Rausch *et al.*, 1990a). A minimum estimate of the cost of hospitalization and wage losses in USA (a non-endemic country) for human neurocysticercosis was US$ 8.8 million annually (Roberts *et al.*, 1994). These authors also estimated that treatment costs in 1992 for neurocysticercosis in Mexico were US $89 million and US $85 million for Brazil.

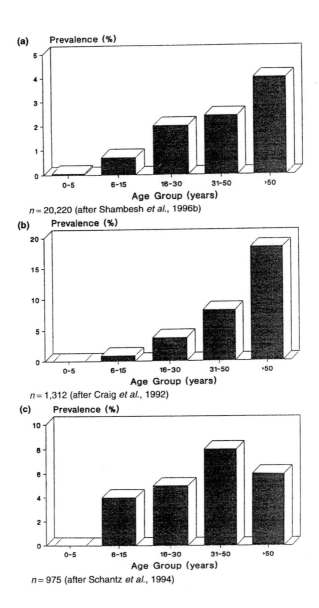

Figure 4 Age prevalence data for (a) human cystic echinococcosis (Libya), (b) alveolar echinococcosis (China) and (c) *T. solium* cysticercosis (Mexico) from mass screening of endemic communities. Data for (a) and (b) based on ultrasound with serological confirmation; (c) data based on glycoprotein immunoblot only.

1.2. Pathology, Diagnosis and Treatment

1.2.1. *Pathology*

The pathological course of metacestode infection due to *E. granulosus, E. multilocularis* or *T. solium* is variable but relates primarily to pressure effects of a space-occupying cyst or lesion in an organ or tissue. In humans, the liver is the main site of cyst growth for *Echinococcus* species, and muscle and the central nervous system (CNS) for *T. solium.* The clinical literature on these three cestode zoonoses is extensive and has been reviewed in detail by others (Pawlowski, 1993; Flisser, 1994; Ammann and Eckert, 1995). However, it is useful to summarize some information if only to enable an appreciation of the difficulties in accurate diagnosis and therefore in undertaking detailed epidemiological or community studies.

Initially, there is always an asymptomatic period after oral egg infection and some individuals may also remain asymptomatic for many years or even for their lifetimes. Incubation periods before symptoms become evident, will depend on the site of cyst growth, especially with respect to effects on adjacent organs or structures. For cystic or alveolar echinococcosis, upper abdominal pain and jaundice are relatively common though the signs are not pathognomonic (Pawlowski, 1993; Ammann and Eckert, 1995). Human alveolar echinococcosis is a severe disease with a calculated mean time between infection and symptoms of 5–15 years (WHO, 1996), and a >90% fatality rate in untreated patients (Wilson *et al.*, 1992; Ammann and Eckert, 1995). In contrast, mortality rates in human cystic echinococcosis are much lower at 1–5% of cases, with increased risk associated with multiple surgical intervention. For neurocysticercosis, time between infection and symptoms may be shorter than for hydatidosis because of the shorter metacestode development period (i.e. approx. 10 weeks versus several years) and the frequent brain location. Never the less, the most common symptom of neurocysticercosis, which is seizures or convulsions, was reported to occur in a large group of expatriate British soldiers (on return from India) at various times, between < 1 year to 30 years, after infection (MacArthur, 1934). Pathology of neurocysticercosis relates primarily to number, size, type (cellulose or racemose), condition and site of cysticerci in the brain (Flisser, 1994).

The natural history of all three human taeniid zoonoses is variable and ill defined, although the basic infection and early vesicular developmental processes are essentially similar (see Section 2). It is apparent, however, that in a proportion of exposed humans in endemic regions spontaneous metacestode destruction resulting in necrosis and ultimately calcification may occur for all three species. This appears as abortive early metacestode stages in alveolar echinococcosis (Rausch *et al.*, 1987; Bresson-Hadni *et al.*, 1994), and degenerate mature hydatid cysts of *E. granulosus* (Romig, 1990) or cysticerci of *T. solium* (Garcia-Noval *et al.*, 1996). The ability of *Echinococcus* spp., but not *T. solium*, to undergo asexual multiplication in the metacestode stage has important pathological consequences. Primarily the regenerative capacity of germinal epithelium and protoscoleces results in endogenous daughter cyst formation in cystic echinococcosis or exogenous budding and metastases in alveolar echinococcosis. Risk of secondary echinococcosis is significant during surgery or following traumatic cyst rupture for cystic hydatidosis, and of distant metastatic foci (frequently lung and brain) in advanced alveolar echinococcosis. Destruction of liver tissue can be very extensive in human alveolar echinococcosis, the lesion(s) typically being a multivesiculated gelatinous matrix (grossly resembling a carcinoma) of 5–200 mm in size often with central necrotic cavitation and peripheral nodular lesions, and vesicles are supported by a thin laminated layer. Typically hepatic cysts of *E. granulosus* are thick walled, unilocular and fluid filled with a thick, parasite-derived, laminated layer (up to 3 mm thick) with cyst size ranging from <1 cm to >20 cm. Cysticerci due to *T. solium* are usually in the range of 0.5–1.5 cm and fluid filled, bearing a single invaginated scolex. The intraventricular acephalic "racemose form", however, may grow to 20 cm in the human brain. *T. solium* cysts in the CNS are usually located in parenchymal, subarachnoid, intraventricular or spinal sites. Death of cerebral cysticerci may cause significant pathology due to inflammatory responses.

1.2.2. *Diagnosis*

Diagnosis of human cystic and alveolar echinococcosis and neurocysticercosis is difficult due to the tissue/organ location of cystic lesions and lack of direct parasitological evidence of infection (Craig *et al.*, 1995a). The most important diagnostic approach utilizes imaging methods for detection of space-occupying lesions. Laboratory methods, especially serological antibody detection, are also very useful and can be used to screen large numbers of serum samples such as from community surveys or blood donor stocks (Gottstein *et al.*, 1987a). There is usually a reasonably good correlation between imaging and serological methods for clinically advanced cases of larval cestodiasis, especially multiple cyst infection (Schantz and Gottstein, 1986; Richards and Schantz, 1991; Babba *et al.*, 1994). Computerized tomography (CT), magnetic resonance imaging (MRI), ultrasound (US) and radiography are currently the most useful imaging methods. CT scan is the best overall technique for human larval cestode detection in terms of resolution and applicability to full body scan. It is especially useful for brain and abdominal organs and it is also good for calcified cysts (Richards *et al.*, 1985; Kramer *et al.*, 1989; Choji *et al.*,

1992; Von Sinner, 1993). MRI is not suitable for calcifications but is excellent for detection of actively metabolizing cysts in soft tissue and therefore also where parasite tissue degeneration in response to chemotherapy has occurred (Jena *et al.*, 1988; Von Sinner, 1993). MRI may also detect viable cysts of *T. solium* not imaged by CT scan (M. Cruz, personal communication). Ultrasound is useful in imaging abdominal cysts especially hepatic lesions and can enable identification of pathognomonic images, for example presence of laminations and/or daughter cysts in cystic echinococcosis (Gharbi *et al.*, 1981; Von Sinner, 1991). X-radiography is particularly important in the diagnosis of pulmonary cystic echinococcosis and may be useful in detection of disseminated cysticercosis with multiple calcified cysts (Bourke and Petana, 1994). There is usually no difficulty in differentiating hepatic alveolar echinococcosis and cystic echinococcosis images in ultrasound (or CT and MRI), though differentiation from hepatic neoplasia especially for early alveolar echinococcosis may be problematic (Didier *et al.*, 1985; Von Sinner, 1991). Antigen-specific histopathology or DNA-based confirmation of the specific parasite, especially *Echinococcus*, may be important after surgical biopsy, laparotomy or fine-needle or percutaneous aspiration (Craig *et al.*, 1986a; Condon *et al.*, 1988; Pogacnik *et al.*, 1990; Gottstein and Mowatt, 1991; Choji *et al.*, 1992).

Immunodiagnostic tests for specific *Echinococcus* or *Taenia* antibodies have an important role in confirmation of clinical diagnosis and in sero-epidemiological studies (reviewed by Gottstein, 1992a; Craig, 1993; Flisser, 1994). Sensitive immunological techniques have been applied to detect specific antibodies to partially purified or electrophoretically separated (Western blot) native antigen preparations or to recombinant antigens. The important goal of specific serological differentiation between these three taeniid zoonoses has, however, not yet been completely achieved. Currently, native or recombinant *E. multilocularis* antigens enable highly species-specific immunodiagnosis (~98% specificity) of human alveolar echinococcosis (Gottstein, 1992b; Gottstein *et al.*, 1983, 1993; Helbig *et al.*, 1993) and *T. solium* glycoprotein antigens for specific immunodiagnosis (100% specific) of neurocysticercosis (Tsang *et al.*, 1989; Wilson *et al.*, 1991). For human cystic echinococcosis the greatest species specificity in sensitive immunodiagnostic tests, such as ELISA, is around 70%, achieved using native hydatid fluid antigens or recombinant antigen B, with approximately 30% of sera from alveolar echinococcosis patients and 5–15% of sera from cysticercosis patients cross-reactive (Maddison *et al.*, 1989; Leggatt *et al.*, 1992; Helbig *et al.*, 1993). The sensitivity of immunodiagnostic antigens (or tests) is generally >70% for the three infections with higher seroreactivity in patients with complicated and/or multiple cyst presentation. The highest sensitivity occurs for alveolar echinococcosis

Table 2 Summary of efficacy of ELISA and immunoblot serological tests for immunodiagnosis of cystic echinococcosis, alveolar echinococcosis or neurocysticercosis.

Disease	Test	Native antigen	Sensitivity	Specificity	Field validate
Cystic echinococcosis	ELISA	antigen B	~70%	~80%	Yes
	Immunoblot	antigen B subunits	~50–70%	>90%	Yes
Alveolar echinococcosis	ELISA	Em2	~85%	>95%	Yes
	Immunoblot	Em18/16	50–90%	>90%	No
Neurocysticercosis	ELISA	cyst fluid	~70%	~70%	Yes
	Immunoblot	GP-EITB	~70%	100%	Yes

Date summarized from: Rogen *et al.* (1991); Leggatt *et al.* (1992); Verastegui *et al.* (1992); Gottstein *et al.* (1987a,b); Bresson-Hadni *et al.* (1994); Craig *et al.* (1992); Ito *et al.* (1993); Wen *et al.* (1995); Schantz and Sarti (1989); Tsang *et al.* (1989); Wilson *et al.* (1991).

because the majority (>95%) of infected persons mount a florid antibody response to the infiltrative parasite mass (see also Section 3). Table 2 summarizes the efficacy of ELISA and Western blot (immunoblot) tests for the detection of antibodies in the three cestode zoonoses.

Diagnosis of the adult intestinal *T. solium* tapeworm has also been constrained in terms of both sensitivity and specificity. The standard approaches have been coproparasitological detection of eggs by microscopy or proglottides in the faeces, and questioning to determine whether individuals are aware of passing proglottides (Denham and Suswilo, 1995). These techniques, although simple and relatively inexpensive, are known to lack specificity and sensitivity, as taeniid eggs appear identical under light microscopy, and eggs or proglottides are absent from the faeces for much of the period of the infection (Schantz and Sarti, 1989). Specific differentiation of the intestinal taeniid species of humans, *T. solium* and *T. saginata*, has been achieved by analysis of the uterine branch structure or the scolex morphology of the parasite after treatment (Verster, 1967). As these parasites often occur sympatrically, the speciation of the infections is important for an understanding of their epidemiology. It is common for reports to be made that do not differentiate the *Taenia* species present, in addition a significant number of infections cannot be differentiated due to the absence of material in suitable condition for analysis.

A number of alternative approaches have been developed for the specific differentiation of adult taeniid cestode infections based on the molecular characteristics of the parasites, either at the DNA or protein level (Le Riche and Sewell, 1978; Bursey *et al.*, 1980; Flisser *et al.*, 1988; Harrison *et al.*, 1990). Some of these techniques, particularly those involving DNA probes or polymerase chain reaction (PCR), are, at least theoretically, capable of differentiating very small quantities of eggs or other somatic material (Gottstein *et al.*, 1991; Chapman *et al.*, 1995). However, these approaches may not fully overcome the problem of diagnostic sensitivity because eggs or proglottides are frequently absent from faeces during infection. Studies have been undertaken to determine whether serum antibody detection might prove a reliable diagnostic method for human taeniasis (Machnicka-Roguska and Zwierz, 1970; Flentje and Padelt, 1981). Generally this has been shown not to be the case, although the recently developed immunoblot for *T. solium* cysticercosis has indicated that it may also detect a significant number of cases of intestinal *T. solium*, but not *T. saginata*, infections (Tsang *et al.*, 1989; Diaz *et al.*, 1992).

Recent studies have involved the use of capture ELISA for coproantigen detection based on polyclonal rabbit antisera raised against either adult somatic (Allan *et al.*, 1990, 1992, 1993, 1996; Maass *et al.*, 1991) or excretory/secretory products (Deplazes *et al.*, 1990, 1991). The results have indicated that coproantigen assays are more sensitive than microscopy, can detect infection early during the prepatent period, are highly specific (at the genus level) and become negative shortly after successful treatment of the infection. Antigen can also be detected in 5% formalin fixed faecal samples.

1.2.3. *Treatment*

Treatment for human cystic and alveolar echinococcosis is primarily surgical (drainage, cystectomy or liver resection), whereas that for uncomplicated neurocysticercosis is primarily chemotherapy using albendazole or praziquantel, or neurosurgery for intraventricular cysts (Flisser, 1994; Amman and Eckert, 1995). Benzimidazole chemotherapy (largely albendazole) has also been increasingly used for treatment of human cystic echinococcosis (Horton, 1989); efficacy is around 30% with up to 70% of patients improving (Wen *et al.*, 1993). Benzimidazole chemotherapy for alveolar echinococcosis is essentially parasitostatic but can result in a significant increase in mean survival time, for instance from an average of 5.3 years in untreated cases to >14 years for patients given benzimidazole chemotherapy (Wilson *et al.*, 1992). A recent report now indicates that long-term (up to 5 years) continuous albendazole treatment of some alveolar echinococcosis patients can result in parasitological cure with "death" of the parasite (Liu *et al.*, 1993a). Neurocysticercosis chemotherapy using praziquantel or albendazole results in approximately 80%

efficacy in eliminating cysts or reduction in cyst numbers and in >90% reduction in numbers of seizures (Vazquez and Sotelo, 1992; Flisser, 1994). The new technique of percutaneous drainage and sterilization of abdominal cystic hydatids under ultrasound guidance, with albendazole cover, appears highly efficacious and offers a new conservative surgico-medical approach for treatment (Filice *et al.*, 1990). Guidelines for treatment of cystic and alveolar echinococcosis have recently been published (WHO, 1996).

1.3. Problems in Community-based Studies

In many endemic regions there is a dichotomy in the perception of the importance of *E. granulosus, E. multilocularis* and *T. solium*. On the one hand it is perceived as a medical problem largely for a few specialized surgeons or other clinicians, many of whom may be unsure about the life cycle and zoonotic basis of the infection. On the other hand, it is perceived as a veterinary problem particularly in the case of *E. granulosus* and *T. solium* where domestic animals, e.g. sheep and pigs, are hosts and the parasitic infection is only commonly encountered during slaughter. Government veterinarians usually have a better understanding of life cycles and transmission but may fail to appreciate the public health importance. Coupled with the asymptomatic nature of animal infection and the relatively little economic importance in livestock intermediate hosts, as well as the chronicity of infection in humans, then it is not surprising that reporting surveillance and control efforts are difficult to set up and co-ordinate. These problems are common with other zoonotic helminth diseases but perhaps most acute in the taeniid cestode zoonoses (Schwabe, 1991). Intersectorial co-operation is therefore vital for effective control programmes — ideally co-ordinated by a veterinary public health authority (Schwabe, 1991). In the case of *E. multilocularis* which is cycled sylvatically rather than synanthropically, the additional role of zoological ecologists becomes important. Intermediate host infection rates, primarily in sheep for *E. granulosus*, voles for *E. multilocularis* and pigs for *T. solium*, are usually the most practically obtained indicators of transmission in a given region but require skilled personnel for specific parasitological identification of the parasites. Infection rates in domestic dogs (*E. granulosus, E. multilocularis*) or foxes (*E. multilocularis*) are also very useful but more difficult to obtain as these hosts are not routinely culled (see Section 4). The relatively recent introduction of modern imaging methods especially in many developing countries has resulted in much more effective detection of human disease particularly coupled with improvement in laboratory diagnostics.

The main reasons for undertaking community screening are the following.

1. To enable early detection and treatment to prevent development of advanced disease. In human alveolar echinococcosis early detection and subsequent surgical removal of small lesions can significantly reduce morbidity and mortality (Vuitton, 1990). Similarly small cysts of *E. granulosus* are more amenable to surgical removal or effective chemotherapy (Wen *et al.*, 1993).
2. To define the public health problem.
3. To understand the epidemiology and ecology of parasite transmission.
4. To provide a rationale for intervention strategies, i.e. selection of control methods will be influenced by local epidemiological data.
5. To evaluate progress of control programmes, i.e. provide baseline data against which future progress can be measured (Schantz, 1993; Gemmell and Roberts, 1995).

2. LIFE-CYCLE BIOLOGY AND TRANSMISSION DYNAMICS

2.1. *Echinococcus* spp.

Echinococcus like all taeniid cestodes is transmitted cyclically between carnivore and herbivorous hosts. The small hermaphroditic adult tapeworms (1.5–11 mm) are essentially non-pathogenic and exist in the small intestine (attached in the crypts of Lieberkühn) of a range of primarily canid definitive hosts. When fully developed, these worms will each possess only three or four proglottides, the most terminal of which will contain approximately 200–600 mature eggs which will ultimately be released in the faeces either as free eggs or within a gravid proglottis.

The eggs (~40 μm) contaminate the environment and the fur of the definitive host. If they are consumed by a suitable intermediate host, the eggs hatch in the gut releasing oncosphere larvae which penetrate the intestinal epithelium gaining access to the circulatory system (Lethbridge, 1980). The oncospheres of *E. granulosus* most frequently become lodged in the capilliaries of the liver or lungs of ungulates (varies with host species) where they will ultimately develop into the mature metacestode stage or hydatid cyst. After several months the hydatid cyst will be fluid filled and may produce thousands of protoscoleces internally. Larval metacestodes of *E. multilocularis* develop more rapidly in the rodent liver and may produce protoscoleces by 60 days post-infection. Protoscoleces of *E. granulosus* released within the intermediate host either by cyst rupture or spillage during surgery; or germinal cells by metastasis for

E. multilocularis, have the ability to develop into further cysts or cystic lesions giving rise to secondary hydatidosis/echinococcosis.

If the hydatid cyst is eaten by a canid following host predation, scavenging (or deliberate feeding), the protoscoleces attach to the intestinal wall and develop into adult worms. In this way it is not uncommon for definitive hosts to harbour hundreds or thousands of adult worms.

2.2. *Taenia solium*

The adult tapeworm of *Taenia solium* strobilates in the human small intestine following ingestion of a viable metacestode (cysticercus) usually in improperly cooked pork. The armed rostellum of the single larval scolex present within the cysticercus, evaginates in the human gut, stimulated by bile and digestive enzymes, and attaches to the mucosa with the aid of four muscular suckers. Strobilation is rapid with development of a mature gravid tapeworm (2–7 metres in length) after 2–3 months post-infection (Flisser, 1994). Gravid segments may contain 50–60000 eggs and are released passively in faeces in small groups daily, two or three times per week, or less regularly (Yoshino, 1934; P.S. Craig and Y. Wang, unpublished observations). Interestingly more than 50% of human *T. solium* carriers may be unaware of their infection (J.C. Allan, unpublished observations). The eggs, like those of all taeniids are relatively resistant to environmental conditions. However, in most endemic areas, transmission to pigs occurs rapidly due to human stools being actively sought out and eaten by pigs with access to defaecation sites or latrines (Flisser, 1988). Eggs hatch and activate in the pig small intestine and develop primarily in striated muscle and the central nervous system. The larval scolex and cyst (8–15 mm) is viable by about 10 weeks after infection. In humans exposed to infective eggs, cysticerci can develop, in general, in the same way as in the pig host but in contrast show both cellulose and racemose morphological types in the human brain.

2.3. Transmission Dynamics

The transmission dynamics of *Echinococcus* or *Taenia* in a carnivore/ herbivore cycle, within a particular area, will depend on a number of factors. At any one time the parasite population consists of three subpopulations. These are the adults in the definitive host, larvae in the intermediate host and eggs in the environment. Transmission will therefore depend on, (i) the biology of the parasite, (ii) the biology of the hosts, and (iii) environmental factors. In addition, for *E. granulosus* and *T. solium* socio-ecological factors such as feeding behaviour of hosts, livestock practices and levels of human awareness are also important (Gemmell and Roberts, 1995), and for *E. multilocularis* ecological factors such as temporal fluctuations in intermediate host population densities (Giraudoux, 1991).

With regards to the biology of the parasite, it is its biotic potential which is important. This is the potential number of viable cysts that can be established in an intermediate host by an individual definitive host per day. On the basis that the mean number of *E. granulosus* worms per definitive host is around 200–400 (although variation, as discussed later, does occur); that the number of eggs shed from an average infected dog per day is 8470, and that the proportion of eggs transforming into viable cysts is 0.0033, a biotic potential of 28 viable hydatid cysts per infected dog per day has been calculated (Gemmell *et al.*, 1987a; Gemmell and Roberts, 1995). This is about 1/100 of the potential of the much larger related canine species, *Taenia hydatigena*.

One of the most important features of host biology for taeniid cestodes is the existence of acquired immunity in the intermediate host. This is the only density-dependent constraint on transmission assuming there is little or no "crowding effect" nor parasite-induced host mortality. Acquired immunity to *E. granulosus* in dogs has been shown to be weak (Gemmell *et al.*, 1986) whereas immunity in the intermediate host is a much stronger regulator of parasite numbers. A strong immunity against the oncosphere/ early post-oncosphere stages exists for most taeniid cestodes and in natural conditions may by acquired by the ingestion of as few as 10 eggs, lasts between 3 and 12 months in the absence of subsequent egg exposure and may be lifelong in the presence of eggs. The result of this is an over-dispersed distribution within the definitive host, with only a few animals harbouring large numbers of cysts and a greater proportion possessing smaller numbers (Gemmell, 1993).

Environmental factors which are important in transmission include temperature and desiccation which can have a lethal effect on the survival of eggs outside of the host. Extremes of temperature of +40°C and −70°C generally kill taeniid eggs within hours (Gemmell, 1990). Between these two extremes, temperature regulates the maturation–ageing process. Eggs released into the environment will therefore be of heterogeneous infectivity to the intermediate host. Mature forms will be able to produce viable larvae whereas senescent ones will be killed off by the immune system or not hatch at all (Gemmell *et al.*, 1987a). Eggs of *E. multilocularis* are well adapted to low temperatures of the holarctic and may survive northern European winters (Veit *et al.*, 1995) and even the sub-arctic winter (Schiller, 1955). Environmental factors are also important in dispersion of the eggs. Most taeniid eggs remain within 180 m of the site of deposition but there is evidence to suggest that some may be dispersed over a much larger area possibly being transported by dipteran flies (Lawson and Gemmell, 1990; Torgersen *et al.*, 1995).

In studying the transmission dynamics of taeniid cestodes in different regions it is necessary to relate the above-mentioned biological and environmental factors to the stability of the parasite population. Sustained transmission will only occur if there are sufficient numbers of viable infective stages available to the hosts and if the immunological status of the intermediate host population includes a proportion of individuals which are susceptible to infection. The stability of a parasite population is dependent on the basic reproductive ratio (R_o). This can be defined as the ratio of the number of adult parasites in "one generation" to the number in the "next generation" and is the reproduction ratio in the absence of density-dependent constraints such as acquired immunity (Anderson and May, 1992). If each parasite replaced itself, on average, by less than one adult parasite then the parasite population would gradually die out. If, on average, each parasite were able to replace itself with one or more adult parasites then the population would increase in size (until regulated by density-dependent constraints). If a parasite population is neither increasing nor decreasing with time, then it is in a steady state. With cestode populations, a parasite population is said to be in an endemic steady state if the population size is constant and the effects of density-dependent constraints, primarily immunity, are insignificant (R_o is equal to or slightly greater than 1). If the population size is constant and is strongly regulated by acquired immunity, R_o is much greater than one and the population is said to be in the hyperendemic steady state. If R_o is less than 1 then the parasite population is in the extinction steady state and will become extinct if R_o is maintained below 1 (reviewed by Gemmell, 1990).

Estimating R_o and the effects of environmental and immunological factors on the parasite population for a particular area is difficult. However, Roberts *et al.* (1986, 1987) and Gemmell and Roberts (1995) have developed a series of mathematical models, based on estimates of infection pressure, host susceptibility and the rate of loss of parasites from the system, for the population dynamics of some taeniid cestodes in domestic animal populations. Studies of the transmission dynamics of the natural endemic synanthropic cycles of *E. granulosus* in northwest China (Ming *et al.*, 1992) and in Uruguay (Cabrera *et al.*, 1995) showed an increase in age intensity and prevalence of ovine hydatidosis not sufficiently regulated by acquired immunity and therefore were described to be in the endemic steady state.

Transmission of *E. multilocularis* or *E. granulosus* in sylvatic cycles occurs through natural predator–prey relationships. However, transmission in synanthropic cycles (i.e. those associated with humans) including that for *T. solium* is considerably modified by human behaviour. The extent and maintenance of such cycles, however, is often poorly understood. Analysis of transmission patterns within human communities requires a knowledge of the levels of human, livestock and dog infection as well as an appraisal of behaviour which may be important in transmission.

The transmission dynamics of *T. solium* in human and pig populations has not yet been described, but requires further data on the adult tapeworm in humans (longevity, reinfection rates, biotic potential), and the metacestode in pigs (effect of immunity, age specific intensity and prevalence rates), as well as human–pig interaction studies.

3. MEASUREMENT OF HUMAN INFECTION WITH TAENIID CESTODES IN ENDEMIC COMMUNITIES

Determination of whether *E. granulosus*, *E. multilocularis* or *T. solium* is present in a given region or community may be indicated by the presence of parasites in domestic or wild animal hosts and/or in the human population. Surprisingly, quantification of human infection is not always easy or obvious, particularly when prevalence is low, e.g. in a region adjacent to or outside a known transmission zone. Prior indication may be available often in the form of sporadic or clustered hospital admissions from one region, district or community. For example, the identification of the Turkana cystic echinococcosis focus in northwest Kenya was initially through follow-up of advanced hydatid cases in Kitale Hospital which was outside the District (Wray, 1958; Schwabe, 1964; French *et al.*, 1982). Likewise the disproportionately high alveolar echinococcosis case admission rate for Chinese peasants in Lanzhou City but originating from Zhang County, 250 km away in south Gansu (Jiang, 1981; Craig *et al.*, 1992). An epidemic of serious burns (head and/or limbs) in the Ekari people from the Wissal Lakes area of Irian Jaya was eventually linked to convulsive episodes close to open fires and the identification of a new endemic focus of neurocysticercosis (Gajdusek, 1978). Autopsy data have also been important in indicating hitherto undefined problems, for example neurocysticercosis in Mexico where >6% prevalence rate was determined at routine autopsies (Rabiela *et al.*, 1972).

Due to the chronicity of larval cystic infection, and therefore long periods between exposure and symptoms, and also the lack of rapid or effective treatment, true incidence cannot be measured. Rather point prevalence or period prevalence are determined. For human taeniasis, however, the true incidence could be determined (but not yet initiated) by undertaking mass treatment with longitudinal follow-up over at least 12 months.

Essentially there are two main approaches to quantify human infection with taeniid parasites in a given area or specific community: (i) collation and analysis of retrospective hospital cases/admissions and autopsy surveys; (ii) active case finding by survey or screening methods.

3.1. Retrospective Case Finding

This has been the most widely used approach to quantify human infection rates with cystic echinococcosis, alveolar echinococcosis or neurocysticercosis. Data have usually been collected over a fixed period of 5–10 years and an annual incidence rate (i.e. new hospital admissions) usually expressed by age and sex per 100000 population determined (e.g. Kamhawi, 1995). Examples of hospital (surgical) case rates in endemic countries for the three zoonoses are shown in Table 1. Records from thoracic, abdominal and neurology surgical departments particularly require assessment, together with general wards and out-patient clinics for chemotherapy cases, the latter for cystic echinococcosis and neurocysticercosis/taeniasis. Other sources of human infection data may also be valuable. For example a recent retrospective study of cystic echinococcosis in Italy was based on data collected between 1988 and 1990 from 19 ultrasonography departments or clinics from persons attending for other complaints not associated with echinococcosis (Caremani *et al.*, 1993).

Although useful, particularly from a medical impact viewpoint, this kind of data, however, does not represent the most accurate description of human infection rates. This is because hospital cases are almost always symptomatic and therefore asymptomatics will be excluded (unless cases were previously identified by screening or by chance diagnosis). Also, not all infected persons will have the same access to medical care, and there may also be a bias for treatment in certain age groups not necessarily related to risk factors, for example most surgery for cystic echinococcosis in developing countries occurs in the 20–40 years age group with relatively few operations for young and old patients. Sex rates for surgical or chemotherapeutic treatment are usually about equal for cystic or alveolar echinococcosis patients or neurocysticercosis patients, though significantly higher female infection rates have been recorded for cystic echinococcosis in Turkana, Kenya (French and Nelson, 1982) and for alveolar echinococcosis in Gansu, China (Craig *et al.*, 1992).

Inaccurate identification of the specific parasite species causing disease as well as poor record keeping are also problems in retrospective surveys. Differentiation between cystic and alveolar echinococcosis is not always clear as histological confirmation may be lacking and patients may be counted more than once if they require multiple surgical interventions over a long period. Misdiagnosis of alveolar echinococcosis as hepatic carcinoma has occurred prior to histological examination (Jiang, 1981; Sasaki *et al.*, 1994). Neurocysticercosis cases may be identified from neurosurgical departments or private CT clinics but many cases whose only symptoms are epileptic fits remain incorrectly classified. Human taeniasis due to *T. solium* adult tapeworm infection is often not specifically identified in hospital out-patient clinics and may be routinely classified as *T. saginata*, even when based only on the presence of eggs in stool, simply because this parasite is usually thought to be more common in many developing countries.

Despite the above drawbacks, retrospective hospital surveys have proved very useful. This is especially so in highlighting the public health problem of cystic echinococcosis in order to mobilize legislative and health authorities prior to implementation of ultimately effective control programmes, for example in Tasmania (Beard, 1987) and in New Zealand (reviewed by Gemmell, 1990). Similarly, hospital morbidity and mortality data due to human alveolar echinococcosis in Alaskan Eskimos ultimately resulted in implementation of a government-funded dog control programme on St Lawrence Island (Rausch *et al.*, 1990a)

3.2. Active Screening

Active screening to detect human cystic and alveolar echinococcosis, and neurocysticercosis can currently be carried out by three approaches: (i) mass radiological imaging surveys, (ii) immunodiagnostic surveys; and (iii) autopsy surveys. Autopsy surveys are the least useful due to the problems of organizing hospitals to undertake a routine but specific search for larval taeniid infections (Schantz, 1993). The first two approaches are the most important especially if used in combination (Schantz and Gottstein, 1986; Craig, 1993; Flisser, 1994).

Radiological imaging methods notably X-ray, ultrasound, CT and MRI are important; however, only X-ray and ultrasound are at present available as transportable units, especially the latter (Macpherson, 1992). Both mass miniature X-ray and portable ultrasound have been used to screen for pulmonary or abdominal cystic echinococcosis, respectively, in Argentina and Uruguay (Schantz *et al.*, 1973; Perdomo *et al.*, 1988). X-radiography was used in Argentina in association with tuberculosis screening (Schantz *et al.*, 1973).

Active screening for human cystic or alveolar echinococcosis has been approached in two main ways. First, a specific serological test was used on the population sample and the presence of seroreactors confirmed by an imaging technique, usually after transportation of individuals to a central medical facility. The second approach, has been to undertake primary screening by portable ultrasound (or less commonly X-ray) and then to carry out confirmation of image positives by serology.

3.3.1. *Immunodiagnostic Screening*

Primary screening using an immunodiagnostic test(s) has been widely used to screen for human cystic echinococcosis. For example use of crude hydatid fluid antigens in ELISA or IHA with confirmation of seropositives by abdominal ultrasound and chest radiograph in Tunisia (Mlika *et al.*, 1984), Argentina (Coltorti *et al.*, 1988a) and in northern Israel (Nahmias *et al.*, 1992).

A similar approach using crude *E. multilocularis* antigens in ELISA has been carried out to screen for human alveolar echinococcosis in Hokkaido, Japan (Sato *et al.*, 1993), in Gansu Province, China (Craig *et al.*, 1992) and the Doubs region of France (Bresson-Hadni *et al.*, 1994). Seroreactors for both types of hydatidosis were followed-up by ultrasound or CT scan. For human neurocysticercosis a highly specific immunoblot test, incorporating *T. solium* glycoprotein antigens, has been used to screen endemic communities in Mexico (Schantz *et al.*, 1994) and Guatemala (Garcia-Noval *et al.*, 1996) with CT scan follow-up for seroreactors.

Serological tests for anti-*Echinococcus* or anti-*T. solium* cysticercus antibodies should have high specificity so that the predictive value remains high despite the low absolute prevalence of these larval cestode infections (Gottstein *et al.*, 1987a; Schantz, 1993). Currently, only native antigen preparations from *E. granulosus, E. multilocularis* or *T. solium* have been used in immunodiagnostic screening at the community level. Recombinant antigens have been produced and clinically assessed for *E. multilocularis* and *E. granulosus* (Lightowlers and Gottstein, 1995) but not yet reported for use in community studies. The positive predictive value for alveolar echinococcosis using the immunoaffinity purified native antigen Em2 in ELISA was 99% (Gottstein *et al.*, 1987a), and ranged between 60 and 90% for crude hydatid fluid extracts in ELISA for cystic echinococcosis (Schantz, 1993). Gottstein *et al.* (1993) have now combined native Em2 with a recombinant antigen 11/3-10 and the reported specificity remained at 99% with respect to sera from cystic echinococcosis or *T. solium* cysticercosis patients. The species specificity of native or recombinant antigen B, an *E. granulosus* cyst fluid lipoprotein (Lightowlers *et al.*, 1989), with respect to alveolar echinococcosis and *T. solium* cysticercosis patient sera is much lower ranging from 60–70% in ELISA (Rogan *et al.* 1991; McVie *et al.*, 1996) to approximately 70–95% for immunoblot identification of the lowest sub-unit (8–12 kDa) of *E. granulosus* antigen B (Maddison *et al.*, 1989). For *E. granulosus* antigen B (native or recombinant) approximately 40% of alveolar echinococcosis sera cross-react (Maddison *et al.*, 1989; Leggatt *et al.*, 1992; Helbig *et al.*, 1993) and 4–12% of sera from *T. solium* cysticercosis patients (Leggatt *et al.*, 1992; Verastegui *et al.*, 1992) (see Table 2).

Clearly species specificity, rather than genus or family specificity, becomes important when more than one of the taeniid zoonoses are co-endemic. *T. solium* metacestode somatic extracts or cysticercus cyst fluid behave quite well as clinical diagnostic antigens in ELISA with up to 80% sensitivity and specificity for neurocysticercosis particularly when cerebrospinal fluid (CSF) is tested (Richards and Schantz, 1991). However, their poor predictive value in community screening studies has limited their use (Schantz and Sarti, 1989). A number of highly sensitive and specific immunodiagnostic assays based on immunoblot have shown improvement over ELISA tests (Gottstein *et al.*, 1987b; Tsang *et al.*, 1989; Wilson *et al.*, 1991). The current bench-mark for immunodiagnosis of human neurocysticercosis is an immunoblot test (using sera or CSF) in which antibody binding to a group of seven specific glycoprotein antigens (50–13 kDa) are identified from an SDS-PAGE separated lentil lectin purified fraction of a *T. solium* metacestode extract (Tsang *et al.*, 1989). This test is also reported to be 100% specific. Table 2 summarizes the efficacy of ELISA and immunoblot tests for the immunodiagnosis of the three cestode zoonoses.

Serological tests for antibody detection used in mass screening should also be as sensitive as possible, this being a function of the technique as well as the antigen employed. For the taeniid zoonoses most variability in sensitivity, i.e. ability to detect antibodies (predominantly IgG isotype) in infected persons, occurs for immunodiagnosis of cystic echinococcosis and neurocysticercosis. For these two larval cestode infections patient scro-reactivity tends to be greater (approx. 70–95%) with multiple viable cyst infection in liver tissue or brain parenchyma, respectively, and lower (approx. 30–60%) with single intact cysts (in non-liver sites for cystic echinococcosis) especially if calcified or partially calcified (Garcia *et al.*, 1991; Verastegui *et al.*, 1992). CSF antibody testing is usually more sensitive than for serum in neurocysticercosis (Cho *et al.*, 1986). The infiltrative nature of human alveolar echinococcosis in contrast, usually results in a detectable antibody response in >95% of infected persons (Gottstein *et al.*, 1983; Helbig *et al.*, 1992).

The natural history of human cystic and alveolar echinococcosis and cysticercosis has not been very well characterized, and there is, therefore, a lack of data on the nature of early versus late antibody responses during the course of infection. This raises an important problem as almost all immunodiagnostic tests based on antibody detection for the taeniid zoonoses have been standardized with clinically defined advanced hospital patient sera. Such sera are more likely to be antibody positive, particularly for cystic echinococcosis and neurocysticercosis patients. Selection of high titred seroreactors would therefore tend to bias a test toward apparent higher sensitivity. For asymptomatic cystic echinococcosis detected by ultrasound screening, a proportion of cysts are more likely to be small (young?) and healthy with intact membranes and therefore less seroreactive than hospitalized cases (Nahmias *et al.*, 1992; Shambesh *et al.*, 1996b). In a recent study, Libyan cystic echinococcosis patients were divided into asymptomatics (diagnosed by ultrasound during community screening) and advanced hospital cases, and the latter group exhibited higher IgG and IgG$_4$ antibody levels (Shambesh *et al.*, 1996a). Immunodiagnostic tests used in screening for human neurocysticercosis may also exhibit reduced sensitivity compared to hospital standardized data but this is probably due to a higher proportion of persons with calcified cysts in the community rather than early asymptomatic cyst development (Garcia-Noval *et al.*, 1996).

The use of crude or only partially purified native antigen preparations in immunodiagnostic tests tends to be criticised, but for all their disadvantages, notably poor specificity and batch variability, crude antigens have been and continue to be useful in immunologic screening for larval taeniid zoonoses particularly alveolar and cystic echinococcosis. Highest possible immunodiagnostic sensitivity is important in alveolar echinococcosis because early diagnosis and treatment can significantly reduce mortality (Uchino *et al.*, 1991). Use of a crude extract of protoscoleces, or a total metacestode extract of *E. multilocularis* in ELISA for mass screening of human alveolar echinococcosis in China (Craig *et al.*, 1992) and France (Bresson-Hadni *et al.*, 1994) resulted in the detection of an additional four and two cases, respectively (confirmed by ultrasound or CT scan) in comparison to use of the highly specific purified Em2 antigen. The recombinant *E. multilocularis* antigens II/3-10 and Em60 offer potential advantages but have not yet been field tested (Gottstein *et al.*, 1993; Helbig *et al.*, 1993).

In contrast, for neurocysticercosis, which arguably has a lower overall pathogenicity, confirmation of a seropositive can only effectively be done by CT scan or MRI (Flisser, 1994). The consequences of producing false positives in a serological survey for neurocysticercosis may therefore be both economically and ethically great and therefore the onus should be on the highest obtainable serological specificity even at the risk of reduced sensitivity. For human cystic echinococcosis the situation regarding serological screening lies somewhere between alveolar echinococcosis and neurocysticercosis. A small proportion of persons with cystic echinococcosis will develop serious life-threatening or fatal disease (Pawlowski, 1993). Confirmation of seropositives could be achieved relatively easily by ultrasound for 60–70% of cases and by radiography for approximately 10% of cases. However, a greater number of false seropositives will be expected when crude hydatid cyst fluid antigens are employed (Schantz and Gottstein, 1986). One approach is to undertake serological screening for cystic echinococcosis using two tests, the first highly sensitive (e.g. ELISA with crude cyst fluid antigen) to maximize case detection, and the second, more specific, e.g. immunoblot of 8–12 kDa sub-units for antigen B recognition, or use of recombinant antigen B, to confirm seropositives for clinical follow-up. However, a small number of cystic echinococcosis cases which are ELISA positive but immunoblot negative could still be missed (Paolillo et al., 1996).

In addition to their use in mass serological screening for case detection, immunodiagnostic tests have a role in seroepidemiology especially in determination of exposure. Exposure will not always result in cystic infection for the three taeniid species. Reasons for this may relate to infection with non-viable or senescent eggs (Gemmell et al., 1987a), presence of effective anti-oncosphere (possibly also cross-protective?) immunity (Craig, 1993), presence of post-oncosphere immunity (Rausch et al., 1987), and/or genetic predisposition in relation to infection suscept-ibility (Gottstein and Bettens, 1994). Never the less antibody responses may occur in the absence of detectable active infection, such as abortive calcified hepatic alveolar echinococcosis lesions described in Alaskan Eskimos (Rausch et al., 1987) and in French farmers (Bresson-Hadni et al., 1994). Measurement of antibody responses to oncosphere rather than metacestode antigens may better reflect exposure (Craig et al., 1986b), higher rates being reported in endemic versus non-endemic cystic echino-coccosis areas (Craig, 1993). High oncosphere antibody levels in E. granulosus endemic sera could be important in mediating acquired immu-nity as increased oncosphere killing rates have been observed in vitro (Rogan et al., 1992). It is known that under natural conditions acquired immunity in livestock taeniid species is maintained by repeated egg exposure (Gemmell et al., 1987a).

The use of filter-paper blood spots has been reported to overcome many of the logistic, cost and sociocultural problems associated with mass serology based on venous blood samples (Craig et al., 1992; Kenny, 1993). Immunoglobulin G antibodies against E. granulosus remained stable in dried blood spots for up to one week at room temperature (~20°C) and were still seropositive after storage at −20°C for 120 days (Coltorti et al., 1988a; M. Morley and P.S Craig, unpublished). Develop-ment of a rapid dot-ELISA for cystic echinococcosis in which 50 μl of heparinized whole blood was tested also provided advantages in commu-nity screening especially in remote regions with mobile populations (Rogan et al., 1991).

Circulating parasite antigen detection has also been proposed as a possible immunodiagnostic approach for cystic echinococcosis and neurocysticercosis, particularly as the presence of parasite antigen in sera or CSF would indicate active viable cysts (Craig and Nelson, 1984; Gottstein, 1984; Correa *et al.*, 1989; Estrada *et al.*, 1989). However, such assays have shown poor or variable sensitivity in human cystic echinococcosis (Craig, 1993; Liu *et al.*, 1993b) and neurocysticercosis (Wang *et al.*, 1992), and therefore their role is probably more suitable for post-treatment surveillance than clinical diagnosis or mass screening (Craig, 1993; Flisser, 1994).

Coproantigen detection using *Taenia* specific antibodies in a capture ELISA (Allan *et al.*, 1990; 1992; 1993) has revolutionized the ability to identify tapeworm carriers in endemic communities (Allan *et al.*, 1996). A test based on a capture antibody against a proglottis somatic extract derived from *T. solium* grown in immunosuppressed hamsters, has enabled detection of >50% more *Taenia* carriers than microscopy alone in two Guatemalan communities (Allan *et al.*, 1996). Coproantigen positive persons were treated with the anthelminthic niclosamide followed by a purgative to remove the tapeworm intact and on this basis the test was 98% sensitive and 99% specific. Faecal samples can be fixed in 5% formal saline without any reduction in test sensitivity (Allan *et al.*, 1992) and because the test detects tapeworm products it does not rely on the presence of eggs in stool samples.

3.3.2. *Imaging Based Screening*

Mass chest X-ray has been used to screen for human cystic echinococcosis (CE) in South America (Schantz *et al.*, 1973; Purriel *et al.*, 1974). However, pulmonary infection accounted for only about 10% of CE cases according to reviews of surgical records involving >1500 patients (Menghebat *et al.*, 1993; Ammann and Eckert, 1995). The advent of portable ultrasound scanners in the late 1970s (Vicary *et al.*, 1977) has made possible effective community-based mass screening for abdominal cystic echinococcosis (e.g. Mlika *et al.*, 1986; Macpherson *et al.*, 1987) and hepatic alveolar echinococcosis (Craig *et al.*, 1992). Ultrasound has no use in diagnosis of neurocysticercosis, though ocular and intramuscular cysts are detectable (Macpherson, 1992). CT scan and MRI remain the only appropriate imaging techniques for neurocysticercosis and therefore a direct community-based radiological approach is not a realistic option as a primary screening tool for neurocysticercosis. CT scan, however, can be employed effectively as a secondary screening method, for instance, after identification of late onset epileptics using a "neurologic" questionnaire (International League Against Epilepsy 1993) and/or to confirm primary serology-based screening tests (Garcia-Noval *et al.*, 1996).

To-date ultrasound has been used in a number of mass-screening surveys for human cystic echinococcosis including studies in Uruguay, Tunisia, Libya, East Africa and China, and now also for human alveolar echinococcosis (Craig et al., 1992). Ultrasound is a non-invasive technique which is always well accepted by the population screened. Macpherson et al. (1987) have stressed the general advantages of ultrasound as a screening tool in their study of cystic echinococcosis in nomadic pastoralists in northwest Kenya. Unlike a specific laboratory test an ultrasound scanner has a number of other medical applications which are useful at the local hospital/clinic level (Doehring-Schwerdtfegen et al., 1992). The technique has been shown to be quite specific (>80%) in screening for cystic or alveolar echinococcosis (Craig et al., 1992; Paolillo et al., 1996) and has a high overall sensitivity (70% to >80%) especially for alveolar echinococcosis where almost all primary lesions are hepatic and therefore amenable to detection by ultrasound (Craig et al., 1992). Cysts of E. granulosus or E. multilocularis lesions as small as 0.5–1 cm can be detected but serological confirmation is required for such small images. Diagnosis of larger cysts or lesions also benefits from serological confirmation, though some characteristic images will be directly pathognomonic for the infection.

The ultrasound image classification for cystic echinococcosis described by Gharbi et al. (1981) is very useful and may be summarized as: Type I, simple univesicular fluid-filled cysts with well defined border; Type II, presence of membrane (laminated or split wall); Type III, presence of septa and or daughter cysts; Type IV, solid mass which is hyper- or hypoechoic; and Type V, presence of calcifications and hyperechoic. In general, Types II and III are considered by most hydatid experienced sonographers to be pathognomonic for cystic echinococcosis (Gharbi et al., 1981; Paolillo et al., 1996). In a recent large Libyan community-based study 56% (169/303) of cystic echinococcosis cases detected were characteristic of Types II and III (Shambesh et al., 1996b). Simple univesicular cystic images (i.e. Type I) may account for 18–>40% of hepatic cystic echinococcosis cases detected by portable ultrasound (Romig, 1990; Shambesh et al., 1996b) but require serological confirmation to differentiate from single non-parasitic epithelial cysts which tend to occur in liver and kidneys (Harris et al., 1986). A solid cystic mass (Type IV) also requires serological testing for differential diagnosis of liver abscess or carcinoma (Von Sinner, 1992). Cystic echinococcosis presented as Types I and IV, however, tend to exhibit lower or variable seroreactivity, for example, 38–70% and 63–100%, respectively (Paolillo et al., 1996; Shambesh et al., 1996b). Non-liver abdominal location of E. granulosus hydatid cysts was reported to occur in 18% of Libyan cases detected with ultrasound, the majority in the kidneys (45%) and spleen (14%) (Shambesh et al., 1996b). In the large mass ultrasound study of Shambesh et al. (1996b) simple non-parasitic

Table 3 Abdominal cysts detected by ultrasound with serologic testing in a cystic echinococcosis mass screening programme in Libya (*n* = 530 with cystic images; population screened, 20 220) (Shambesh *et al.*, 1996b).

Organ location/cyst type	Number and prevalence of cyst type (%)	AgB-ELISA sero-reactivity (%)
Liver		
Type I — univesicular	53 (17.5)	20 (37.7)
Type II — laminations	104 (34.3)	89 (85.6)
Type III — daughter cysts	65 (21.5)	62 (95.4)
Type IV — solid mass	17 (5.6)	17 (100)
Type V — calcifications	40 (13.2)	9 (22.5)
Type VI — multiple cysts	24 (7.9)	23 (95.8)
Total	303	220 (72.6)
Kidney	191 (84.1)	29 (15.2)
Spleen	9 (4.0)	6 (66.7)
Abdominal wall	7 (3.1)	1 (14.3)
Ovarian cysts	10 (4.4)	4 (40.0)
Uterine cysts	4 (1.8)	2 (50.0)
Muscle cysts	3 (1.3)	0 (0)
Polycystic	3 (1.3)	0 (0)
Total	227	42 (18.5)

kidney cysts were recorded in a significant proportion of people screened, 0.8% (162/20 220), mostly in the >50 years age group. Table 3 shows the proportion of ultrasound detected cysts of different pathological presentation together with seroprevalence (after Shambesh *et al.*, 1996b).

Ultrasound based primary screening for hepatic alveolar echinococcosis has also proved highly effective. One study in central China recorded ultrasound as exhibiting 84% sensitivity and specificity for diagnosis of hepatic alveolar echinococcosis (Craig *et al.*, 1992). Advanced lesions of *E. multilocularis* are usually located in the right liver lobe and on ultrasound are hypodense with irregular contours and peripheral calcifications often also with central necrotic cavitation (Didier *et al.*, 1985; Choji *et al.*, 1992). "Early" small hypoechoic lesions due to alveolar echinococcosis may resemble haemangioma or hepatoma. Furthermore, small focal hyperechoic calcifications and clusters of microcalcifications could be confused with tuberculosis or may in fact be due to abortive *E. multilocularis* infection (Bresson-Hadni *et al.*, 1994; B. Bartholomot, D. Vuitton, P.S. Craig, unpublished observations). In these cases serological confirmation for specific *E. multilocularis* antibodies has been shown to be important (Rausch *et al.*, 1987; Bresson-Hadni *et al.*, 1994). In a recent community study of alveolar echinococcosis in China 17.2% (110/640) of persons scanned by ultrasound had images showing microcalcifications of which 11.8% (13/110) of sera gave some degree of seroreactivity suggestive of abortive *E. multilocularis* infection; in the same study small hyperechoic lesions (1.5–2 cm) with specific seroreactivity were identified in 0.8% (5/640) of persons compared with 5% (32/640) for confirmed cases of alveolar echinococcosis (Giraudoux *et al.*, 1995).

Serological confirmation of a positive radiological image should be carried out with a test of high specificity (>90%) and acceptable sensitivity (>70%). In mass radiological screening for cystic echinococcosis serological confirmation has been successfully employed using the DD5 gel diffusion test (Frider *et al.*, 1988), crude cyst fluid in ELISA (Mlika *et al.*, 1986; Bchir *et al.*, 1991), purified antigen B in ELISA (Shambesh *et al.*, 1996b) and also the immunoblot test for the low molecular weight antigen B subunits (Paolillo *et al.*, 1996). It should be emphasized, however, that ultrasound-based screening may miss a number of infected persons, e.g. with pulmonary cystic echinococcosis (or with other non-abdominal site locations) or very small cystic or alveolar echinococcosis lesions <1 cm. It is therefore important to test serologically all individuals screened/registered during a community study and any ultrasound negative seropositives followed up by repeat ultrasonography, radiography and/or CT scan. In this way during a community study in Uruguay two additional cases of cystic echinococcosis (one pulmonary and one hepatic), who were ultrasound negative, but filter-paper blood seropositive were identified after follow-up by radiography and CT scan (Paolillo *et al.*, 1996).

Under-reporting of *T. solium* neurocysticercosis cases from rural areas may be due to the lack of diagnostic imaging facilities in these areas, and the failure to include the disease in differential diagnosis (Gracia *et al.*, 1990). As greatly increased reporting has been found to occur after the introduction of accurate new image diagnostic techniques, such as CT scan to the USA (Richards *et al.*, 1985), this may also occur as these imaging techniques begin to be more widely applied. Modern imaging techniques, such as CT and MRI scanning are, however, generally relatively expensive and therefore inaccessible to much of the population of rural endemic areas. Perhaps more appropriate, therefore, would be the wider application of simple and relatively inexpensive serological techniques for the immunodiagnosis of neurocysticercosis. However, methods have only recently been developed with characteristics suitable for application in the general population.

High rates of epilepsy in Latin America have been explained by the local high prevalence of neurocysticercosis (Medina *et al.*, 1990; Botero *et al.*, 1993). These studies have been conducted in hospital neurological wards with access to both trained neurologists and neuroimaging facilities. One major handicap for studies on *T. solium* has been that characterization of neurological disease in general, and epilepsy in particular, has been difficult due to the multiplicity of aetiologies. The lack of staff and facilities in rural, high risk, areas also limits the assessment of the occurrence of such conditions. This has often meant that studies of neurological disease carried out by different groups or in different countries have been difficult to compare due to differences in methodology and the results have appeared discordant.

If a picture of the neurological impact of *T. solium* is to be built up in rural populations using simple and easily applied methodologies the use of a standardized approach for the characterization of these neurological conditions would appear to be necessary in order that studies in different areas can be directly compared. A simple and reliable screening protocol, probably based initially around detection of epileptic cases by questionnaire with some follow-up by a neurologist would appear to be an appropriate method for screening open populations. For instance, a classification of seizures and epileptic syndromes has been proposed by the International League Against Epilepsy (1993) and the use of such standardized classifications and simple screening methods for the detection of such conditions, coupled with the immunological and imaging diagnostic technologies now available may add considerably to our understanding of the neurological disease impact in the epidemiology of *T. solium*.

4. DETECTION OF TAENIID INFECTION IN ANIMAL DEFINITIVE AND INTERMEDIATE HOSTS

In addition to human data, any detailed community study of taeniid zoonoses also requires detection and quantification of the parasite (adult and larval stages) in animal hosts. For *T. solium* the domestic pig is the only significant intermediate host. For *E. granulosus* a number of ungulate livestock species may be infected though not all will take part in a true transmission cycle; worldwide the sheep is the most important intermediate host (Rausch, 1995). A number of arvicolid rodent species are important natural intermediate hosts for *E. multilocularis* (Rausch, 1995). Foxes of the genera *Vulpes* and *Alopex* form the main definitive hosts for *E. multilocularis* in its sylvatic cycle, but the dog is an important domestic definitive host for this parasite and is probably responsible for the bulk of human infections. The domestic dog is the most important definitive host of *E. granulosus* in its domestic cycle, though wild canids may be susceptible to infection from the synanthropic cycle.

The dynamics of transmission in domestic animal cycles has now been described mathematically for *E. granulosus, Taenia hydatigena* and *T. ovis*

(Gemmell *et al.*, 1986, 1987a; Gemmell and Roberts, 1995). A preliminary model to describe the transmission of *E. multilocularis* between foxes and rodents has now been published (Roberts and Aubert, 1994) (also Section 2). As yet the transmission dynamics of *T. solium* in its human–pig cycle has not been described.

4.1. Animal Intermediate Hosts

Cystic echinococcosis in sheep (or other domestic ungulates, e.g. cattle, camels, reindeer, etc.) and cysticercosis in pigs are most accurately quantified by examination at necropsy. In many countries these infections are notifiable and slaughterhouse records are kept by law, however in other endemic areas slaughter records are collected voluntarily at meat inspection. Mature ovine hydatid cysts (approx. 1–8 cm) are located primarily in liver and/or lungs, whereas in porcine cysticercosis cysts (0.5–1 cm) are located primarily in striated muscle especially heart, tongue and limbs (Harrison and Sewell, 1991). Both *E. granulosus* and *T. solium* cysts are easily identified macroscopically by a trained meat inspector. Small hydatid cysts (<5 mm) however, require histological identification (Lloyd *et al.*, 1991) especially to differentiate from other larval taeniids such as *T. hydatigena* (which is sometimes even confused with larger hydatid cysts) and *T. ovis*, and also caseous lymphadenitis (due to *Corynebacterium pseudotuberculosis*) (Cabrera *et al.*, 1995). Differential diagnosis of porcine cysticercosis at post mortem is really only problematic for caseated or necrotic cysts when infection intensity is low.

In ovine hydatidosis due to *E. granulosus*, cyst numbers are important as age intensity data may be used to calculate R_o (basic reproductive ratio) and therefore to describe the endemic steady state which is important in developing rational control intervention. R_o has now been calculated for *E. granulosus* transmission in specific regions of Xinjiang, China (Ming *et al.*, 1992) and Uruguay (Cabrera *et al.*, 1995).

Pre-mortem detection of porcine cysticercosis can be achieved by the ancient and still widely used method of tongue palpation which was accurately described around 400BC (Viljoen, 1937; Grove, 1990). This method is reported to be around 70% sensitive (Gonzalez *et al.*, 1990) and highly specific (Onah and Chiejina, 1995). It is also practised by pig owners in rural areas and leads to removal of animals from the formal inspection process (Sarti *et al.*, 1992b; CWG Peru; 1993; Boa *et al.*, 1995). Palpation is not applicable to ovine hydatidosis due to the organ location of hydatid cysts. Radiological diagnosis of ovine hydatidosis by radiography has been used successfully on a very small scale (Wyn-Jones and Clarkson,

1984), but the potential (if any) for ultrasound scanning has yet to be assessed.

Immunodiagnostic tests offer possible specific and sensitive indirect detection of ovine hydatidosis and porcine cysticercosis. In ovine hydatidosis specific serum antibody reactivity is almost always less reliably detected than when the same tests are applied to human cystic hydatidosis (Lightowlers, 1990; Craig, 1993). The frequent presentation of high intensity infection for porcine cysticercosis appears to stimulate a better antibody response than occurs in ovine hydatidosis. Serum antibodies to crude *T. solium* metacestode extracts can be detected by ELISA, though identification of low molecular weight antigens by immunoblot appears more sensitive (Pathak *et al.*, 1994). The immunoblot test using glycoprotein-purified *T. solium* antigen has been shown to be 70% sensitive and 100% specific for porcine cysticercosis but in practice may not be significantly better than the method of tongue palpation (Gonzalez *et al.*, 1990). Detection of circulating antigen for immunodiagnosis of ovine hydatidosis lacks sufficient sensitivity (Judson *et al.*, 1985), but has been successful in detecting *T. saginata* infections in cattle when numbers of cysticerci exceed 100 (Brandt *et al.*, 1992). Preliminary data on serum antigen detection look promising for porcine cysticercosis due to *T. solium* but requires further research (Rodriguez *et al.*, 1989).

The need for pre-mortem identification of rodent alveolar echinococcosis is not really an important priority, however conservation considerations could warrant development of specific tests in the future. Both experimental and natural infections of *E. multilocularis* elicit putative specific antibody responses to *E. multilocularis* antigens (Ito *et al.*, 1994). Specific confirmation of small *E. multilocularis* lesions could be achieved at rodent autopsy using DNA hybridization after PCR amplification, for example using the pAL1 DNA probe described by Vogel *et al.* (1991).

4.2. Animal Definitive Hosts

Detection of *E. granulosus* or *E. multilocularis* infection in dogs, foxes or other canids is most reliably undertaken at necropsy by careful examination of the small intestine for the presence of the small tapeworms (2–11 mm in length for *E. granulosus* and 1.2–4.5 mm for *E. multilocularis*). Examination should include gut washing/incubation and mucosal scrapings for microscope slide examination. Worm burdens can be counted, however, for heavy infections, i.e. >1000 worms. An estimation is usually made based on sectional counts along the gut, but low intensity infections, e.g. <20 worms, can be missed.

Parasitological detection of *Echinococcus* in domestic dogs pre-mortem is currently routinely carried out by purgation using arecoline hydrobromide. This method is 100% specific but lacks sensitivity, is biohazardous, laborious and up to 20% of dogs fail to purge (Wachira *et al.*, 1990; Craig *et al.*, 1995b). Despite these drawbacks arecoline purgation has provided important surveillance data in hydatid control programmes (e.g. McConnell and Green, 1979) and for epidemiological studies (e.g. Walters and Clarkson, 1980).

Recent developments in immunodiagnosis based on serum antibody or faecal antigen detection (reviewed by Craig, 1993) have provided potential alternatives to arecoline diagnosis of canine echinococcosis. Coproantigen detection by ELISA in particular offers a reliable and sensitive alternative to arecoline purgation (PAHO, 1994). Current coproantigen tests, based on polyclonal anti-somatic or anti-ES adult *E. granulosus* antibodies, are about 97% specific and 98% sensitive when worm burdens or purge counts are >50 worms and is applicable to 5% formalin fixed stools (Allan *et al.*, 1992; Deplazes *et al.*, 1992; Craig *et al.*, 1995b). Despite a general increase in ELISA absorbance values with increasing worm burdens the coproantigen test is not yet however sufficiently quantitative to estimate worm burdens (Craig *et al.*, 1995b). Serum antibody detection by ELISA using *E. granulosus* protoscolex antigens (native or recombinant) does not differentiate between current and previous infections (Gasser *et al.*, 1990; Jenkins *et al.*, 1990). Antibody detection, however, can be useful in epidemiological studies where exposure in dogs or wild canid populations needs to be identified (Gasser *et al.*, 1993; Reichel *et al.*, 1996).

Current coproantigen or serum antibody tests are highly genus (*Echinococcus*) specific but not species specific (Deplazes *et al.*, 1992; Craig, 1993). Differential diagnosis of *E. granulosus* and *E. multilocularis* infections in definitive hosts could be achieved by specific DNA hybridization of PCR-amplified worm products in faeces. However, significant progress is required if "non-egg" DNA targets are to be detected in faeces (Bretagne *et al.*, 1993). This is because egg output in adult taeniid infections in dogs is not constant (Deplazes and Eckert, 1988; Allan *et al.*, 1992; Kinder *et al.*, 1992).

5. CYSTIC ECHINOCOCCOSIS IN ENDEMIC COMMUNITIES

Patterns of transmission of *E. granulosus* within human communities are influenced not only by parasite biotic factors (see Section 2), but also by patterns of human behaviour and life style. In particular the migratory behaviour of people and their association with dogs is important. The disease is most frequently associated with pastoral people where the infection is naturally transmitted between livestock and domestic dogs. The periodic movement of people and livestock can influence where and when transmission to humans takes place and is particularly important in transhumant and nomadic lifestyles (Macpherson, 1994). Other modes of transmission, however, also occur, including urban transmission.

In general, the pastoral communities affected can either be (i) settled, such as sheep farming communities in Wales and South America; (ii) transhumant, involving regular seasonal population movements to defined areas and typified by herding communities in northwest China, (iii) nomadic, involving frequent movement of people and livestock to different areas for varying periods of time, e.g. Nilotic tribes of East Africa; (iv) urban, where transmission occurs within a town or city, e.g. Kathmandu.

Within these communities transmission to man generally occurs where there is a close association of dogs, domestic livestock and people. For infection to occur, dogs must get access to infected livestock and humans must ingest eggs in food, water or through direct dog contact. In virtually all human cases, where parasite isolates have undergone DNA typing analysis, it is the sheep strain which seeems to be most important (Bowles and McManus, 1993). Some evidence now also suggests that the cattle strain may also be infective to humans (Bowles *et al.*, 1992b). Communities where sheep are numerous are therefore more likely to be at risk, although it should be remembered that the sheep strain may also infect other livestock such as cattle.

Clinical, epidemiological and DNA analysis of the cervid strain of the parasite suggests that this too is infective to humans (Wilson *et al.*, 1968; Bowles *et al.*, 1994). Transmission of this strain to man is one of the few situations where wild intermediate hosts may be important. In Lappland and Siberia infection rates of 50–70 per thousand have been recorded among reindeer herders, hunters and fur-farmers (Nemurovskaia *et al.*, 1980; Ageyeva, 1989). Transmission to humans in these communities is likely to be associated with domestic dogs rather than wild canids.

5.1. Cystic Echinococcosis in a Highly Endemic Nomadic Community in East Africa

The prevalence of cystic echinococcosis is significant in several groups of nomadic pastoralists, particularly in East Africa (Macpherson and Craig, 1991). The close association and dependence of the people on their livestock, and unregulated home slaughter of animals, provides a situation where, if dogs are also present, the parasite can be readily transmitted. Since the mid 1960s it has been recognized that the Turkana tribe of northwest Kenya have a very high incidence of cystic echinococcosis. The exact extent of the disease has been hard to assess due to the remote nature of the location. Early evidence from hospitals outside of this area indicated an annual surgical incidence rate of 40 cases per 100000 (Schwabe, 1964). Later hospital reports from inside the region revealed a much higher incidence at around 220 per 100000 (French and Nelson, 1982). Since then other studies using a combination of ultrasound and serology have indicated a prevalence of between 6% and 9% for people in the north west of the district (Macpherson et al., 1987; Romig, 1990; Rogan et al., 1991). These people are essentially nomadic and tend to have many dogs associated with the community. It is now recognized that this focus of disease probably extends to neighbouring Nilotic pastoralists such as the Toposa in Sudan, the Karamajong in Uganda and the Dassanetch and Hamar in Ethiopia (Eisa et al., 1962; Owor and Bitakamire, 1975; Macpherson et al., 1989; Klungsoyr et al., 1993). It is noteworthy, however, that although E. granulosus is endemic in Somalia, there are no reported cases of hydatid disease among Somali pastoralists, in part due to the fact that the Somali are Muslims having very limited contact with dogs (though this does not preclude the massive transmission problem in Islamic North Africa and the Middle East), but also the possibility that a dominant dog–camel strain is of low infectivity to humans (McManus et al., 1987; Wachira et al., 1993).

Typical nomadic behaviour of the Turkana involves the communities setting up a series of temporary huts (akai) made from sticks, and grouped together as extended family units into manyattas, which are surrounded by thorn fences. Several manyattas may exist in one area for a period of weeks or months. They are usually situated in areas where there is adequate grazing for the livestock and supplies of water for both people and animals. This region is, however, very hot and arid (<200 mm annual rainfall), and grazing and water do not last long. The people must therefore continually move to where conditions are better. Water is most frequently obtained from temporary wells which are dug in dry river beds (French et al., 1982; Macpherson, 1994).

The Turkana people have a very close association with dogs which they use as guard animals to warn of frequent livestock raids from neighbouring tribes. Additionally, because, water is scarce, dogs are allowed to clean both babies and cooking utensils by licking (French et al., 1982). Such behaviour means that there is adequate opportunity for transmission of eggs to people, since the prevalence of adult worms in dogs is high at 39–70% (Macpherson et al., 1985). The mean number of E. granulosus worms in infected dogs is generally considered to be around 200 for most areas of the world (Gemmell et al., 1987a) but in Turkana 36% of dogs had mean burdens >1000 (Macpherson et al., 1985). It may, therefore, be possible that the type(s) (breed?) of dog present in Turkana may be more susceptible to the infection than others as appears to be the case for dingoes in Australia (Gemmell and Roberts, 1995). Differences in maturation rates of *E. granulosus* after experimental infection of Turkana versus Nairobi dogs supports this notion (Wachira, 1993). The routes of transmission from dogs to humans may be varied. Due to the climate in this area *Echinococcus* eggs can only survive for a short time and so it is likely that for infection to occur people (and animals) must come into contact with freshly deposited eggs (Wachira *et al.*, 1991). As mentioned previously close direct contact with dogs is important as eggs may stick to fur around the mouth and anus. In addition *Echinococcus* eggs have been found to be present in and around water holes (Craig *et al.*, 1988) and here, due to the more moist and cooler conditions, egg survival is prolonged (Wachira *et al.*, 1991). Unfortunately, as yet the transmission dynamics of *E. granulosus* in the domestic animal populations in Turkana have not been described.

Analysis of ultrasound detection rates indicates that there is a general pattern of increasing human prevalence with age, with people under 20 years having the lowest rates of abdominal infection and the over 40s having the highest (T. Romig and E. Zeyhle, unpublished; Craig, 1993) (see also Figure 4). This pattern may be due to increased chance of exposure with time, increased contact with dogs, and/or a lack of acquired immunity. Some individuals in the older age groups will have acquired the infection earlier in life but, if seroprevalence data are analysed, results produce a similar age-dependent trend. Since this type of diagnostic procedure will detect small cysts it suggests a true increasing prevalence with age. Whether this is due to increasing exposure with time is questionable. Dogs appear to spend more time in the "home area" which is more frequented by women and children than men, and this may explain a higher incidence of infection in women (French and Nelson, 1982; Watson-Jones and Macpherson, 1988). It does not, however, explain why there is a lower incidence in children. It is possible that children are less exposed to the eggs but, if anti-oncospheral antibody levels are studied results indicate that the highest response is in the under 10 age group. The significance of this is unclear but suggests that the younger people are heavily exposed to the parasite. Whether high anti-oncosphere titres reflect a greater chance of resistance to hydatid disease is unclear but it has been shown that oncosphere antibody positive sera from clinically normal people in the endemic area of Turkana has a greater ability to kill oncospheres *in vitro* than non-endemic sera (Rogan *et al.*, 1992). Further investigations into this are required but it may suggest that certain age groups with a high exposure to eggs become resistant due to acquired immunity (Craig, 1993).

Natural transmission from dogs to livestock intermediate hosts is likely to occur either by dogs accompanying their owners during grazing of the animals, or from water holes where dogs and livestock are present, or within the manyattas. This latter situation is evident in the evenings when dogs and animals are brought inside the enclosing manyatta fence for protection. Evidence from occasional slaughter data indicates that infection rates in sheep and goats are on the low side ~5%, with the highest prevalence rates in camels (33%) (Macpherson, 1981; Macpherson *et al.*, 1983). As mentioned above, the camel strain of the parasite is probably of low infectivity to man. Hydatid cysts can also be found in a variety of wild intermediate hosts such as wildebeest (*Gorgon taurinus*), cape buffalo (*Syncerus caffer*) and warthogs (*Phacochoerus aethiopicus*) but, as numbers of these animals are low in Turkana, it is unlikely that these are significantly involved in transmission of the parasite (Macpherson and Craig, 1991). They may, however, form part of a sylvatic cycle involving hyenas (*Crocuta crocuta*), cape-hunting dogs (*Lycaon pictus*), jackals (*Canis* spp.) and lions (the only felid to acquire adult *E. granulosus* infections) as definitive hosts in central and southern Africa (Nelson and Rausch, 1963; Macpherson and Craig, 1991).

The infection of domestic dogs takes place primarily during home slaughter of livestock. Organs containing hydatid cysts are frequently thrown to the dogs. In addition to this route dogs can become infected by scavenging on carcasses of animals which have died outside of the manyattas. This is particularly evident during periods of drought (Wachira *et al.*, 1990). As the Turkana people do not bury their dead it has also been postulated that dogs may be infected by eating human remains containing hydatid cysts (French *et al.*, 1982; Macpherson, 1983).

In 1983 a pilot control programme for cystic hydatid disease was initiated in an area of northwest Turkana by the African Medical and Research Foundation (Macpherson *et al.*, 1984). It was initially aimed at (i) reducing the population of stray or unwanted dogs, (ii) mass treatment of wanted dogs with praziquantel every six weeks to remove adult worms, and (iii) to establish community education programmes to heighten people's awareness of disease transmission. The control programme has been hindered over the years by several problems. First, the nomadic nature of the community makes it difficult to monitor people and dogs successfully. Migration into and out of the control area can also cause difficulties. Secondly, the reduced literacy levels of the community and frequent famines have resulted in breakdown of community education principles. Thirdly, lack of government involvement and therefore continued funding have resulted in discontinuous dog chemotherapy and contact with the community. Despite these constraints it has been estimated that about half of the population recognizes dogs as the source of infection. The prevalence of the parasite in dogs has been reduced from 63% to 25% and the dog population has been reduced from a 1:4 dog/man ratio to 1:25

(Wachira and Zeyhle, 1993). Although this represents a significant decrease in the parasite population, if control measures are not effective enough or sustained then levels of the parasite in the community could rise quickly.

5.2. Cystic Echinococcosis among Transhumant and Settled Communities in Northwest China

Human cystic echinococcosis is highly endemic in China occurring in 21 of the 31 provinces and autonomous regions (Craig *et al.*, 1991). Within the Xinjiang Uygur Autonomous Region of northwest China the population (total approximately 15 million) is composed of 13 nationalities of which Uygur (7 million), Han (5.5 million), Kazakh (1 million), Hui (674000) and Mongolian (140000) are dominant (Chai, 1993a). The Uygur, Han and Hui populations are mainly engaged in industrial and agricultural production whereas the Kazakhs and Mongolians tend to be involved with animal husbandry involving grassland grazing. Xinjiang is China's largest administrative region (approx. 1.65 million km^2). The geography of the region includes groups of mountains which rise steeply from flat basins. Temperature extremes range from around 30°C in summer to -20°C in winter and on average temperatures are below 0°C for 4 months of the year. The variation in topography and large annual temperature range are the main reasons why people migrate at different times of the year. Herding communities in Xinjiang are typically semi-settled in communes during the winter months. These are located in the low-lying basins and are adjacent to "winter pastures" where livestock can be grazed. Once the temperatures start to rise in spring groups of people and animals will move towards the foothills and eventually up into the mountains where the "summer pastures" are located. Here they live in tents and periodically move location. Younger men (18–40 years) tend to move to the summer pastures first whereas women and older men remain in the communes for longer and sometimes permanently.

The problem of cystic hydatidosis in Xinjiang has only come to light in recent years and in 1988 the National Hydatid Disease Centre was established in Urumqi. A retrospective survey of surgical cases between 1951 and 1990 revealed 16663 cases with most cases originating in the north of the provence in the Tacheng, Bole and Altai Prefectures (Menghebat *et al.*, 1993). However, these figures are likely to be underestimates of the true surgical incidence because of incomplete records. The surgical data indicated that children in the 6–11 year age group were the most frequently operated on but there was little difference in surgical rates between nationalities.

Since surgical incidence only reflects the number of symptomatic cases who have access to medical treatment it is likely that the true prevalence in these communities is much higher. Several sero-epidemiological surveys have been carried out in different areas of Xinjiang. Most of these have been done using the ELISA technique with crude hydatid fluid as an antigen. These have produced seroprevalence values of between 21% and 32% in areas around the Tianshan, Altai and Kunlun mountains with the majority of positive cases coming from Kazakh, Uygur and Mongolian herdsmen and also Han peasants (Chai, 1993b). Matched radiological or ultrasonographic examination of these groups showed lower prevalences of between 2.3% and 3.1%. These data suggest that only around 10% of the seropositive individuals have ultrasound-detectable cystic lesions. The significance of the other 90% is unclear. It is certainly possible that some of these people are false positives and have antibodies which cross react with the crude hydatid fluid antigen. In other studies where more specific purified antigens (i.e. antigen B) have been used the proportion of antibody-positive people having no detectable cysts dropped to around 67% (Rogan et al., 1991). Other possible explanations for this seropositive group are that they have cysts in locations which are not detected by abdominal ultrasound or that they have been exposed to the parasite but it has not survived for long enough to form large cysts. Whatever the explanation, the use of serology, especially in association with portable ultrasound, can be confirmatory and indicative of parasite exposure.

Settled mixed Han and Hui agricultural communities in north central Xinjiang were investigated for transmission of E. granulosus by Chi et al. (1990). Arecoline treatment of dogs (n = 390) gave a prevalence of 16.2%, and hydatid cysts were detected at post mortem in 88% of sheep (n = 1797), 56% of goats (n = 123) and 94% of cattle (n = 50). The annual surgical case rate for CE was 43.8 per 100000. Though people are primarily agriculturalists in this region, they also commonly keep small numbers of livestock for each household. Transmission of E. granulosus here was associated with home slaughter, which was practised by 84% of villagers, as was feeding offal to dogs, ownership of which was >80% (Chi et al., 1990; Andersen et al., 1991). Despite very high prevalence rates of ovine hydatidosis in this community, recent analysis of age-specific intensity and prevalence for sheep infection, in a similar community in the region, showed the parasite to be in an endemic steady state in which infection pressure (i.e. egg contamination) was too low to stimulate effective acquired immunity in the sheep population (Ming et al., 1992). Consequently, a well-organized regular dog dosing programme in such communities, perhaps in combination with health education, could drive R_o below unity and break transmission (Andersen et al., 1991; Ming et al., 1992).

The cycle of transmission of the parasite which affects transhumant

herding communities in Xinjiang is unclear. The principal animal intermediate hosts are sheep with abattoir-detected infection levels of between
3% and 90%, although cattle (46–73%), yaks (5–81%), goats (4–41%), pigs
(1–37%) and camels (up to 80%) are also commonly infected (Craig *et al.*,
1991; Liu *et al.*, 1993b). The limited DNA analyses carried out on these
human and animal isolates indicated that they are all of the sheep strain
(McManus *et al.*, 1994). The animals are often infected at an early age
(possibly in the summer pastures) by grazing on land contaminated with
dog faeces. The prevalence of adult worms in dogs is relatively high (mean,
22%) and they will frequently go with their owners to the summer pasture
in early spring. At this time the temperatures are around 5–18°C and
moisture is abundant due to snow melt. Under these conditions survival
of *Echinococcus* eggs in the environment is prolonged and, with a dog to
sheep ratio of between 1:50 and 1:150, livestock may be exposed for a
considerable period of time in both the spring and autumn months.

The prevalence of the parasite in dogs varies with the social background
of the owners. Dogs owned by Kazakh herdsmen had a higher infection rate
than those owned by Han and Mongolian peasants (Liu, 1993). Probably
the most significant feature contributing to dog infection is the low numbers of official abattoirs and the tendency for people to carry out home
slaughter. In some areas 95% of the inhabitants routinely discard sheep
viscera or purposely feed raw offal to dogs (Liu, 1993).

Risk factors for human CE in transhumant communities are still unclear
though sero-epidemiological data indicate that predominant pastoral
groups such as Kazakh and Mongol have higher exposure (Chai, 1993b).
In recent years immunological investigations coupled with detailed ques-
tionnaires designed to deduce aspects of the life style of individual people
involved have been undertaken (Andersen, 1993). Results have suggested
that the majority of families owned dogs and that they were frequently fed
raw offal. Generally less than half the people knew what hydatid disease
was or how it was transmitted but more than 75% recognized hydatid cysts
in sheep viscera. Human infection in transhumant communities is likely to
take place by several routes. The younger adult males will spend considerable time with the dogs in the summer pastures and frequent handling may
lead to significant exposure to *Echinococcus* eggs. A similar situation
exists for children who frequently fondle "pet" dogs in the winter communes. The transhumant life style of the people means that for significant
periods of time there is a close association of humans, dogs and livestock; a
situation which is highly conducive to transmission of cystic echinococcosis. It is known that the distribution and prevalence of *E. granulosus* in
southwest USA was strongly associated with such transhumant life styles
(Crellin *et al.*, 1982).

5.3. Cystic Echinococcosis in the British Isles — Low Endemicity

Echinococcus granulosus is endemic in the British Isles, transmission occurring in England, Wales, Scotland and Ireland. Human CE cases, however, predominantly occur in only two foci, both characterized by hill sheep farming, i.e. mid/south Wales (principally Powys county) and northwest Scotland (principally the Hebridean Islands) (Walters, 1978; Chisholm *et al.*, 1983; Stallbaumer *et al.*, 1986). The majority of autochthonous cases occur in Wales (10–16 cases per year) which had an annual incidence rate of 0.4 per 100000 for the period 1974–1983, compared to 0.02 per 100000 for England (Palmer and Biffin, 1987). Before 1934 no local human CE cases were recorded for northwest Scotland (Hill, 1989), however 17 cases were recorded from the islands of Lewis and Skye between 1966 and 1982 (Chisholm *et al.*, 1983). The small number of non-imported CE cases (<20 cases) from England in 1974–83 occurred in the Herefordshire and Birmingham areas close to the Welsh borders, and a proportion, but not all of these cases previously resided in or visited Wales (Stallbaumer *et al.*, 1986; Palmer and Biffin, 1987). Human CE appears to be very rare in Ireland, mostly imported cases, though apparent autochthonous cases have been reported (O'Rourke, 1969).

In Wales no community mass screening has been implemented though sheep farming communities, particularly those in the Usk and Wye valleys, appear most at risk. The highest annual surgical incidence was 7 per 100000 in Brecknock District of Powys in mid-Wales (Palmer and Biffin, 1987). Age-specific CE rates (1984–90) also indicate recent transmission occurring in children (three cases of 3, 8 and 12 years of age) in the south Wales counties of mid-Glamorgan and Gwent, but not in south Powys in which a hydatid control programme, based on supervised dog dosing and health education, was initiated in 1983 (Palmer *et al.*, 1996). Sheep post-mortem infection rates for south Powys region were 21–37% for the period 1973–84 (Walters, 1984) and fell to 10% in 1988/89. This decline was considered to be a direct result of the South Powys Hydatid Control Scheme (Palmer *et al.*, 1996). Arecoline purge studies of farm dogs in mid-Wales gave an *E. granulosus* prevalence rate of 25–28% in a total of 937 dogs (Walters and Clarkson, 1980; Hackett and Walters, 1980). Another study found more than 50% of farms with an infected dog (Walters, 1978). The coproantigen ELISA was also recently used to monitor canine echinococcosis rates in the south Powys control area. None of the dogs were positive ($n = 107$) from farms within the control zone, whereas 3.4% ($n = 267$) of dogs were coproantigen positive from areas bordering the control area (Palmer *et al.*, 1996). As home slaughter is very rare in Wales, transmission of the parasite from sheep to dogs is probably associated with scavenging on carcasses of sheep which have died on the hills. This is also the case for foxes. The red fox (*Vulpes vulpes*) is also naturally infected in south and mid Wales (1–7%), but mean worm burdens were <50 (Hackett and Walters, 1980; Jones and Walters, 1992). The red fox appears to be a poor host for *E. granulosus* and therefore may not be important in parasite transmission (Clarkson and Walters, 1991).

The human infections in Wales and Scotland are almost certainly due to the dog–sheep strain of *E. granulosus*. The dog–horse strain is also considered to occur in the British Isles principally involving foxhounds which become infected after deliberate feeding with raw horse offal and carcasses (Hatch, 1975; Thompson and Smyth, 1975). However, the apparent absence of autochthonous human cases in regions of UK and Ireland, where equine hydatidosis but not ovine hydatidosis occurs (or is common), suggests that the horse strain is of low infectivity for humans (Nelson, 1972; Thompson and Smyth, 1975). Morphological and isoenzyme studies and DNA hybridization analyses have consistently shown differences between UK horse and sheep hydatid isolates (reviewed by Thompson and Lymbery, 1988; Bowles *et al.*, 1992a). The genetic variation is now considered significant enough for recommendation of separate species status for the horse strain of *E. granulosus* namely *E. equinus* (Bowles *et al.*, 1995; Thompson *et al.*, 1995). Never the less some cross-infection experiments have indicated that horse isolates, via dog infection, were infective to sheep and *vice versa* and have led to the suggestion of a single horse strain in the UK (Cook, 1989). Unfortunately, the parasite material used in the study by Cook (1989) was not available for DNA analysis (McManus *et al.*, 1989). Another anomaly is why the sheep strain of *E. granulosus* is restricted to south and mid-Wales and northwest Scotland when similar hill sheep farming practices occur in other parts of northern Wales, northern England and mainland Scotland. It may be that the parasite does occur in some other regions but prevalence rates are very low, and coupled with the fact that hydatidosis is not a notifiable disease and the difficulties in trace-back of sheep at slaughter means that the significance of infected sheep from outside the main endemic areas may go unnoticed. *E. granulosus* has been recorded in 3.6% of farm dogs from the Lake District (northwest England) and 3.4% from the northern English Pennines (Cook and Clarkson, 1971). Also, transmission to other herbivores appears to occur in parts of England where ovine hydatidosis is rare, this is suggested by the finding that 35% of deer are apparently infected with hydatid on some forestry commission estates (Adams *et al.*, 1989). If confirmed, these probably represent infection from the sheep or horse strain rather than an undescribed sylvatic cycle. It is clear, however, that further studies are required on the epidemiology and ecology of *E. granulosus* in Britain.

5.4. Cystic Echinococcosis Transmission in an Urban Setting — Uruguayan Town and Kathmandu City

5.4.1. *CE in a Rural Uruguayan Town*

Echinococcus granulosus is highly endemic in Uruguay as in other countries of the southern-cone of South America (Figure 1). Ovine hydatidosis has been reported in all 19 Departments of the Country with a mean slaughter prevalence rate of 34%; dog infection rates based on arecoline purgation average around 11% (PAHO, 1994). Retrospective surgical data analysis has shown annual CE incidence rates as high as 105 per 100000 for some Departments, the national rate being around 18 per 100000 (Purriel *et al.*, 1974). A more recent ultrasound survey (*n* = 6035) recorded an abdominal CE rate of 1.4% (Perdomo *et al.*, 1988), which is also similar to the seroprevalence rate (Bonifacino *et al.*, 1991). Although Uruguay is a small country (177500 km^2) with 44% of the population in Montevideo, there are 24 million sheep distributed in large ranches (*estancias*) throughout the country. A large number of small rural towns and villages are also scattered across the grasslands.

Transmission of *E. granulosus* occurs in the main dog–sheep ("European" form) cycle in which dogs are infected primarily from deliberate feeding of offal during "home" slaughter on the sheep ranches. On the *estancia* dogs are used to round up sheep as in other parts of South America and Europe. Sheep are infected from contaminated pasture, and as both intensity and prevalence of infection increased with age this suggested that *E. granulosus* in central Uruguay was only relatively stable (i.e. endemic rather than hyperendemic) with a R_0 value of 1.2 and therefore amenable to effective control (Cabrera *et al.*, 1995).

Human CE infection occurs in people living on sheep farms or in the rural towns themselves. A recent epidemiological study in one town in the Department of Durazno included mass ultrasound and serological screening of the human population (*n*=1120) as well as mass arecoline purgation, coproantigen and serum antibody analysis of the dog population (*n*=210) (Paolillo *et al.*, 1996). The prevalence rate in the town for confirmed asymptomatic abdominal CE was 3.6% (40/1120) with a further 29 persons having a previous surgical history of CE (total prevalence was therefore 6.2%).

The age specific prevalence of ultrasound detected human CE increased from 1.3% in the 5–9 years age group to 14% in the >60 years age group (mean 46 years) with no significant difference between males and females; 74% of cases were seroreactive for hydatid cyst fluid antibodies. Interestingly, a 3.9% ultrasound rate was obtained for individuals (*n*=514) residing on farms in the area surrounding the town of study. It was also known that some people, especially males, owned a house in the town but worked on an *estancia*, and therefore dogs could move between the sheep ranches and the town. Although infection of townspeople from such dogs probably occurs, a positive correlation was found for ownership of a currently infected town dog (which had a prevalence of 18%) and the presence of a human CE case in a household cluster. This suggested that a long-term history of dog ownership within the town (or of living close to a household with such history) was a risk factor for CE. Human CE probably takes 1–2 years to become detectable by ultrasound (1–2 cm cyst size resolution), but much longer in many cases with large cysts. Longevity of *E. granulosus* in dogs is probably 1–2 years (M.A. Gemmell, personal communication) and infection intensity (arecoline counts) was overdispersed, range 1–2384 (mean=146) in the Uruguay study. Dogs may live for 5–10 years. Therefore, the risk of human exposure at any one time is variable but probably relatively low. Human infection would appear, therefore, to require long periods (years) of constant contact with high intensity infected dog(s). This was further suggested by the fact that 20% of CE cases resided in just five household clusters in the town where at least one heavily infected dog (>100 worms) was present (Paolillo *et al.*, 1996).

5.4.2. E.granulosus *in Kathmandu City*

Human cystic echinococcosis is known to occur in Nepal though data are sparse. At least 47 cases were recorded from 1985 to 1990 in hospital records in Kathmandu (D.D. Joshi, unpublished). It was, however, not clear whether transmission occurred within the city itself or merely reflected imported rural cases.

Livestock, principally buffaloes (*Bos bubalis*), but also goats, sheep and pigs, are openly slaughtered daily in the city on the banks of the Bagmali River without regulation or inspection. Only pigs are allowed to forage permanently, the other livestock species are imported from outside the city and usually held for 1–2 days before slaughter. A recent survey showed that CE infection rates were 5% in buffaloes ($n = 3065$), 8% in sheep ($n = 150$), 3% in goats ($n = 1783$), and 7% in pigs ($n = 143$) (D.D. Joshi, unpublished observations). Buffaloes are almost exclusively imported for slaughter from northern India where *E. granulosus* is known to be endemic (Singh and Dar, 1988), and sheep from Tibet where the parasite is also endemic (Schantz *et al.*, 1995). Human CE is in fact endemic throughout much of India and a dog–buffalo cycle appears to predominate (Gill and Venkateswara, 1967; Irshadullah *et al.*, 1989). Protoscoleces from isolates of Indian buffalo hydatid cysts, however, showed a similar DNA profile to sheep strain material (Bowles and McManus, 1993), despite morphological characteristics of the adult tapeworm similar to the cattle strain (Gill and

Rao, 1967; Thompson and Lymbery, 1988). Adult *E. granulosus* recovered at autopsy from Kathmandu dogs resembled closely the morphological description of the Swiss cattle strain (Eckert and Thompson, 1988) in length of terminal segment and numbers of testes (P.S. Craig and D. Baronet, unpublished observations). A total of 15% (*n* = 20) of Kathmandu street dogs were found infected at autopsy in a study which also included coproantigen screening of dogs from the clandestine livestock slaughtering areas or from city veterinary clinics (Baronet *et al.*, 1994). In the city ward in which livestock slaughter occurred 5.7% of dogs (*n* = 88) were coproantigen positive compared to 1.8% (*n* = 171) in the low risk area. Dogs from the high risk area were also more likely to be fed offal and wander and defecate in the streets. No human mass screening data for CE are available at present.

It would appear that urban transmission of human CE probably occurs in Kathmandu, most likely associated with street dogs living near the slaughter areas. Construction of dog-proof slaughterhouses could, in theory, break the transmission to dogs in Kathmandu. However, the presence of abattoirs or slaughterhouses, which occurs in many Indian cities, will not necessarily prevent dogs gaining access to infected livestock offal (Irshadullah *et al.*, 1989).

6. ALVEOLAR ECHINOCOCCOSIS IN ENDEMIC COMMUNITIES

Of the three important cestode zoonoses discussed in this review, *Echinococcus multilocularis* is the only species for which the bulk of transmission of the parasite is restricted to cycles involving sylvatic hosts. For this reason exposure of humans in general, is less likely than for the predominant domestic animal cycles of *E. granulosus* and *T. solium*. It is usually assumed that human alveolar echinococcosis can be contracted from exposure to infected fox (or other wild canid) faeces or the contaminated environment though it is difficult to obtain direct evidence of this. Indeed, the few studies of clinical data or from serological screening of fox trappers in the known endemic areas of St Lawrence Island (Alaska), north central USA, Japan and southern Germany indicate such activity was not apparently a risk factor for exposure or clinical disease (Stehr-Green *et al.*, 1988; Doi *et al.*, 1990; Saileela *et al.*, 1992; E. Zeyhle, personal communication). Of course the occurrence of individuals negative for serum antibodies against *E. multilocularis* metacestode antigens (e.g. Em2-ELISA) does not necessarily imply absence of exposure as this may occur without development of infection and subsequent metacestode antibodies. Detection of stage-specific oncosphere antibodies would be useful in investigating exposure versus disease as suggested for *E. granulosus* (Craig *et al.*, 1986b).

In the North American continent human alveolar echinococcosis is currently only a public health problem in Yupik Eskimos in western Alaska, despite the enzootic transmission of the parasite over huge areas of Canada and central north America (Rausch *et al.*, 1990a; Schantz *et al.*, 1995). Low rural human population densities and diversity of fox prey (rodent) species, and/or possible strain differences, may partly explain this but the potential role of the domestic dog as a known definitive host of *E. multilocularis* is probably a neglected but critical factor in human transmission. Perhaps not surprisingly, however, for ethical reasons and lack of accurate pre-mortem diagnosis, very few studies have been directed to quantifying the prevalence of *E. multilocularis* in domestic dogs in endemic areas. There is also a lack of data on comparative susceptibility to infection for dogs, coyotes and vulpine fox species. Such studies are difficult (and biohazardous), however the recent development of coproantigen detection of *Echinococcus* infection in definitive hosts should enable at least the start of epidemiological studies directed toward the role of domestic dogs in transmission (Allan *et al.*, 1992; Deplazes *et al.*, 1992).

The other important area which has been somewhat neglected in studies of *E. multilocularis* transmission is the quantitative ecology of the organism within its common sylvatic cycle(s). Mathematical models of transmission dynamics have been described for *E. granulosus* which may aid control (see Section 2.3) and have been proposed for *E. multilocularis* (Roberts and Aubert, 1994; Roberts, 1994). These models assumed, however, that red fox and vole population densities were constant in order not to complicate the analysis. This is not the case in nature and definitive and intermediate host populations vary significantly in both time and space (Giraudoux, 1991). Other parameters which are important in transmission but for which data are lacking, are the mean life-span of the adult tapeworm in the fox gut, the possibility for super-infection of the fox and the duration of immunity in the rodent intermediate host and whether it acts as a density-dependent constraint (Roberts, 1994; Gemmell and Roberts, 1995). Land-use variables in endemic rural areas also have a significant effect on rodent species densities and thus potential impact on transmission of *E. multilocularis* (Delattre *et al.*, 1989; Giraudoux, 1991).

6.1. Alveolar Echinococcosis on St Lawrence Island, Alaska

St Lawrence Island is the largest island (5000 km^2) in the Bering Sea, it possesses a treeless tundra biotope and has an approximate population of 1000 Yupik Eskimos confined to two coastal villages (Savoonga and

Gambell). Human alveolar echinococcosis is endemic on the island. Studies on the natural history of *E. multilocularis* on St Lawrence began in 1950 with the discovery of the parasite in northern voles (*Microtus oeconomus*). This finding and the confirmation that this was also the same cause of human alveolar echinococcosis (AE) in Germany, resulted in the recognition and designation of formal species status for the organism (Rausch and Schiller, 1951; Vogel, 1955). Human cases of AE on St Lawrence Island have been clinically diagnosed since the 1950s at an annual incidence rate of around 65 per 100000, one of the highest in the world (Rausch *et al.*, 1990a).

On St Lawrence Island the parasite is maintained by an active sylvatic predator–prey cycle involving the only indigenous carnivorous mammal the arctic fox (*Alopex lagopus*) as definitive host and the northern vole (*M. oeconomus*), as intermediate host (Fay and Stephenson, 1986; Fay and Rausch, 1992). Although the diet of arctic foxes is variable and includes birds and marine mammal carrion, they feed mainly on voles throughout the summer resulting in up to 100% infection rates in foxes by autumn. Infection and viability rates in voles is also very high (average around 50%) especially in spring and summer in part as a consequence of coincidence of onset of (previously delayed) protoscolex development of the parasite and the rapid sexual maturation of the microtine host. As a further adaption to arctic transmission, eggs of *E. multilocularis* are tolerant of low temperatures (to $-50°C$) and therefore voles may become infected as soon as winter snow cover is lost (Schiller, 1955), this is also the time when arctic foxes are again able to predate actively on voles with up to 80% of young foxes infected (Rausch *et al.*, 1990b).

Direct human contact with arctic foxes through trapping and skinning or associated with collecting potentially contaminated wild berries or vegetation, was not however a significant risk factor for alveolar echinococcosis. Rather, a case control study on St Lawrence Island and the west coast of Alaska indicated that AE patients were more likely than controls to have owned dogs (for their entire lives), tethered dogs near the house and lived in houses built directly on the tundra (i.e. boards rather than gravel foundation) (Stehr-Green *et al.*, 1988). The prevalence of *E. multilocularis* in populations of voles directly associated with villages, i.e. in or under buildings and boardwalks, was high up to 35% compared to 20–83% away from villages (Rausch *et al.*, 1990a). Domestic dogs kept chained or allowed to wander within the village were known to catch voles and thus become infected — up to 12% in the only autopsy study (Rausch and Schiller, 1956). A 30-day interval dog-dosing programme using praziquantel in Savoonga village over a 10-year period resulted in a reduction in vole prevalence rates from 29% to 1–5% with concomitant reduction in risk of human infection (Rausch *et al.*, 1990a).

Mass serological screening for human AE was carried out in endemic areas of Alaska using the highly specific Em2-ELISA (Gottstein *et al.*, 1985). Of 1103 persons screened 35 were seropositive for Em2 antibodies and seven were confirmed by CT scan to have active hepatic AE lesions. The remaining 28 seroreactors had negative image findings and may represent an immunologically or genetically resistant group in which only limited post-oncospheral development occurs (Gottstein and Felleisen, 1995). Abortive AE lesions manifested as small hepatic calcifications in Em2 seropositive persons were also observed in this Eskimo population (Rausch *et al.*, 1987).

The Alaskan study was important because it showed, for the first time, that in an endemic community transmission of AE was most likely to be associated with domestic dogs rather than the sylvatic fox host. Furthermore, human exposure and early infection is probably much higher than clinical disease rates. Similar observations are now being made in endemic areas of China and France (Section 6.2).

6.2. Alveolar Echinococcosis in Northwest China

E. multilocularis is endemic in northern China and it has a more or less contiguous distribution involving parts of Kazakstan, Mongolia and Russia. From the human AE hospital case data it appears that there are two major Chinese foci of public health importance. The most serious occurs in an area of central China involving adjacent regions of south Gansu Province, eastern Qinghai Province, northern Sichuan Province and Ningxia Hui Autonomous Region. The other focus is situated in north Xinjiang Uygur Autonomous Region especially along the line of the Altai and Tian mountain ranges (Craig *et al.*, 1991; Schantz *et al.*, 1995). Since the first clinical case of human AE was reported in Xinjiang (Yao, 1965) approximately 500 cases have been documented from these five Provinces or autonomous regions.

Transmission appears to occur among both rural agriculturalists and transhumant pastoralists. Most of the cases from Qinghai and Sichuan provinces are ethnic Tibetan (mostly pastoralist), the majority from Gansu are Han Chinese, those from Ningxia are Hui (i.e. Chinese Moslems), and in Xinjiang AE occurs predominantly in Han, Kazakh and Mongol ethnic groups, the latter two also mainly pastoralists and the other group primarily agriculturalists. AE cases appear to have a slight male bias in Qinghai and Xinjiang but in contrast a significantly higher ratio of female to male cases (>2.5:1) has been observed in Ningxia and Gansu (Lin and Hong, 1991; Craig, *et al.*, 1992). This sex difference in clinical case rates may reflect different transmission patterns associated with the male-dominated transhumance movement of livestock (with accompanying dogs) to the high altitude summer pastures for Tibetan (in Qinghai) and Kazakh/ Mongol (in Xinjiang) pastoralists. In contrast, in Gansu and Ningxia poor peasant communities are characterized by permanent villages surrounded by varying degrees of crop production and land use. Domestic dogs are usually common in both the pastoral and settled agricultural type communities and in the latter spend most time in the home area. As in the St Lawrence Island communities (Section 6.1), dogs are likely to play a significant role in transmission of E. multilocularis to humans in these endemic areas of China (Craig et al., 1992). The average age of AE patients (n = 150) was 41 years in Gansu and Ningxia (Lin and Hong, 1991; Craig et al., 1992); which was similar to the 43 years (n = 124) in Hokkaido, Japan (Sasaki et al., 1994). This compared to 53 years in Alaska (Wilson and Rausch, 1980) and 55 years in Switzerland (Gloor, 1988). The reason(s) for this difference between mean age of AE patients in oriental and western foci is not clear, but may in part relate to more active mass-screening programmes in Japan (e.g. Sato et al., 1993) and China (Craig et al., 1992).

An ongoing epidemiological study of human AE in south Gansu (Zhang County) began in 1990 (Craig et al., 1992). Initially only serological screening was undertaken using crude protoscolex (EmP) or purified Em2 antigens in ELISA, when an overall seroprevalence rate of 8.8% (53/606) was recorded, with village rates from 0 to 20%. Subsequent follow-up with portable ultrasound scanning for hepatic AE resulted in confirmation of disease in seroreactors and a combined (EmP plus Em2 positivity) serologic sensitivity of 93%. Interestingly, the crude protoscolex antigen preparation in ELISA identified four additional asymptomatic AE cases that were seronegative for Em2 antibodies (Craig et al., 1992). A similar situation of seronegative AE cases with purified Em2 antigen but seropositive for crude EmC (cyst homogenate) was also recently described in a mass seroepidemiological study for AE in the Doubs region of France (Bresson-Hadni et al., 1994). Mass ultrasound-based screening is currently being used in the south Gansu study with confirmation testing by serology (i.e. EmP, EmC and Em2-ELISAs). The overall human AE prevalence rate for Zhang County is around 5% (n = >1900) with village AE rates ranging from 0 to 16% (P.S Craig, D. Shi, B. Bartholomot, G. Barnish, D. Vuitton, unpublished observations). As also observed in Alaska and France (Rausch et al., 1987; Bresson-Hadni et al., 1994) a small proportion of persons screened exhibited small calcified hepatic lesions and were also seroreactive for E. multilocularis antibodies suggesting abortive infection (Giraudoux et al., 1995).

The data from this community indicate the major public health importance of alveolar echinococcosis in this region of China. Other studies, primarily based on retrospective hospital case data, suggest that equally serious foci (or with even greater transmission) exist in the southern counties of the neighbouring Ningxia Hui Autonomous Region (Li, 1986). The epidemiology of human AE in either region is, however, not yet defined. Such high human AE infection rates point to a role for the domestic dog as a definitive host able to act as a source of infection at the village and household level as described on St Lawrence Island, Alaska (Rausch *et al.*, 1990a). Dog autopsy studies have shown 10–15% infection rates in south Gansu and north Sichuan (Qiu, 1989; Craig *et al.*, 1992). *E. multilocularis* prevalence rates for a small sample ($n = 20$) of red foxes (*V. vulpes*) in Ningxia were 15% (Li, 1985) and 30% in Xinjiang (Wang and Ding, 1989) but no data are available for Gansu. Other potential definitive hosts in western China are the sand fox (*V. corsac*), the Tibetan fox (*V. ferrilata*) (Tang, 1988; Z. Guo, personal communication) and the wolf (*Canis lupus*) (Wang and Ding, 1989).

Similarly, information on intermediate hosts of *E. multilocularis* in China is sparse, with only those species that could be caught easily by locals tending to be examined (summarized in Table 4). In the highly endemic Gansu and Ningxia foci potentially important data on microtine rodents have not yet been collected. The ground squirrel (*Citellus* spp.) has, however, been found with a low infection rate (0.6%) in Ningxia. Recent observations in the endemic focus in Zhang County (Gansu) surprisingly showed that *Microtus oeconomus* (*ratticeps* complex) was one of the most abundant rodent species living in scrubland adjacent to ploughed fields (P. Giraudoux and G. Bao, unpublished observations). This was unexpected

Table 4 Intermediate hosts of *E. multilocularis* in China (modified from Lin and Hong, 1991).

Host species	Prevalence (%)		Location	Reference
Microtus brandti	2.3	(64/2635)	Inner Mongolia	Tang (1988)
Ochotona sp.	4.3	(9/214)	Sichuan	Qiu (1989); Guo (1987)
Citellus sp.	0.6	(12/2156)	Ningxia	Li (1985); Li (1986)
	0	(0/7068)	Gansu	D. Liu personal communication
Myospalax fontanieri	0.3	(1/321)	Ningxia	Lin and Hong (1991)
	0	(0/870)	Gansu	D. Liu personal communication
Meriones unguiculatus	16.7 (1/6)		Inner Mongolia	Tang (1988)
Citellus erythrogenys	0.1 (2/2211)		Xinjiang	Lin *et al.* (1993)
Mus musculus	0.01 (1/6890)		Xinjiang	Lin *et al.* (1993)

because the accepted southern limit to the main range of *M. oeconomus* was 1000 km north of the study area (Allen, 1938; Ognev, 1964). This microtine is the main intermediate host of *E. multilocularis* on St Lawrence Island, Alaska (Rausch *et al.*, 1990a). Although no infected rodents have yet been trapped in this area, there was a highly significant correlation between abundance of *M. oeconomus* (based on transect analysis) and a "land-use" index (Giraudoux *et al.*, 1995). This indicated that villages close to areas with a high ratio of scrub or rough pasture had significantly higher human AE prevalence rates, whereas those surrounded by larger areas of ploughed field had much lower human AE prevalence rates. A landscape analysis in the endemic Doub region of France showed a similar trend with *M.arvalis*, in that rodent density was positively correlated with a permanent grassland landscape (Delattre *et al.*, 1989). We hypothesize that currently in the Zhang County region of China, human AE infection rates (mean age 40 years) reflect transmission 10–15 years in the past, when landscape under wood or scrub cover was greater and red fox and domestic dog population densities were higher. Furthermore, known cyclical changes (3–10 years) in microtine rodent densities which coincide with increased canid populations would enable local increases in transmission within both the sylvatic and the semi-domestic (dog–rodent) cycles (Craig *et al.*, 1992; Giraudoux *et al.*, 1995).

6.3. Alveolar Echinococcosis in Japan — a Newly Endemic Country

Human alveolar echinococcosis is highly endemic in Hokkaido the north island of Japan where more than 250 cases have been diagnosed (Schantz *et al.*, 1995). Before 1966, however, the disease was not recorded. *E. multilocularis* appears to have been introduced to Hokkaido from infected red foxes (*V. vulpes*) from the Kuriles islands, northeast of Japan, between 1924 and 1926. The cestode was probably always highly endemic in eastern Russia including the Kamchatka peninsula, Sakhalin Island and the Kuriles where it is transmitted between the red fox (or arctic fox in the north) and a number of arvicolid intermediate hosts including *Microtus oeconomus, Clethrionymus rufocanus* and *C. rutilus* (Rausch, 1995). In the 1920s a small number of red foxes were translocated from the Kuriles to Reuben Island northwest of Hokkaido in order to help control the vole population (probably also *C. rufocanus*). Approximately 10 years later the first human AE cases were diagnosed on Reuben Island, followed by Hokkaido itself about 25 years later (Suzuki *et al.*, 1993). Public health records showed that 129 cases of AE occurred on Reuben Island between 1937 and 1967 and that approximately 1% of the population was infected (Furuya *et al.*, 1990; Schantz *et al.*, 1995). Also, 19% of red foxes (*n* = 2026) and 1.6% of dogs (n = 3224) were infected with *E. multilocularis* (Yamashita, 1973). A 5-year fox and dog control programme, in which large numbers of animals were killed, appeared to eliminate the parasite on Reuben Island by 1955 (Yamashita, 1973).

The first AE patient in Hokkaido University Hospital (Sapporo) was initially misdiagnosed as heptocellular carcinoma but confirmed as AE after histological examination (Sasaki *et al.*, 1994). A mass serological screening programme was implemented based on ELISA with crude *E. multilocularis* metacestode antigen. Between 1987 and 1988, 154636 people were screened with a seroprevalence rate of 0.29%, of which 116 were seropositive by Western blot for antigens of 29–205 kDa especially 55–65 kDa (complete profile) indicative of long-term antigen exposure, and an incomplete pattern with few bands of 30–35 kDa or >90 kDa suggestive of more recent infection (Furuya *et al.*, 1990). However, only 50% of all immunoblot seropositive cases were confirmed to have AE by hepatic ultrasound, the majority showing the "complete" Western blot pattern. More recently, a Western blot test using an *E. multilocularis* protoscolex antigen preparation (rather than a whole cyst extract) has indicated that two low-molecular-weight antigens designated Em18 and Em16 were highly specific for confirmed AE (Ito *et al.*, 1993). Studies with chinese AE patients, however, suggest the Em16 antigen is also recognized by a small proportion of cystic echinococcosis patients (Wen *et al.*, 1995). Mass serological screening has identified most of the AE cases in Hokkaido with the resulting benefit of early resection and reduced morbidity (Sato *et al.*, 1993). This may also explain the relatively young average age (43 years) of patients compared to some other endemic areas of the world.

The main intermediate hosts on Hokkaido are red–grey voles (*C. rufocanus*) and the northern red-backed vole (*C. rutilus*), with prevalence rates, determined at autopsy, of 4–22%, with the highest prevalence occurring in voles caught close to fox dens (Takahashi *et al.*, 1989). The prevalence of *E. multilocularis* in red foxes was 19% (n = 1724) around the northeastern districts of Nemuro and Kashiro during the first years of the introduction of the parasites to Hokkaido (1966–73), and subsequently has spread throughout the island with reports of 30% of red foxes infected (Kamiya and Ohbayashi, 1975; Kamiya, 1988). Sero-epidemiological data indicate that farmers were more likely to be seropositive than other groups, including fox hunters (Doi *et al.*, 1990). This may be due to contamination by fox faeces around farms but is also likely to occur from contact with infected domestic dogs.

There is growing concern that *E. multilocularis* is now spreading south to Honshu Island where apparent autochthonous cases have been reported, but no animal studies have as yet been described (Schantz *et al.*, 1995).

7. *T. SOLIUM* CYSTICERCOSIS AND TAENIASIS IN COMMUNITIES

The epidemiological picture in endemic countries probably varies considerably both between countries and within them. *Taenia solium* transmission is generally associated with areas where pig husbandry techniques and sanitary facilities allow pigs access to human faeces. Therefore, although any one country or region may be considered endemic, it is clear that active transmission of the parasite from pigs to man may be quite focal and related specifically to areas within endemic countries where these practices remain (Pawlowski, 1990).

As a corollary of this, the pattern of infection in many of the larger towns and cities in endemic countries (where pigs are generally not present in any numbers and sanitation is generally better than that available in more rural areas) may be more like that which occurs in developed countries. That is, the majority of cases of cysticercosis being imported from surrounding rural areas, or due to human to human transmission from tapeworm carriers. Exceptions to this probably occur in peri-urban areas where sanitation is poor and pig husbandry continues to some extent. Conditions like this exist in a number of endemic countries including Mexico, Guatemala and India (J.C. Allan and K. Pathak, unpublished observations). As the proportion of the population living in large urban areas in endemic developing countries is often high and rising, this has important consequences for the epidemiology of *T. solium*. Studies in Peru have indicated that the seroprevalence of cysticercus antibodies was higher in individuals living outside the capital, Lima (Garcia *et al.*, 1991, 1995), and in Guatemala the prevalence of intestinal taeniasis was reported to be lowest in the capital (Acha and Aguillar, 1964; Gonzalez, 1989).

Some other endemic areas show considerable variations in other factors that may influence the transmission of the parasite. For instance in Indonesia cultural and religious differences within the archipelago are probably reflected in the levels of parasite transmission. Although most of the country is Muslim with the concomitant constraints on pork consumption and pig husbandry, other areas do not have such constraints and indeed pork makes up an important part of the diet. Pig to human transmission of *T. solium* is known to take place in Bali and Irian Jaya but cysticercotic cases detected in other, mainly Muslim, islands, such as Surabaya, appear to originate elsewhere (Giri, 1978). In Peru, on the other hand, studies have indicated that the seroprevalence of the infection varies with agricultural practices, being higher in areas where pigs are common and lower in high altitude areas where they are absent (Moro *et al.*, 1994). Within endemic countries, therefore, there is a need to understand the distribution of the parasite and what factors influence that distribution. There has generally been a lack of surveys at the national level of endemic countries to identify areas of active *T. solium* transmission. In those areas where surveys have been carried out there have been limitations imposed upon them by the techniques employed. Nevertheless some studies may, accepting their limitations, provide useful data.

7.1. *T. solium* in Rural Latin America

In Mexico, where cysticercosis was made an officially reportable disease in 1979 (Schantz and Sarti, 1989) the system of reporting would appear to indicate that most cases originate from large urban areas (Schenone *et al.*, 1982). This is likewise the case in most of the endemic countries of Latin America. Since it is generally accepted that most cases are actually present in rural areas this pattern of reporting probably more closely reflects the distribution of facilities where accurate diagnosis of cysticercosis can be made and the willingness of institutions, such as hospitals, to participate in the system (Schantz and Sarti, 1989). Trace-back of cases and case histories indicated that in fact many cases of human cysticercosis diagnosed in Lima (Peru) actually corresponded to individuals either resident or born outside the city (Garcia *et al.*, 1995). There has recently been increased interest in the epidemiological study of *T. solium* in rural communities of endemic countries, which are thought to be where greatest transmission occurs, particularly in Latin America. These studies have coincided with the development of many of the tools described above for the diagnosis of the parasite and in some cases have actively led to the development of new and improved techniques whereas in others have been promoted by an increasing awareness of the parasite due to improvements in its clinical diagnosis. With the advent of these studies we have begun to build a detailed picture of the transmission of the parasite and it is hoped that these findings will be used in improvements in its control.

One problem that remains in the assessment of much of the data has been the range of immunodiagnostic techniques employed to identify potential cases of cysticercosis. This has meant that the results of these studies have been hard to compare (Schantz *et al.*, 1994). It has recently been demonstrated that the results of an ELISA based on *T. solium* vesicular fluid antigen showed no correlation with cases of neurocysticercosis detected in a community study (Schantz *et al.*, 1994). The same study did, however, indicate that a Western blot (EITB) assay using purified *T. solium* metacestode glycoprotein extract (Tsang *et al.*, 1989) did show correlation with cases of neurocysticercosis. For this reason the results discussed here with respect to serology will be limited to those where a test of high specificity has been used. At the present time immunodiagnostic tests for human cysticercosis appear to be limited to one of a number of Western blot assays (Gottstein *et al.*, 1987b; Tsang *et al.*, 1989; Michault *et al.*, 1990), but that of Tsang *et al.* (1989) has been most extensively assessed and currently used (Tsang and Wilson, 1995). Studies have indicated that the prevalence of intestinal *T. solium* taeniasis detected by microscopy is generally below 5%, although figures above this have occasionally been recorded (Kaminsky, 1991). Data from the use of coproantigen tests in rural communities, however, have indicated that these microscopy-based figures may underestimate the true prevalence by as much as two to three times (Allan *et al.*, 1996). In Latin America preliminary data indicate that intestinal taeniasis rates appear higher in Central America (Moore *et al.*, 1995) although this may reflect the limited number of studies that have been carried out and the fact that these studies have been restricted to only a few countries within the region.

Very little is known about the epidemiology of the intestinal tapeworm infection from studies carried out on high risk rural populations. This is probably for several reasons. First, the clinically more severe neurocystic infection has received most of the attention. Secondly, the low prevalence of intestinal taeniasis has generally meant that the size of the field epidemiological studies conducted to date has resulted in the detection of very few tapeworm carriers (in many studies less than 10). Thirdly, many studies fail to identify the species of *Taenia* responsible for the infection. This means that very few risk factors for intestinal taeniasis have been identified. There are for instance no good data on such important factors as the average life span of the parasites in the gut (data are largely anecdotal). This type of information is important for any quantitative transmission modelling and if the parasite is to be controlled through mass treatment, which has been proposed as a potential intervention strategy (Cruz *et al.*, 1989). Data from some studies have shown females to be at higher risk of infection than males (Richards *et al.*, 1985; Cruz *et al.*, 1989; Kaminsky, 1991) but this has not been a consistent finding. None the less this is an area which bears further investigation.

A number of studies in rural *T. solium* endemic communities have indicated that a significant risk factor for positive antibody serology against *T. solium* cyst antigens using the glycoprotein immunoblot assay (Tsang *et al.*, 1989) is the presence of a current or past case of intestinal *Taenia* in the household (Sarti *et al.*, 1992a; Garcia Noval *et al.*, 1996). This association has also been shown using the less specific ELISA assays (Diaz Camacho *et al.*, 1990, 1991). Intestinal *T. solium* carriers or individuals with a history of passing proglottides are also significantly more likely to be metacestode antibody positive than members of the general population (Sarti *et al.*, 1992a; Garcia Noval *et al.*, 1996). A laboratory-based study indicated that all intestinal *T. solium* carriers tested for metacestode antibodies were seropositive (Diaz *et al.*, 1992). However, recent field studies in Mexico and Guatemala have shown that this appears not to be the case with all carriers (Sarti *et al.*, 1994; Garcia Noval *et al.*, 1996). This is an important point; if the existing serological test for cysticercosis which is being applied for epidemiological and clinical work cross-reacts with the adult *T. solium* tapeworm this means that individuals who are cysticercus antibody seropositive must be examined to ensure that a seropositive result is not due to intestinal infection.

As the presence of an intestinal *T. solium* tapeworm carrier in the immediate environment appears to significantly increase the likelihood of human cystic infection the accurate identification and treatment of such cases has been identified as a priority in the control of the parasite (Gemmell *et al.*, 1983). Whether this is best done by passive or active case finding or mass treatment of high risk populations has not been assessed. The availability and low cost of highly efficacious drugs for treating the intestinal infection have meant that mass treatment has been proposed as a potential control strategy (Pawlowski, 1990; Schantz *et al.*, 1993). Initial studies involving mass treatment with praziquantel have shown that the approach may be a useful one (Cruz *et al.*, 1989; Diaz Camacho *et al.*, 1991). More work remains to be done, however, particularly as these studies were relatively short term (one year). One potential problem area is the possibility that low dose praziquantel treatment may induce neurological symptoms in individuals with previously asymptomatic neurocysticercosis (Flisser *et al.*, 1993).

Neuroepidemiological studies on *T. solium* neurocysticercosis in Mexico have begun to show that outside the hospital setting, where neurocysticercosis has been recognized for some time as a major cause of neurological disease (Medina *et al.*, 1990; Botero *et al.*, 1993), in rural communities an association of a history of seizures with metacestode antibody seropositivity against *T. solium* has now become apparent (Sarti *et al.*, 1992a, 1994). In a larger study in Guatemala this association was not significant (Garcia Noval *et al.*, 1996). The high proportion of probable neurocysticercotics, however, had calcified (dead) lesions which may explain why they were seronegative in the study of Garcia Noval *et al.* (1996). Other studies have also indicated that neurological symptoms, such as chronic headache, also appear to be significantly associated with neurocysticercosis in rural endemic communities (Sarti *et al.*, 1994; Cruz *et al.*, 1994b). Neurocysticercosis may therefore be a significant cause of morbidity due to neurological disease in *T. solium* endemic communities. Rates of migraine headache, histories of convulsions or epilepsy in the communities studied to date in Mexico, Ecuador and Guatemala are high compared to those encountered in developed countries (Sarti *et al.*, 1992a, 1994; Cruz *et al.*, 1994b; Garcia Noval *et al.*, 1996).

A further significant impact of *T.solium* at the level of the rural community is its effect on pig husbandry. The prevalence of infected pigs in rural areas appears to be much higher than that detected in abattoirs (Aluja, 1982). Studies from Mexico, Guatemala and Peru in rural areas have indicated pre-mortem prevalence rates of infected pigs by tongue palpation of up to 30% (Diaz Camacho *et al.*, 1991; Sarti *et al.*, 1992a,b; Diaz *et al.*, 1992; CWGP, 1993; Garcia Noval *et al.*, 1996).

In one of the largest community studies in Latin America two adjacent villages (total *n* = 2365) in southeast Guatemala were screened for *T. solium* cysticercus antibodies (glycoprotein immunoblot assay), history of convulsive attacks (neurological questionnaire) and *Taenia* carriers (coproantigen ELISA and microscopy). CT scan was used to follow-up individuals with convulsive history suggestive of neurocysticercosis (Garcia-Noval *et al.*, 1996). In addition, a knowledge, attitudes and practices questionnaire was used on female heads of households, and a sample of pigs from both villages (total *n* = 471) were screened by tongue palpation. One of the villages was significantly more "developed" than the other so that for example 40% of persons defecated in the open in the poor village compared to only 4% in the other, and only 9% of households in the poor village had proper water drainage compared to 79% in the better developed village. Although *T. solium* transmission occurred in both villages there was a significantly higher cysticercus antibody seroprevalence rate in individuals from the poor village (15% versus 10%) and a greater prevalence of taeniasis (2.8% versus 1%). All coproantigen and/or *Taenia* egg positive persons were confirmed by identification of a tapeworm after anthelminthic and purgative treatment. Tongue cysts were present in 15% of pigs in the poor village versus 4% in the richer village. Interestingly, the epileptic rate was nearly the same in both villages 2.8% and 2.9%. CT scan follow-up showed that 47% of individuals with a history of convulsions and 24% of non-convulsive controls had images suggestive of neurocysticercosis, but there was no relationship between an abnormal image case and seropositivity in the immunoblot assay (Garcia-Noval *et al.*, 1996). Other interesting findings included a greater risk of cysticercus antibody seropositivity and intestinal taeniasis in females in both villages. There was also a higher risk of positive immunoblot for *Taenia* carriers. The latter observation could be due to autoinfection or antibody stage cross-reactivity. The study also showed that seroprevalence was higher in family members from houses where a *Taenia* carrier was identified. This household clustering was also observed in community studies in Mexico (Sarti *et al.*, 1994) and Ecuador (Cruz *et al.*, 1994a).

7.2. *T. solium* and Neurocysticercosis in a Non-endemic Community

The recent reported increase in the numbers of cases of *T. solium* cysticercosis detected in some developed countries may be due to the increasing levels of international travel, either related to business or tourism to endemic regions, or altered patterns of population movement from these endemic areas.

One country where increasing numbers of neurocysticercosis case reports has led to a consequent interest in *T. solium* has been the United States of America. Recent epidemiological investigations have implicated both importation of cases and a degree of local human to human transmission of the parasite as the source of these cases. These studies have probably led to the best understanding of the epidemiology of *T. solium* in any developed country although this may also be a reflection of the relative importance of cysticercosis in the US in comparison with other similarly developed areas.

In the USA the majority of cases of human cysticercosis appear to be in immigrants from *T. solium* endemic areas, especially Latin America. This is particularly apparent in southwest and southern USA where large numbers of immigrants choose to live and work (Richards *et al.*, 1985; Earnest *et al.*, 1987; Sorvillo *et al.*, 1992; Shandera *et al.*, 1994). There have, however, been locally acquired infections, involving transmission from individuals with *T. solium* taeniasis resulting in cases of cysticercosis in individuals with no history of travel to endemic areas (Sorvillo *et al.*, 1992; Schantz *et al.*, 1992). Based on US abattoir data, active human to pig transmission in the US appears to be very infrequent for example only three of 83 million pigs were infected in 1990 and this probably reflects the relatively good levels of hygiene and pig husbandry practices (Schantz and McAuley, 1991). Consequently there is little if any local transmission of *T. solium* from pigs to man.

Recently, however, an interesting epidemiological study in New York identified significant numbers of human neurocysticercosis cases within a "closed" community (6000–8000 families) where there was little or no history of travel to endemic *T. solium* areas (Schantz *et al.*, 1992). Even more surprising was that all the patients were Orthodox Jews having a strict taboo against eating pork, and so these individuals were in an extremely low-risk group. Initially four patients (6–39 years), all from the same community developed symptoms of seizures, and were admitted for CT or MRI brain scans which indicated lesions. The glycoprotein immunoblot test for specific metacestode antibodies was positive in three patients and brain biopsy confirmed larval taeniid infection in two of the cases. Stool examinations for eggs or proglottides were negative in the four cases.

Seven relatives of two of the patients were also seropositive by immunoblot test and two were confirmed by MRI to have neurocysticercosis. An
epidemiological investigation directed by the Division of Parasitic Diseases, Centers for Disease Control, identified that the employment of
recently emigrated Latin American housekeepers was a common factor
for the four index cases one of which was a confirmed *Taenia* carrier. Also
around 80% of households employed such individuals (Schantz *et al.*,
1992). A follow-up seroprevalence survey (*n*=1789) in this community
found 1.3% of persons seropositive for cysticercosis antibodies by glycoprotein immunoblot test. All were asymptomatic with no intracerebral
lesions. Seropositivity was, however, highly statistically significantly associated with hiring a domestic worker from Central America (Moore *et al.*,
1995). *T. solium* is highly endemic in central America at least in Guatemala
and Honduras for which prevalence data are beginning to be collected
(Kaminksy, 1991; Allan *et al.*, 1996). Screening of emigrants from *T.
solium* endemic countries, especially Central America, who seek employment as housekeepers or food handlers has been recommended (Schantz *et
al.*, 1992) and a *Taenia* coproantigen-ELISA would be useful for this
purpose (Allan *et al.*, 1990, 1996).

8. CONCLUSIONS

The public health importance of taeniid cestode zoonoses especially cystic
echinococcosis, alveolar echinococcosis and *T. solium* cysticercosis was
highlighted in the 1970s by J.D. Smyth and colleagues (Matossian *et al.*,
1977; Smyth, 1979). Transmission of *E. granulosus, E. multilocularis* and
T. solium together covers almost every region on the globe, and community
studies in a few selected countries have been described above. New control
programmes have been implemented in some countries, notably for cystic
echinococcosis, but success is difficult to achieve and requires years of
effort, support and appropriate surveillance (Gemmell, 1978; Gemmell *et
al.*, 1987b). Concern is also growing about the major, but mostly unrecog-
nized, problem of neurocysticercosis and epilepsy in developing countries
particularly Latin America and also its public health consequences in the
USA and elsewhere (Shandera *et al.*, 1994; Tsang and Wilson, 1995). *E.
multilocularis* which is the cause of one of the most pathogenic of all
human parasitic infections, appears to be spreading in parts of western
Europe, USA and Japan (Suzuki *et al.*, 1993; Storandt and Kazacos, 1993;
Lucius and Bilger, 1995).

The remarkable development in the last 20 years or so of high resolution imaging technologies such as CT scan and ultrasound have enabled

precision detection of taeniid larval cystic infection in humans. Coupled with improvements in immunodiagnostic test sensitivity and antigen specificity, diagnosis of cestode zoonoses is now more accurate than ever before. Although further developments in immunodiagnosis are required, notably standardization with species specific recombinant taeniid antigens for serum antibody detection, laboratory tests have provided improved capability for screening populations in both epidemiological and community studies (Gottstein, 1992a; Craig, 1993; Flisser, 1994). Parallel, and recent, development of genus specific coproantigen tests for human taeniasis and canine echinococcosis have also provided major new tools for epidemiological and surveillance programmes (Allan *et al.*, 1992). Epidemiological studies, particularly on *E. granulosus*, have also been significantly advanced by the development of DNA typing approaches for identification of subspecific variants within parasite–host assemblages (Bowles and McManus, 1993).

Mathematical approaches to modelling the transmission dynamics of taeniid cestodes in domestic or sylvatic animal host cycles have begun to provide a rational approach to consider options for *Echinococcus* control intervention (Gemmell and Roberts, 1995), but in addition, the transmission ecology of *E. multilocularis* will require special consideration (Giraudoux, 1991). How, basic models can now be applied to the transmission of *T. solium* between humans and pigs is an important question. Furthermore, can mathematical approaches be used to describe zoonotic transmission where the human host is not an inherent part of the parasite's (i.e. *Echinococcus*) life cycle?

Combination of all the above approaches together with traditional methods (e.g. hospital records analysis), for application to community studies concerning cestode zoonoses, will undoubtedly lead to further detailed description and understanding of the patterns of transmission, risk factors and appropriate control strategies.

ACKNOWLEDGEMENTS

We are grateful for the major support of our research from the Wellcome Trust and the European Commission. Thanks also to Beverley Butler for skilful production of the manuscript.

REFERENCES

Acha, P. and Aguillar, F.J. (1964). Studies on cysticercosis in Central America and Panama. *American Journal of Tropical Medicine and Hygiene* **13**, 48–53.

Adams, J.C., Bode, R. and Dannatt, N. (1989). The public health, animal health and health of culling wild deer. *State Veterinary Journal* **43**, 53–65.

Ageyeva, N.G. (1989). Pecularities of epidemic and epizootological processes in foci of echinococosis in the European north of USSR. In: *Epidemiological Control of Echinococosis*. Proceedings of IV All-Union Scientific-Practical Conference, 17–20 October 1989, Chimkent, Moscow. pp. 3–11.

Allan J.C., Avila, G., Garcia Noval, J., Flisser, A. and Craig, P.S. (1990). Immunodiagnosis of taeniasis by coproantigen detection. *Parasitology* **101**, 473–477.

Allan, J.C., Craig, P.S., Garcia Noval, J., Mencos, F., Liu, D., Wang, Y., Wen, H., Zhou, P., Stringer, R., Rogan, M. and Zeyhle, E. (1992). Coproantigen detection for the immunodiagnosis of echinococcosis and taeniasis in dogs and humans. *Parasitology* **104**, 347–355.

Allan, J.C., Mencos, F., Garcia Noval, J., Sarti, E., Flisser, A., Wang, Y., Liu, D. and Craig, P.S. (1993). Dipstick dot ELISA for detection of *Taenia* coproantigens in humans. *Parasitology* **107**, 79–85.

Allan, J.C., Velasquez-Tohom, M., Torres-Alvarez, R., Yurrita, P. and Garcia-Noval, J. (1996). Field trial of the coproantigen based diagnosis of *Taenia solium* taeniasis by enzyme linked immunosorbent assay. *American Journal of Tropical Medicine and Hygiene* (In Press).

Allen, G.M. (1938). *Mammals of China and Mongolia X1*. New York: Museum of Natural History.

Aluja, A.S. (1982). Frequency of porcine cysticercosis in Mexico. In: *Cysticercosis: Present State of Knowledge and Perspectives* (A. Flisser, K. Wilms, J.P. Laclette, C. Larralde, C. Ridaura and F. Beltran, eds), pp. 53–62. New York: Academic Press.

Ammann, R. and Eckert, J. (1995). Clinical diagnosis and treatment of echinococcosis in humans. In: *Echinococcus and Hydatid Disease* (R.C.A. Thompson and A.J. Lymbery, eds), pp. 411–463. Wallingford: CAB International.

Andersen, F.L. (1993). General introduction to cystic echinococcosis and description of Cooperative research efforts in the Xinjiang Uyger Autonomous region, PRC. In: *Compendium on Cystic Echinococcosis with Special Reference to the Xinjiang Uyger Autonomous Region, The Peoples Republic of China* (F.L. Andersen, J.J. Chai and F.J. Liu, eds). Utah: Brigham Young University.

Andersen, F.L., Tolley, H.D., Schantz, P.M., Chai, P., Liu, F. and Ding, Z. (1991). Cystic echinococcosis in Xinjiang Uyger Autonomous Region, PRC. II. Comparison of three levels of a local preventive and control program. *Tropical Medicine and Parasitology* **42**, 1–10.

Anderson, R.M. and May, R.M. (1992). *Infectious Disease of Humans, Dynamics and Control*. Oxford: Oxford University Press.

Babba, H., Messedi, A., Masmoudi, S., Zribi, M., Grillot, R., Ambrose-Thomas, P., Beyrouti, I. and Sahnoun, Y. (1994). Diagnosis of human hydatidosis: comparison between imagery and six serologic techniques. *American Journal of Tropical Medicine and Hygiene* **50**, 64–68.

Baronet, D., Waltner-Toews, D., Craig, P.S. and Joshi, D.D. (1994). *Echinococcus granulosus* infections in the dogs of Kathmandu, Nepal. *Annals of Tropical Medicine and Parasitology* **88**, 485–492.

Bchir, A., Larouze, B., Soltani, M.A., Hamdi, A., Bouhaouala, H., Duci, S., Bouden, L., Ganouni, A., Achour, A., Gandebout, C., Rousset, J.J. and Jemmali, M. (1991). Echotomographic and serological population based study of hydatidosis in central Tunisia. *Acta Tropica* **49**, 149–153.

Beard, T.C. (1987). Human hydatid disease in Tasmania. In: *Epidemiology in Tasmania* (H. King, ed.), pp. 77–88. Canberra: Brologa Press.

Boa, M.E., Bogh, H.O., Kassuku, A.A. and Nansen, P. (1996). The prevalence of *Taenia solium* metacestodes in northern Tanzania. *Journal of Helminthology* **69**, 113–117.

Bonifacino, R., Malgor, R., Barbeito, R., Balleste, R., Rodriguez, M.J., Botto, C. and Klug, F. (1991). Seroprevalence of *Echinococcus granulosus* infection in a Uruguayan rural human population. *Transactions of the Royal Society of Tropical Medicine and Hygiene*, **85**, 769–772.

Botero, D., Tanowitz, H.B. and Weiss, L.M. (1993). Taeniasis and cysticercosis. *Infectious Diseases Clinics of North America* **7**, 683–697.

Bourke, G.J. and Petana, W.B. (1994). Human *Taenia* cysticercosis: a bizarre mode of transmission. *Transactions of the Royal Society of Tropical Medicine and Hygiene*, **88**, 680.

Bowles, J. and McManus, D.P. (1993). Molecular variation in *Echinococcus. Acta Tropica* **53**, 291–305.

Bowles, J. and McManus, D.P. (1994). Genetic characterisation of the Asian *Taenia*, a newly described taeniid cestode of humans. *American Journal of Tropical Medicine and Hygiene* **50**, 33–44.

Bowles, J., Blair, D. and McManus, D.P. (1992a). Genetic variants within the genus *Echinococcus* identified by mitochondrial DNA sequencing. *Molecular and Biochemical Parasitology* **54**, 165–174.

Bowles, J., van Knapen, F. and McManus, D.P. (1992b). Cattle strain of *Echinococcus granulosus* and human infection. *Lancet* **339**, 1358.

Bowles, J., Blair, D. and McManus, D.P. (1994). Molecular genetic characterization of the cervid strain ("northern form") of *Echinococcus granulosus. Parasitology* **109**, 215–221.

Bowles, J., Blair, D. and McManus, D.P. (1995). A molecular phylogeny of the genus *Echinococcus. Parasitology* **110**, 317–328.

Brandt, J.R.A., Geerts, S., DeDeken, R., Kumar, V., Ceulemans, F., Brijs, L. and Falla, N. (1992). A monoclonal antibody based ELISA for the detection of circulating excretory–secretory antigens in *Taenia saginata* cysticercosis. *International Journal for Parasitology* **22**, 471–477.

Bresson-Hadni, S., Laplante, J.J., Lenys, D., Rohmer, P., Gottstein, B., Jacquier, P., Mercet, P., Meyer, J.P., Miguet, J.P. and Vuitton, D.A. (1994). Seroepidemiologic screening of *Echinococcus multilocularis* infection in a European area endemic for alveolar echinococcosis. *American Journal of Tropical Medicine and Hygiene* **51**, 837–846.

Bretagne, S., Guillou, J.P., Morand, M. and Houin, R. (1993). Detection of *Echinococcus multilocularis* DNA in fox faeces using DNA amplification. *Parasitology* **106**, 193–199.

Bursey, C.C., McKenzie, J.A. and Burt, M.D.B. (1980). Polyacrilamide gel electrophoresis in differentiation of *Taenia* (Cestoda) by total protein. *International Journal for Parasitology* **10**, 167–174.

Cabrera, P.A., Haran, G., Benavidez, V., Valledor, S., Perera, G., Lloyd, S., Gemmell, M.A., Baraibar, M., Morana, A., Maissonave, J. and Carballo, M. (1995). Transmission dynamics of *Echinococcus granulosus, Taenia hydatigena* and *Taenia ovis* in sheep in Uruguay. *International Journal for Parasitology* **25**, 807–813.

Caremani, M., Maestrini, R., Occhini, U., Sassoli, S., Accorsi, A., Giogio, A. and Filice, C. (1993). Echographic epidemiology of cystic hydatid disease in Italy. *European Journal of Epidemiology* **9**, 401–404.

Chai, J.J. (1993a). Brief overview of the Xinjiang Uyger Autonomous Region, The Peoples Republic of China. In: *Compendium on Cystic Echinococcosis with Special Reference to the Xinjiang Uyger Autonomous Region, The Peoples Republic of China* (F.L. Andersen, ed.), pp. 16–21. Provo: Brigham Young University.

Chai, J.J. (1993b). Sero-epidemiological Surveys for Cystic Echinococcosis in the Xinjiang Uyger Autonomous region, PRC. In: *Compendium on Cystic Echinococcosis with Special Reference to the Xinjiang Uyger Autonomous Region, The Peoples Republic of China* (F.L. Andersen, ed.), pp. 153–162. Provo: Brigham Young University.

Chapman, A., Vallejo, V., Mossie, K.G., Ortiz, D., Agabian, N. and Flisser, A. (1995). Isolation and characterization of species-specific DNA probes from *Taenia solium* and *Taenia saginata* and their use in an egg detection assay. *Journal of Clinical Microbiology* **33**, 1283–1288.

Chi, P., Zhang, W., Zhang, Z., Hasyet, M., Liu, F., Ding, Z., Anderson, F.L., Tolley, H.D. and Schantz, P.M. (1990). Cystic echinococcosis in the Xinjiang Uygur Autonomous Region, Peoples Republic of China. Demographic and epidemiologic data. *Tropical Medicine and Parasitology* **41**, 157–162.

Chisholm, I.L., MacVicar, M.J. and Williams, H. (1983). Hydatid disease in the Western Isles. *Journal of Hygiene (Cambridge)* **90**, 19–25.

Cho, S.Y., Kim, S.I., Kang, S.Y., Choi, D.Y., Suk, J.S., Choi, K.S., Ha, Y.S., Chung, C.S. and Myong, H. (1986). Evaluation of enzyme linked immunosorbent assay in serological diagnosis of human neurocysticercosis using paired samples of serum and cerebrospinal fluid. *Korean Journal of Parasitology* **24**, 25–41.

Choji, K., Fujita, N., Chen, M., Spiers, A.S.D., Morita, Y., Schinohara, M., Norima, T. and Irie, G. (1992). Alveolar hydatid disease of the liver: computed tomography and transabdominal ultrasound with histopathological correlation. *Clinical Radiology* **46**, 97–103.

Clarkson, M.J. and Walters, T.M.H. (1991). The growth and development of *Echinococcus granulosus* of sheep origin in dogs and foxes in Britain. *Annals of Tropical Medicine and Parasitology* **85**, 53–61.

Coltorti, E.A., Fernandez, E., Guarnera, E., Lago, J. and Iriarte, I. (1988a). Field evaluation of an enzyme immunoassay for detection of asymptomatic patients in a hydatid control programme. *American Journal of Tropical Medicine and Hygiene* **38**, 603–607.

Coltorti, E.A., Guarnera, E., Larrieu, E., Santillan, G. and Aquino, A. (1988b). Seroepidemiology of human hydatidosis. Use of dried blood samples on filter paper. *Transactions of the Royal Society of Tropical Medicine and Hygiene* **82**, 607–610.

Condon, J., Rausch, R.L. and Wilson, J.F. (1988). Application of the avidin–biotin immunohistochemical method for the diagnosis of alveolar hydatid disease from tissue sections. *Transactions of the Royal Society of Tropical Medicine and Hygiene* **82**, 731–735.

Cook, B.R. (1989). The epidemiology of *Echinococcus granulosus* in Great

Britain. V. The status of subspecies of *Echinococcus granulosus* in Great Britain. *Annals of Tropical Medicine and Parasitology* **83**, 51–61.

Cook, B.R. and Clarkson, M.J. (1971). The epidemiology of *Echinococcus* infection in Great Britain III. *Echinococcus granulosus* and cestodes of the genus *Taenia* in farm dogs in England in the Lake District, northern Pennines and East Anglia. *Annals of Tropical Medicine and Parasitology* **65**, 71–79.

Correa, D., Sandoval M., Harrison L., Parkhouse M., Plancarte A., Meza-Lucas, A. and Flisser, A. (1989). Human neurocysticercosis: comparison of enzyme immunoassay capture techniques based on monoclonal and polyclonal antibodies for the detection of parasite products in cerebrospinal fluid. *Transactions of the Royal Society of Tropical Medicine and Hygiene* **83**, 814–816.

Craig, P.S. (1993). Immunodiagnosis of *Echinococcus granulosus*. In: *Compendium on Cystic Echinococcosis with Special Reference to the Xinjiang Uygur Autonomous Region, The Peoples Republic of China* (F.L. Andersen, ed.), pp. 85–118. Provo: Brigham Young University.

Craig, P.S. and Nelson, G.S. (1984). The detection of circulating antigen in human hydatid disease. *Annals of Tropical Medicine and Parasitology* **78**, 219–227.

Craig, P.S., Bailey, W. and Nelson, G.S. (1986a). A specific test for the identification of cyst fluid samples from suspected human hydatid infections. *Transactions of the Royal Society of Tropical Medicine and Hygiene* **80,** 256–257.

Craig, P.S., Zeyhle, E. and Romig, T. (1986b). Hydatid disease: research and control in Turkana. II. The role of immunological techniques for the diagnosis of hydatid disease. *Transactions of the Royal Society of Tropical Medicine and Hygiene* **80**, 183–192.

Craig, P.S., Macpherson, C.N.L., Watson-Jones, D. and Nelson, G.S. (1988). Immunodetection of *Echinococcus* eggs from naturally infected dogs and from environmental contamination sites in settlements in Turkana, Kenya. *Transactions of the Royal Society of Tropical Medicine and Hygiene* **82**, 268–274.

Craig, P.S., Liu, D. and Ding, Z. (1991). Hydatid disease in China. *Parasitology Today* **7**, 46–50.

Craig, P.S., Liu, D., Macpherson, C.N.L., Shi, D., Reynolds, D., Barnish, G., Gottstein, B. and Wang, Z. (1992). A large focus of alveolar echinococcosis in central China. *Lancet* **340**, 826–831.

Craig, P.S., Rogan, M.T. and Allan, J.C. (1995a). Hydatidosis and cysticercosis — larval cestodes. In: *Medical Parasitology a Practical Approach* (S.H. Gillespie and P.M. Hawkey, eds), pp. 209–237. Oxford: IRL Press.

Craig, P.S., Gasser, R.B., Parada, L., Cabrera, P., Parietti, S., Borgues, C., Acuttis, A., Agulla, J., Snowden, K. and Paolillo, E. (1995b). Diagnosis of canine echinococcosis: comparison of coproantigen and serum antibody tests with arecoline purgation in Uruguay. *Veterinary Parasitology* **56**, 293–301.

Crellin, J.R., Andersen, F.L., Schantz, P.M. and Condie, S.J. (1982). Possible factors influencing distribution and prevalence of *Echinococcus granulosus* in Utah. *American Journal of Epidemiology* **116**, 463–474.

Cruz, M., Davis, A., Dixon, H. and Pawlowski, Z.S. (1989). Operational studies on the control of *Taenia solium* taeniasis/cysticercosis in Ecuador. *Bulletin of the World Health Organisation* **67**, 563–566.

Cruz, M.E., Cruz, I., Preux, P.M., Schantz, P. and Dumas, M. (1994b). Headache and cysticercosis in Ecuador, South America. *Headache* **35**, 93–97.

Cruz, I., Cruz, M., Teran, W., Schantz, P.M., Tsang, V. and Barry, M. (1994a). Human subcutaneous *Taenia solium* cysticercosis in an Andean population with neurocysticercosis. *American Journal of Tropical Medicine and Hygiene* **51**, 405–407.

CWG (Cysticercosis Working Group in Peru) (1993). The marketing of cysticercotic pigs in the Sierra of Peru. *Bulletin of the World Health Organisation* **71**, 223–228

Delattre, P., Pascal, M., Le Pesteur, M.H., Giraudoux, P. and Damange, J.P. (1989). Caracteristiques ecologiques et epidemiologiques de *Echinococcus multilocularis* au cours d'un cycle complet des populations d'un hote intermediare (*Microtus arvalis*). *Canadian Journal of Zoology* **66**, 2740–2750.

Denham, D.A. and Suswilo, R.R. (1995). Diagnosis of intestinal helminth infections. In: *Medical Parasitology a Practical Approach* (S.H. Gillespie and P.M. Hawkey, eds), pp. 253–255. Oxford: IRL Press.

Deplazes, P. and Eckert, J. (1988). Utersuchungen zur infektion des hundes mit *Taenia hydatigena*. *Schweizer Archiv fur Tierheilkunde*, **128**, 307–320.

Deplazes, P., Gottstein, B., Stingelin, Y. and Eckert, J. (1990). Detection of *Taenia hydatigena* coproantigens by ELISA in dogs. *Veterinary Parasitology* **36**, 91–103.

Deplazes, P., Eckert, J., Pawlowski, Z.S., Machowska, L. and Gottstein, B. (1991). An enzyme linked immunosorbent assay for diagnostic detection of *Taenia saginata* copro-antigens in humans. *Transactions of the Royal Society of Tropical Medicine and Hygiene* **85**, 391–396.

Deplazes, P., Gottstein, N., Eckert, J., Jenkins, D.J., Ewald, D. and Jimenez-Palacios, S. (1992). Detection of *Echinococcus* coproantigens by enzyme-linked immunosorbent assay in dogs, dingoes and foxes. *Parasitology Research* **78**, 303–308.

Diaz, J.F., Verastegui, M., Gilman, R.H., Tsang, V.C.W., Pilcher, J.B., Gallo, C., Garcia, H.H., Torres, P., Montenegro, T., Miranda, E. and The Cysticercosis Working Group in Peru. (1992). Immunodiagnosis of human cysticercosis (*Taenia solium*): a field comparison of an antibody enzyme linked immunosorbent assay (ELISA) an antigen ELISA and an enzyme linked immunoelectrotransfer blot (EITB) assay in Peru. *American Journal of Tropical Medicine and Hygiene* **46**, 610–615.

Diaz-Camacho, S., Candil Ruiz, A., Beltran, M.U. and Willms, K. (1990). Serology as an indicator of *Taenia solium* tapeworm infections in a rural community in Mexico. *Transactions of the Royal Society of Tropical Medicine and Hygiene* **84**, 563–566.

Diaz-Camacho, S., Candil Ruiz, A., Suate Peraza, V., Zazueta Ramos, M.L., Felix Medina, M., Lozano, R. and Willms, K. (1991). Epidemiologic study and control of *Taenia solium* infections with praziquantel in a rural village of Mexico. *American Journal of Tropical Medicine and Hygiene* **45**, 522–531.

Didier, D., Weiler, S., Rohmer, R., Lassegue, A., Deschamps, J.P., Vuitton, D., Miguet, J.P. and Weill, F. (1985). Hepatic alveolar echinococcosis correlative US and CT study. *Radiology* **154**, 179–186.

Doehring-Schwerdtfeger, E., Abdel-Rahim, I.M., Dittrich, M., Mohamed-Ali, Q., Franke, D., Kardorff, R., Richter, J. and Ehrich, H.H. (1992). Ultrasonography as a diagnostic and for a district hospital in the tropics. *American Journal of Tropical Medicine and Hygiene* **46**, 727–731.

Doi, R., Deo, H., Fukuyama, H., Nakao, M., Inaoka, T., Kutsumi, H., Onishi, K., Amoh, K. and Ishimaru, O. (1990). Epidemiology of multilocular echinococcosis in Hokkaido. A sero-epidemiological study of hunters. *Japanese Journal of Public Health* **34**, 357–365.

Dumas, M., Grunitzky, E., Deniau, M., Dabis, F., Bouteille, B., Belo, M., Pestre, M., Catanzano, G., Darde, M.L.D. and Almeida, M. (1989). Epidemiological study of neurocysticercosis in northern Togo (West Africa). *Acta Leidensia* **57**, 191–196.

Earnest, M.P., Barth Reller, L., Filley, C.M. and Grek, A.J. (1987). Neurocysticercosis in the United States: 35 cases and a review. *Revue of Infectious Diseases* **9**, 961–979.

Eckert, J. and Thompson, R.C.A. (1988). *Echinococcus* strains in Europe: a review. *Tropical Medicine and Parasitology* **39**, 1–8.

Eisa, A.M., Mustafa, A.A. and Soliman, K.N. (1962). Preliminary report on cysticercosis and hydatidosis in southern Sudan. *Sudan Journal of Veterinary Science* **3**, 97–108.

Estrada, J.J., Estrada, J.A. and Kuhn, R.E. (1989). Identification of *Taenia solium* antigens in cerebrospinal fluid and larval antigens from patients with neurocysticercosis. *American Journal of Tropical Medicine and Hygiene* **41**, 50–55.

Fan, P.C. (1988). Taiwan *Taenia* and taeniasis. *Parasitology Today* **4**, 86–88.

Fay, F.H. and Rausch, R.L. (1992). Dynamics of the arctic fox population on St. Lawrence Island, Bering Sea. *Arctic* **45**, 393–397.

Fay, F.H. and Stephenson, R.O. (1986). Annual, seasonal and habitat-related variation in feeding habits of the arctic fox (*Alopex lagopus*) on St. Lawrence Island, Bering Sea. *Canadian Journal of Zoology* **67**, 1986–1994.

Filice, C., Pirola, F., Brunetti, E., Dughetti, S. and Strosselli, M. (1990). A new therapeutic approach for hydatid liver cysts. Aspiration and alcohol injection under sonographic guidance. *Gastroenterology* **98**, 1366–1368.

Flentje, B. and Padelt, H. (1981). Value of a serological diagnosis in *Taenia saginata* infestations of man. *Angewandte Parasitologie* **22**, 65–68.

Flisser, A. (1988). Neurocysticercosis in Mexico. *Parasitology Today* **4**, 131–134.

Flisser, A. (1994). Taeniasis and cysticercosis due to *Taenia solium*. In: *Progress in Clinical Parasitology Volume 4* (Tsieh Sun, ed.). Boca Raton: CRC Press.

Flisser, A., Reid, A., Garcia Zepeda, E. and McManus, D.P. (1988). Specific detection of *Taenia saginata* eggs by DNA hybridization. *Lancet* **ii**, 1429–1430.

Flisser, A., Madrazo, I., Plancarte, A., Schantz, P.M., Allan, J.C., Craig, P.S. and Sarti, E. (1993). Neurological symptoms in occult neurocysticercosis after single taeniacidal dose of praziquantel. *Lancet* **342**, 748.

French, C.M. and Nelson, G.S. (1982). Hydatid disease in the Turkana region of Kenya. II. A study in medical geography. *Annals of Tropical Medicine and Parasitology* **76**, 439–457.

French, C.M., Nelson, G.S. and Wood, M. (1982). Hydatid disease in the Turkana District of Kenya. I. The background to the problem with hypothesis to account for the remarkably high prevalence of the disease in man. *Annals of Tropical Medicine and Parasitology* **76**, 425–437.

Frider, B., Losada, C.A., Larrieu, E. and de Zavaleta, O. (1988). Asymptomatic abdominal hydatidosis detected by ultrasonography. *Acta Radiologica* **29**, 431–434.

Furuya, K., Nishizuka, M., Honma, H., Kumagai, M., Sato, N., Takahashi, M. and Uchino, J. (1990). Prevalence of human alveolar echinococcosis in Hokkaido as evaluated by western blotting. *Japanese Journal of Medical Science and Biology* **43**, 43–49.

Gajdusek, D.C. (1978). Introduction of *Taenia solium* into West New Guinea with a note on an epidemic of burns from cysticercus epilepsy in the Ekari people of the Wissel Lakes area. *Papua New Guinea Medical Journal* **21**, 329–342.

Garcia-Albea, E. (1989). Cisticercosis en Espana. Algunos datos epidemiologicos. *Revista Clinica Espanola* **184**, 3–6.

Garcia, H.H., Martinez, M., Gilman, R.H., Herrera, G., Tsang, V.C.W., Pilcher, J.B., Diaz, F., Verastegui, M., Gallo, C., Porras, M., Alvarado, M., Navanjo, J., Miranda, E. and the Cysticercosis Working Group in Peru (1991). Diagnosis of cyticercosis in endemic regions. *Lancet* **338**, 549–551.

Garcia, H., Gilman, R.H., Tovar, M.A., Flores, E.E., Jo, R., Tsang, V.C.W., Diaz, F., Torres, P., Miranda, E. and the Cysticercosis Working Group in Peru (CWG) (1995). Factors associated with *Taenia solium* cysticercosis analysis of nine hundred forty-six Peruvian neurologic patients. *American Journal of Tropical Medicine and Hygiene* **52**, 145–148.

Garcia-Noval, J., Allan, J.C., Fletes, C., Moreno, E., deMata, F., Torres-Alvarez, R., Soto de Alfaro, H., Yurrita. P., Higueros-Morales, H., Mencos, F. and Craig, P.S. (1996). Epidemiology of *Taenia solium* taeniasis and cysticerosis in two rural Guatemalan communities. *American Journal of Tropical Medicine and Hygiene* (In Press).

Gasser, R.B., Lightowlers, M.W. and Rickard, M.D. (1990). A recombinant antigen with potential for serodiagnosis of *Echinococcus granulosus* infection in dogs. *International Journal for Parasitology* **20**, 943–950.

Gasser, R.B., Jenkins, D.J., Paolillo, E., Parada, L., Cabrera, P. and Craig, P.S. (1993). Serum antibodies in canine echinococcosis. *International Journal for Parasitology* **23**, 579–586.

Geerts, S. (1995). Cysticercosis in Africa (letter). *Parasitology Today* **11**, 389.

Gemmell, M.A. (1978). Perspective on options for hydatidosis and cysticercosis control. *Veterinary Medicine Revue* **1**, 3–46.

Gemmell, M.A. (1990). Australasian contributions to an understanding of the epidemiology and control of hydatid disease caused by *Echinococcus granulosus* — past, present and future. *International Journal for Parasitology* **20**, 431–456.

Gemmell, M.A. (1993). Quantifying the transmission dynamics of the family Taeniidae with particular reference to *Echinococcus* spp. In: *Compendium on Cystic Echinococcosis with Special Reference to the Xinjiang Uygur Autonomous Region, The People's Republic of China* (F.L. Andersen, ed.), pp. 57–73. Provo: Brigham Young University.

Gemmell, M.A. and Roberts, M.G. (1995). Modelling *Echinococcus* life cycles. In: *Echinococcus and Hydatid Disease* (R.C.A. Thompson and A.J. Lymbery, eds), pp. 333–354. Wallingford: CAB International.

Gemmell, M.A., Matyas, Z., Pawlowski, Z. and Soulsby, E.J.L. (eds) (1983). *Guidelines for Surveillance Prevention and Control of Taeniasis/Cysticercosis.* Geneva; World Health Organization, 207 pp.

Gemmell, M.A., Lawson, J.R. and Roberts, M.G. (1986). Population dynamics in echinococcosis and cysticercosis: biological parameters of *Echinococcus granulosus* in dogs and sheep. *Parasitology* **92**, 599–620.

Gemmell, M.A., Lawson, J.R. and Roberts, M.G. (1987a). Population dynamics in echinococcosis and cysticercosis: evaluation of the biological parameters of *Taenia hydatigena* and *T. ovis* and comparison with those of *Echinococcus granulosus. Parasitology* **94**, 161–180.

Gemmell, M.A., Lawson, J.R. and Roberts, M.G. (1987b). Towards global control of cystic and alveolar hydatid disease. *Parasitology Today* **3**, 144–155.

Gharbi, H.A., Hassine, B., Brauner, M.W. and Dupuch, K. (1981). Ultrasound examination of hydatid liver. *Radiology* **139**, 459–463.

Gill, H.S. and Rao, B.V. (1967). On the biology and morphology of *Echinococcus granulosus* (Batsch, 1786) of buffalo–dog origin. *Parasitology* **57,** 695–704.

Gill, H.S. and Venkateswara, R. (1967). Incidence and fertility rate of hydatid cysts in Indian buffalo (*Bos bubalis*). *Bulletin de l'office Internationale des Epizootes* **69,** 989–997.

Giraudoux, P. (1991). Utilisation de l'espace par les hotes du énia multiloculcaire (*Echinococcus multilocularis*): conséquences épidémiologiques. PhD Thesis, Université de Bourgogne.

Giraudoux, P., Vuitton, D.A., Bresson-Hadni, S., Craig, P.S., Bartholomot, B., Barnish, G., Laplante, J.J., Zhong, S.D., Wan, Y.H. and Lenys, D. (1995). Mass screening and epidemiology of alveolar echinococcosis in France, Western Europe and in Gansu central China: from epidemiology towards transmission ecology. Abstract, International Symposium on Alveolar Echinococcosis. Sapporo, Japan, 8–9 February.

Giri, W.I. (1978). Cysticercosis in Surabaya, Indonesia. *Southeast Asian Journal of Tropical Medicine and Public Health* **9,** 232–236.

Gloor, G. (1988). Echinokokkose beim Menschen in der Schweiz 1970–83. Medical Dissertation, University of Zurich.

Gonzalez, C.L. (1989). Teniasis/cisticercosis en guatemala. In: *Cistecercosis* (F.J Aguilar, R. Maselli and A. Samayoa, eds), pp. 81–82. Guatemala: Asociacion Guatamalteca de Parasitologia y Medicina Tropical.

Gonzalez, A.E., Cama, V., Gilman, R.H., Tsang, V.C.W., Pilcher, J.B., Chavera, A., Castro, M., Montenegro, T., Verastegui, M., Miranda, E. and Bazalar, H. (1990). Prevalence and comparison of serologic assays, necropsy and tongue examination for the diagnosis of porcine cysticercosis in Peru. *American Journal of Tropical Medicine and Hygiene* **43,** 194–199.

Gottstein, B. (1984). An immunoassay for the detection of circulating antigens in human echinococcosis. *American Journal of Tropical Medicine and Hygiene* **33,** 1185–1191.

Gottstein, B. (1992a). Molecular and immunological diagnosis of echinococcosis. *Clinical Microbiology Reviews* **5,** 248–261.

Gottstein, B. (1992b). *Echinococcus multilocularis* infection: immunology and immunodiagnosis. *Advances in Parasitology* **31,** 321–380.

Gottstein, B. and Bettens, F. (1994). Association between HLA-DR13 and susceptibility to alveolar echinococcosis. *Journal of Infectious Diseases* **169,** 1416–1417.

Gottstein, B. and Felleisen, R. (1995). Protective immune mechanisms against the metacestode of *Echinococcus multilocularis*. *Parasitology Today* **11,** 320–326.

Gottstein, B. and Mowatt, M.R. (1991). Sequencing and characterisation of an *Echinococcus multilocularis* DNA probe and its use in the polymerase chain reaction. *Molecular and Biochemical Parasitology* **44,** 183–194.

Gottstein, B., Eckert, J. and Fey, H. (1983). Serological differentiation between *Echinococcus granulosus* and *E. multilocularis* infections in man. *Zeitschrift fur Parasitenkunde* **69,** 347–356.

Gottstein, B., Schantz, P.M. and Wilson, J.F. (1985). Serologic screening for *Echinococcus multilocularis* infections with ELISA. *Lancet* **1,** 1097–1098.

Gottstein, B., Lengelar, C., Bachmann, P., Hagemann, P., Kocher, P., Brossard, M., Witassek, F. and Eckert, J. (1987a). Sero-epidemiological survey for alveolar echinococcosis (by EM2-ELISA) of blood donors in an endemic area of Switzerland. *Transactions of the Royal Society of Tropical Medicine and Hygiene* **81,** 960–964.

Gottstein, B., Tsang, V.C.W. and Schantz, P.M. (1987b). Demonstration of species specific and cross reactive components of *Taenia solium* metacestode antigens. *American Journal of Tropical Medicine and Hygiene* **35**, 308–313.

Gottstein, B., Deplazes, P., Tanner, I. and Skaggs, J.S. (1991). Diagnostic identification of *Taenia saginata* with the polymerase chain reaction. *Transactions of the Royal Society of Tropical Medicine and Hygiene* **85**, 248–249.

Gottstein, B., Jaquier, P., Bresson-Hadni, S. and Eckert, J. (1993). Improved primary immunodiagnosis of alveolar echinococcosis in humans by an enzyme-linked immunosorbent assay using the Em2^{plus} antigen. *Journal of Clinical Microbiology* **31**, 373–376.

Gracia, F., Chavarria, R., Archbold, C., Larreategui, M., Castillo, L., Schantz, P. and Reeves, W.C. (1990). Neurocysticercosis in Panama: preliminary epidemiologic study in the Azuero region. *American Journal of Tropical Medicine and Hygiene* **42**, 67–69.

Grove, D.I. (1990). *A History of Human Helminthology*. Wallingford: CAB International.

Guo, Z.X. (1987). Hydatidosis in wild-life — *Ochotona* sp. serve as an intermediate host of *Echinococcus. Qinghai Medical Journal* **1**, 9–11.

Guo, Z.X., He, D.L., Li, Y.Q., Zhao, G.Q. and Ma, Y.F. (1993). Research on the natural focus of hydatidosis in the wild-life in Qinghai–Tibet plateau of China. 16th International Congress of Hydatidology, 12–16 October, Beijing.

Hackett, F. and Walters, T.M.H. (1980) Helminths of the red fox in mid-Wales. *Veterinary Parasitology* **7**, 181–184.

Harris, K.M., Morris, D.L., Tudor, R., Toghill, P. and Hardcastle, R. (1986). Clinical and radiographic features of simple and hydatid cysts of the liver. *British Journal of Surgery* **73**, 835–838.

Harrison, L.J.S., Delgado, J. and Parkhouse, R.M.E. (1990). Differential diagnosis of *Taenia saginata* and *Taenia solium* with DNA probes. *Parasitology* **100**, 459–461.

Harrison, L.J.S. and Sewell, M.M.H. (1991). The zoonotic taeniae of Africa. In: *Parasitic Helminths and Zoonoses in Africa* (C.N.L. Macpherson and P.S. Craig, eds), pp. 54–82. London: Unwin Hyman.

Hatch, C. (1975). Observations on the epidemiology of equine hydatidosis in Ireland. *Irish Veterinary Journal* **29**, 155–157.

Helbig, M., Frosch, P., Kern, P. and Frosch, M. (1993). Serological differentiation between cystic and alveolar echinococcosis by use of recombinant antigens. *Journal of Clinical Microbiology* **31**, 3211–3215.

Hill, P.W. (1989). Hydatid disease in the Outer Hebrides. PhD Thesis, University of London.

Horton, R.J. (1989). Chemotherapy of *Echinococcus* infection in man with albendazole. *Transactions of the Royal Society of Tropical Medicine and Hygiene* **83**, 97–102.

International League Against Epilepsy, Commission on Epidemiology and Prognosis (1993). Guidelines for epidemiologic studies on epilepsy. *Epilepsia* **34**, 592–596

Irshadullah, M., Nizami, W.A. and Macpherson, C.N.L. (1989). Observations on the suitability and importance of domestic intermediate hosts of *Echinococcus granulosus* in Uttar Pradesh, India, *Journal of Helminthology* **63**, 39–45.

Ito, A., Nakao, M., Kutsumi, H., Lightowlers, M.W., Itoh, M. and Sato, S. (1993). Serodiagnosis of alveolar hydatid disease by Western blotting. *Transactions of the Royal Society of Tropical Medicine and Hygiene* **87**, 170–172.

Ito, A., Nakao, M., Ito, M., Matsuzaki, T., Kamiya, M. and Kutsumi, H. (1994). Antibody responses in the wild vole, *Clethrionomys rufocanus bedfordiae*, naturally infected with *Echinococcus multilocularis* by Western blotting. *Journal of Helminthology* **68**, 267–269.

Jena, A., Sanchetee, P.C., Gupta, R.K., Klaushu, S., Chandnov, R. and Lakashmipathi, N. (1988). Cysticercosis of the brain shown by magnetic resonance imaging. *Clinical Radiology* **38**, 542–546.

Jenkins, D.J., Gasser, R.B., Zeyhle, E., Romig, T. and Macpherson, C.N.L. (1990). Assessment of a serological test for the detection of *Echinococcus granulosus* infection in dogs in Kenya. *Acta Tropica* **47**, 245–248.

Jiang, C. (1981). Liver alveolar echinococcosis in the northwest: report of 15 patients and collective analysis of 90 cases. *Chinese Medical Journal* **94**, 771–778.

Jones, A. and Walters, T.M.H. (1992). The cestodes of foxhounds and foxes in Powys, mid-Wales. *Annals of Tropical Medicine and Parasitology* **86**, 143–150.

Judson, D.G., Dixon, J.B., Clarkson, M.J. and Pritchard, J. (1985). Ovine hydatidosis: some immunological characteristics of the seronegative host. *Parasitology* **91**, 349–357.

Kamhawi, S. (1995). A retrospective study of human cystic echinococcosis in Jordan. *Annals of Tropical Medicine and Parasitology* **89**, 409–414.

Kaminsky, R. (1991). Taeniasis–cysticercosis in Honduras. *Transactions of the Royal Society of Tropical Medicine and Hygiene* **85**, 531–435.

Kamiya, M. (1988). Infectious diseases transmitted by dogs to humans. *Asian Medical Journal* **31**, 87–93.

Kamiya, H. and Ohbayashi, M. (1975). Some helminths of the red fox, *Vulpes vulpes schrencki* Kishida, in Hokkaido, Japan with a description of a new trematode, *Massaliatrmea yamashita* n.sp. *Japanese Journal of Veterinary Research* **23**, 60–67.

Kenny, J.V. (1993). The use of filter paper in seroepidemiological studies. Community based approach. *Tropical Doctor* **23**, 3.

Kinder, A., Carter, S.D., Allan, J.C., Marshall-Clarke, S. and Craig, P.S. (1992). Salivary and serum antibodies in experimental canine taeniasis. *Veterinary Parasitology* **41**, 321–327.

Klungsoyr, P., Courtright, P. and Hendrickson, T.H. (1993). Hydatid disease in the Hamar of Ethiopia: a public health problem for women. *Transactions of the Royal Society of Tropical Medicine and Hygiene* **87**, 254–255.

Kramer, L.D., Locke, G.E., Byrd, S.E. and Daryabagi, J. (1989). Cerebral cysticercosis: documentation of natural history with CT. *Radiology* **171**, 459–462.

Lawson, J.R. and Gemmell, M.A. (1990). Transmission of taeniid tapeworm eggs via blow flies to intermediate hosts. *Parasitology* **100**, 143–146.

Leggatt, G.R., Yang, W. and McManus, D.P. (1992). Serological evaluation of the 12 kDa subunit of antigen B in *Echinococcus granulosus* cyst fluid by immunoblot analysis. *Transactions of the Royal Society of Tropical Medicine and Hygiene*, **86**, 189–192.

Le Riche, P.D. and Sewell, M.M.H. (1978). Differentiation of taeniid cestodes by enzyme electophoresis. *International Journal for Parasitology* **8**, 479–483.

Lethbridge, R.C. (1980). The biology of the oncosphere of cyclophyllidean cestodes. *Helminthological Abstracts* A**49**, 59–72.

Li, W. (1985). Natural animal hosts of adult *Echinococcus multilocularis* and morphological studies in China. *Zoologica Acta Sinica* **31**, 365–371.

Li, W. (1986). Investigation of echinococcosis in Ningxia. *Endemic Disease Bulletin* **1**, 131–135.

Lightowlers, M.W. (1990). Cestode infections in animals: immunological diagnosis and vaccination. *Revue Scientifique et Technique de L'Office International des Epizooties* **9**, 463–487.

Lightowlers, M.W. and Gottstein, B. (1995). Echinococcosis/hydatidosis: antigens, immunological and molecular diagnosis. In: *Echinococcus and Hydatid Disease* (R.C.A. Thompson and A.J. Lymbery, eds), pp. 355–410. Wallingford: CAB International.

Lightowlers, M.W., Liu, D., Haralambous, A. and Rickard, M.D. (1989). Subunit composition and specificity of the major cyst fluid antigens of *Echinococcus granulosus*. *Molecular and Biochemical Parasitology* **37**, 171–182.

Lin, Y. and Hong, L. (1991). The biology and geographical distribution of *Echinococcus multilocularis* infection in China. *Endemic Diseases Bulletin* **6**, 117–126.

Lin, Y., Hong, L., Yang, W. and Peng, W. (1993). Observations on the natural rodent hosts of alveolar hydatid cyst of *Echinococcus multilocularis* in Tacheng Region, Xinjiang. *Endemic Diseases Bulletin* **8**, 29–33.

Liu, D., Rickard, M.D. and Lightowlers, M.W. (1993). Assessment of monoclonal antibodies to *Echinococcus granulosus* Antigen 5 and Antigen B for detection of human hydatid circulating antigens. *Parasitology* **106**, 75–81.

Liu, F.J. (1993). Prevalence of *Echinococcus granulosus* in dogs in the Xinjiang Uyger Autonomous Region, PRC. In: *Compendium on Cystic Echinococcosis with Special Reference to the Xinjiang Uyger Autonomous Region, The Peoples Republic of China* (F.L. Andersen, ed.), pp. 168–177. Provo: Brigham Young University.

Liu, F.J. Che, X.H. and Chang, Q. (1993b). Prevalence of hydatid cysts in livestock in the Xinjiang Uyger Autonomous Region, PRC. In: *Compendium on Cystic Echinococcosis with Special Reference to the Xinjiang Uyger Autonomous Region, The Peoples Republic of China* (F.L. Andersen, ed.), pp. 177–190. Provo: Brigham Young University.

Liu, Y., Wang, X., Chen, Y. and Yao, Y. (1993a). Computer tomography of liver in alveolar echinococcosis treated with albendazole. *Transactions of the Royal Society of Tropical Medicine and Hygiene* **87**, 319–321.

Lloyd, S., Martin, S.C., Walters, T.M.H. and Soulsby, E.J.L. (1991). Use of sentinal lambs for early monitoring of the South Powys hydatidosis control scheme: prevalence of *Echinococcus granulosus* and some other helminths. *Veterinary Record* **129**, 73–76.

Lucius, R. and Bilger, B. (1995). *Echinococcus multilocularis* in Germany: increased awareness or spreading of a parasite. *Parasitology Today* **11**, 430–434.

Lymbery, A.J. (1992). Interbreeding, monophyly and the genetic yardstick: species concepts in parasites. *Parasitology Today* **8**, 208–211.

Maass, M., Delgado, E. and Knobloch, J. (1991). Detection of *Taenia solium* antigens in merthiolate–formalin preserved stool samples. *Tropical Medicine and Parasitology* **42**, 112–114.

MacArthur, W.P. (1934). Cysticercosis as seen in the British Army, with special reference to the production of epilepsy. *Transactions of the Royal Society of Tropical Medicine and Hygiene* **27**, 343–363.

Machnicka-Roguska, B. and Zwierz, C. (1970). Intradermal test with antigenic fractions in *Taenia saginata* infection. *Acta Parasitologia* **18**, 293–299.

Macpherson, C.N.L. (1981). Epidemiology and strain differentiation of *Echinococcous granulosus* in Kenya. PhD Thesis. University of London.

Macpherson, C.N.L. (1983). An active intermediate host role for man in the life-cycle of *Echinococcus granulosus* in Turkana, Kenya. *American Journal of Tropical Medicine and Hygiene* **32**, 397–404.

Macpherson, C.N.L. (1992). Ultrasound in the diagnosis of parasitic disease. *Tropical Doctor* January, 14–20.

Macpherson, C.N.L. (1994). Epidemiology and control of parasites in nomadic situations. *Veterinary Parasitology* **54**, 87–102.

Macpherson, C.N.L. and Craig, P.S. (1991). Echinococcosis — a plague on pastoralists. In: *Parasitic Helminths and Zoonoses in Africa* (C.N.L. Macpherson and P.S. Craig, eds), pp. 25–53. London: Unwin Hyman.

Macpherson, C.N.L., Karstad, L., Stevenson, P. and Arundel, J.H. (1983). Hydatid disease in the Turkana District of Kenya III. The significance of wild animals in the transmission of *Echinococcus granulosus* with particular reference to Turkana and Maasailand. *Annals of Tropical Medicine and Parasitology* **77**, 61–73.

Macpherson, C.N.L., Zeyhle, E. and Romig, T. (1984). An *Echinococcus* pilot control programme for north-west Turkana, Kenya. *Annals of Tropical Medicine and Parasitology* **78**, 188–192.

Macpherson, C.N.L., French, C.M., Stevenson, P. *et al.* (1985). Hydatid disease in the Turkana district of Kenya. IV. The prevalence of *Echinococcus granulosus* infections in dogs and observations on the role of the dog in the lifestyle of the Turkana. *Annals of Tropical Medicine and Parasitology* **79**, 51–61.

Macpherson, C.N.L., Zeyhle, E., Romig, T. and Rees, P. (1987). Portable ultrasound scanner versus serology in screening for hydatid cysts in a nomadic population. *Lancet* **ii**, 259–262.

Macpherson, C.N.L., Spoerry, A., Zeyhle, E., Romig, T. and Gorfe, M. (1989). Pastoralists and hydatid disease: an ultrasound scanning prevalence survey in East Africa. *Transactions of the Royal Society of Tropical Medicine and Hygiene* **84**, 243–247.

Maddison, S.E., Slemenda, S.B., Schantz, P.M., Fried, J.A., Wilson, M. and Tsang, V.C.W. (1989). A specific diagnostic antigen of *Echinococcus granulosus* with an apparent molecular wt of 8 kDa. *American Journal of Tropical Medicine and Hygiene* **40**, 377–383.

Matossian, R.M., Rickard, M.D. and Smyth, J.D. (1977). Hydatidosis: a global problem of increasing importance. *Bulletin of the World Health Organisation* **55**, 499–507.

McConnell, J.D. and Green, R.J. (1979). The control of hydatid disease in Tasmania. *Australian Veterinary Journal* **55**, 140–145.

McManus, D.P., Simpson, A.J.G. and Rishi, A.K. (1987). Characterisation of the hydatid organism, *Echinococcus granulosus* from Kenya using cloned DNA markers. In: *Helminth Zoonoses with Particular Reference to the Tropics* (S. Geerts, V. Kumar and J. Brandt, eds), pp. 29–36. Dordrecht: Martinus Nijhoff.

McManus, D.P., Thompson, R.C.A. and Lymbery, A.J. (1989). Comment on the status of *Echinococcus granulosus* in the U.K. *Parasitology Today* **5**, 365–367.

McManus, D.P., Ding, Z. and Bowles, J. (1994). A molecular genetic survey indicates the presence of a single, homogeneous strain of *Echinococcus granulosus* in north-western China. *Acta Tropica* **56**, 7–14.

McVie A., Ersfeld, A., Rogan, M.T. and Craig, P.S. (1996). Expression and immunological characterisation of *Echinococcus granulosus* recombinant antigen B for

IgG4 antibody subclass detection in huamn cystic echinococcosis. *Journal of Clinical Microbiology* (in press).

Medina, M., Rosas, E., Rubio, F. and Sotelo, J. (1990). Neurocysticercosis as the main cause of late onset epilepsy in Mexico. *Archives of Internal Medicine* **150**, 325–327.

Menghebat, L., Jiang, L. and Chai, J.J. (1993). A retrospective survey for surgical cases of cystic echinococcosis in the Xinjiang Uyger Autonomous Region, PRC. In: *Compendium on Cystic Echinococcosis with Special Reference to the Xinjiang Uyger Autonomous Region, The Peoples Republic of China* (F.L. Andersen ed.), pp. 135–145. Provo: Brigham Young University.

Michault, A., Riviere, B., Fressy, P., Laporte, J.P., Bertil, G. and Mignard, C. (1990). Apport de l'enzyme-linked immunoelectrotransfer blot assay au diagnostic de la neurocysticercose humaine. *Pathologie Biologie* **38**, 119–125.

Ming, R., Tolley, H.D., Anderson, F.L., Chai, J. and Sultan, Y. (1992). Frequency distribution of *Echinococcus granulosus* hydatid cysts in sheeep populations in the Xinjiang Uygur Antonomous Region, China. *Veterinary Parasitology* **44**, 67–76.

Miyazaki, I. (1991). *Helminthic Zoonoses*. Tokyo: International Medical Foundation of Japan.

Mlika, N., Larouze, B., Dridi, M., Yang, R., Gharbi, S., Jemmali, M., Gaudebout, C. and Rousset, J.J. (1984). Serological survey of human hydatid disease in high risk populations for central Tunisia. *American Journal of Tropical Medicine and Hygiene* **33**, 1182–1184.

Mlika, N., Larouze, B., Gaudebout C., Braham, B., Allegue, M., Dazza, M.C., Dridi, M., Gharbi, S., Gaumer, B., Bchir, A., Rousset, J.J., Delattre, M. and Jemmali, M. (1986). Echotomatographic and serologic screening for hydatidosis in a Tunisian village. *American Journal of Tropical Medicine and Hygiene* **35**, 815–817.

Moore, A.C., Lutwick, L.I., Schantz, P.M., Pilcher, J.B., Wilson, M., Hightower, A.W., Chapnick, E.K., Abter, E.I.M., Grossman, J.R., Fried, J.A., Ware, D.A., Haichou, X., Hyon, S.S., Barbour, R.L., Antar, R. and Hakim, A. (1995). Seroprevalence of cysticercosis in an orthodox Jewish community. *American Journal of Tropical Medicine and Hygiene* **53**, 439–442.

Moro, P.L., Guevara, A., Verastegui, M., Gilman R.H., Poma, H., Tapia, B., Tsang, V.C.W., Garcia, H.H., Pacheco, R., Lapel, C., Miranda, E. and the Cysticercosis Working Group in Peru (1994). Distribution of hydatidosis and cysticercosis in different Peruvian populations as demonstrated by an enzyme-linked immuno-electrotransfer blot (EITB) assay. *American Journal of Tropical Medicine and Hygiene* **51**, 851–855.

Nahmias, J., Goldsmith, R., Schantz, P., Siman, M. and El-On, J. (1992). High prevalence of human hydatid disease (echinococcosis) in communities in north Israel: epidemiological studies in the town of Yirka. *Acta Tropica* **50**, 1–10.

Nelson, G.S. (1972). Human behaviour in the transmission of parasitic disease. In: *Behaviour Aspects of Parasite Transmission* (E.U. Canning and C.A. Wright, eds), pp. 109–122. London: Zoological Journal of the Linnean Society.

Nelson, G.S. and Rausch, R.L. (1963). *Echinococcus* infections in man and animals in Kenya. *Annals of Tropical Medicine and Parasitology* **57**, 136–149.

Nemurovskaia, A.I., Nekipelow, V.Y., Iakovleva, T.A. and Iasinskii, A.A. (1980). Problems of echinococcosis and alveococcosis in the Russian Federation. *Medical Parasitology and Parasitic Disease* **49**, 17–21 (In Russian).

Ognev, S.I. (1964). *Mammals of USSR and Adjacent Countries VI (Rodents)*. Israel Program for Scientific Translation Ltd., Jerusalem.

Onah, D.N. and Chiejina, S.N. (1995). *Taenia solium* cysticercosis and human taeniasis in the Nsukka area of Enugu State, Nigeria. *Annals of Tropical Medicine and Parasitology* **89**, 399–407.

O'Rourke, F.J. (1969). Hydatid disease in Ireland. *Journal of the Irish Medical Association* **62**, 91–93.

Owor, R. and Bitakamire, P.K. (1975). Hydatid disease in Uganda. *East African Medical Journal* **52**, 700–704.

PAHO (Pan American Health Organisation) (1994). Meeting of the scientific working group on the advances in the prevention, control and treatment of hydatidosis. Montevideo, 26–28 October.

Palmer, S.R. and Biffin, A.H.B. (1987). The changing incidence of human disease in England and Wales. *Epidemiology and Infection* **99**, 693–700.

Palmer, S.R., Biffin, A.H., Craig, P.S. and Walters, T.M.H. (1996). The control of hydatid disease in Wales: Evidence of success but a continuing treat. *British Medical Journal* **312**, 674–675.

Paolillo, E., Cohen, H., Bonifacino, R., Botta, B., Parada, L., Cabrera, P., Snowden, K., Gasser, R., Tessier, R., Dibarboure, L., Wen, H., Rogan, M.T., Allan, J.C., Soto, H. and Craig P.S. (1996). Human cystic echinococcosis in a Uruguayan community: a sonograhpic, serologic and epidemiological study. (Submitted).

Pathak, K.M.L., Allan, J.C., Ersfeld, K. and Craig, P.S. (1994). A Western blot and ELISA assay for diagnosis of *Taenia solium* infection in pigs. *Veterinary Parasitology* **53**, 209–217.

Pawlowski, Z.S. (1990). Perspectives on the control of *Taenia solium*. *Parasitology Today* **6**, 371–373

Pawlowski, Z.S. (1993). Critical points in the clinical management of cystic echinococcosis. In: *Compendium on Cystic Echinococcosis with Special Reference to Xinjiang Uygur Autonomous Region, The People's Republic of China* (F.L. Andersen, ed.), pp. 119–134. Provo: Brigham Young University.

Perdomo, R., Alvarez, C., Genninazzi, H., Ferreira, C., Monti, J., Parada, R., Cativelli, D., Barragne, A.D., Rivero, M.E. and Parada, J. (1988). Early diagnosis of hydatidosis by ultrasonography. *Lancet* **i**. 244.

Pogacnik, A., Pohat-Marinsek, Z. and Us-Krasovec, M. (1990). Fine needle aspiration biopsy in the diagnosis of liver echinococcosis. *Acta Cytologica* **34**, 765–766.

Powell, S.Y., Proctor, A.J., Wilmot, B. and MacLeod, N. (1966). Cysticercosis and epilepsy in Africans: a clinical and neurological study. *Annals of Tropical Medicine and Parasitology* **60**, 142–158.

Purriel, P., Schantz, P.M., Beovide, H. and Mendoza, G. (1974). Human echinococcosis (hydatidosis) in Uruguay: a comparison of indices of morbidity and mortality, 1962–71. *Bulletin of the World Health Organisation* **49**, 395–402.

Qiu, G.M. (1989). Sichuan pika naturally infected with *Echinococcus multilocularis*. *Endemic Diseases Bulletin* **4**, 64–65.

Rabiela, M.T., Lombardo, L. and Flores, F. (1972). Cisticercosis cerebral. Estudio de 68 casos de autopsia. *Patologia (Mexico)* **10**, 27–40.

Rausch, R.L. (1995). Life cycle patterns and geographic distribution of *Echinococcus* species. In: *Echinococcus and Hydatid Disease* (R.C.A. Thompson and A.J. Lymbery, eds), pp. 89–134. Wallingford: CAB International.

Rausch, R.L. and Schiller, E.L. (1951). Hydatid disease (echinococcosis) in Alaska and the importance of rodent intermediate hosts. *Science* **113**, 57–58.

Rausch, R.L. and Schiller, E.L. (1956). Studies on the helminth fauna of Alaska. XXV. The ecology and public health significance of *Echinococcus sibiricensis* Rausch and Schiller, 1954, on St. Lawrence Island. *Parasitology* **46**, 395–419.

Rausch, R.L., Wilson, J.F., Schantz, P.M. and McMahon, B.J. (1987). Spontaneous death of *Echinococcus multilocularis*: cases diagnosed serologically (by Em2-ELISA) and clinical significance. *American Journal of Tropical Medicine and Hygiene* **36**, 576–585.

Rausch, R.L., Wilson, J.F. and Schantz, P.M. (1990a). A programme to reduce the risk of infection by *Echinococcus multilocularis*: the use of praziquantel to control the cestode in a village in the hyperendemic region of Alaska. *Annals of Tropical Medicine and Parasitology* **84**, 239–250.

Rausch, R.L., Fay, F.H. and Williamson, F.S.L. (1990b). The ecology of *Echinococcus multilocularis* (*cestoda: Taeniidae*) on St. Lawrence Island, Alaska. II — Helminth populations in the definitive host. *Annals de Parasitologie Humaine et Comparee* **65**, 131–140.

Reichel, M.P., Baber, D.J., Craig, P.S. and Gasser, R.B. (1996). Epidemiological survey of cystic echinococcosis in the Falkland Islands. *Preventive Veterinary Medicine* (In press).

Richards, F. and Schantz, P.M. (1991). Laboratory diagnosis of cysticercosis. *Clinics of Laboratory Medicine* **11**, 1011–1028.

Richards, F.O., Schantz, P.M., Ruiz Tiben, E. and Sorvillo, F.J. (1985). Cysticercosis in Los Angeles county. *Journal of the American Medical Association* **254**, 3444–3448.

Robbana, M., Rachid, B.M.S., Zitouna, M.M. and Hafsia, M. (1981). Première observation d'echinococcose alveolar autochtone en Tunisie. *Archives d'Anatomie et Cytologie Pathologique* **29**, 311–312.

Roberts, M.G. (1994). Modelling of parasitic populations: cestodes. *Veterinary Parasitology* **54**, 145–160.

Roberts, M.G. and Aubert, M.F.A. (1994). A model for the control of *Echinococcus multilocularis* in France. *Veterinary Parasitology* **56**, 67–74.

Roberts, M.G., Lawson, J.R. and Gemmell, M.A. (1986). Population dynamics in echinococcosis and cysticercosis: mathematical model of the lifecycle of *Echinococcus granulosus*. *Parasitology* **92**, 621–641.

Roberts, M.G., Lawson, J.R. and Gemmell, M.A. (1987). Population dynamics in echinococcosis and cysticercosis: mathematical model of the lifecycle of *Taenia hydatigena* and *T. ovis*. *Parasitology* **94**, 181–197.

Roberts, T., Murrell, K.D. and Marks, S. (1994). Economic losses caused by food borne parasitic diseases *Parasitology Today* **11**, 419–423.

Rodriguez del Rosal, E., Correa, D. and Flisser, A. (1989). Swine cysticercosis: Detection of parasite products in serum. *Veterinary Record* **124**, 488.

Rogan, M.T., Craig, P.S., Zeyhle, E., Romig, T., Lubano, G.M. and Liu, D. (1991). Evaluation of a rapid dot ELISA as a field test for the diagnosis of cystic hydatid disease. *Transactions of the Royal Society of Tropical Medicine and Hygiene* **85**, 773–777.

Rogan, M.T., Craig, P.S., Zeyhle, E., Masinde, G., Wen, H. and Zhou, P. (1992). In vitro killing of taeniid oncospheres, mediated by human sera from hydatid endemic areas. *Acta Tropica* **51**, 291–296.

Romig, T. (1990). Beobachtungen zur zystischen Echinokokkose des Minschen in Turkana-Gebeit, Kenia. PhD thesis, University of Hohenheim.

Saileela, S., Hildreth, M.B., Gottstein, B., Wilson, M. and Schantz, P.M. (1992).

Abstracts of the 67th Annual Meeting of the *American Society of Parasitologists*, 4–8 August, Philadelphia, p. 142.

Sarti, E., Schantz, P.M., Plancarte, A., Wilson, M., Gutierrez, I.O., Lopez, A.S., Roberts, J. and Flisser, A. (1992a). Prevalence and risk factors for *Taenia solium* taeniasis and cysticercosis in humans and pigs in a village in Morelos, Mexico. *American Journal of Tropical Medicine and Hygiene* **46**, 677–685.

Sarti, E., Schantz, P.M., Aguilera, J. and Lopez, A. (1992b). Epidemiologic observations on porcine cysticercosis in a rural community of Michoacan State, Mexico. *Veterinary Parasitology* **41**, 195–201.

Sarti, E., Schantz, P.M., Plancarte, A., Wilson, M., Gutierrez, I., Aguilera, J., Roberts, J. and Flisser, A. (1994). Epidemiological investigation of *Taenia solium* and cysticercosis in a rural village of Michoacan State, Mexico. *Transactions of the Royal Society of Tropical Medicine and Hygiene* **88**, 49–52

Sasaki, F., Hata, Y., Sato, N., Hamada, H., Takashashi, H. and Uchino, J. (1994). Alveolar echinococcosis of the liver in children. *Pediatric Surgery International* **9**, 32–34.

Sato, N., Uchino, J., Suzuki, K., Kamiyama, T., Takahashi, T., Shimamura, T., Une, Y. and Nakajima, Y. (1993). IX Mass Screening. In: *Alveolar Echinococcosis of the Liver* (J. Uchino and N. Sato, eds), pp. 121–129. Sapporo: Hokkaido University Medical School.

Schantz, P.M. (1993). Surveillance and surveys for cystic echinococcosis. In: *Compendium of Cystic Echinococcosis with Special Reference to the Xinjiang Uygur Autonomous Region, The People's Republic of China* (F.L. Andersen, ed.), pp. 74–84. Provo: Brigham Young University.

Schantz, P.M. and Gottstein, B. (1986). Echinococcosis (hydatidosis). In: *Immunoserology of Parasitic Diseases, Vol I: Helminthic Diseases* (K.F. Walls and P.M. Schantz, eds), pp. 69–107. New York: Academic Press.

Schantz, P.M. and McAuley, J. (1991). Current status of food borne parasite zoonoses in The United States. *South East Asian Journal of Tropical Medicine and Public Health* **22**, supp. 65–71.

Schantz, P.M. and Sarti, E. (1989). Diagnostic methods and epidemiological surveillance of *Taenia solium* infection. *Acta Leidensia* **57**, 153–163

Schantz, P.M., Williams, J.F. and Riva Posse, C. (1973). The epidemiology of hydatid disease in Southern Argentina. Comparison of morbidity indices, evaluation of immunodiagnostic tests and factors affecting transmission in Southern Rio Negro Province. *American Journal of Tropical Medicine and Hygiene* **22**, 629–641.

Schantz, P.M., Gottstein, B., Ammann, R. and Lanier, A. (1991). Hydatid and the arctic. *Parasitology Today* **7**, 35–36.

Schantz, P.M., Moore, A.C., Muñoz, J.L., Hartman, B.J., Schaeffer, J.A., Aron, A.M., Persaud, D., Sarti, E., Wilson, M. and Flisser, A. (1992). Neurocysticercosis in an Orthodox Jewish community in New York City. *New England Journal of Medicine* **327**, 692–695.

Schantz, P.M., Cruz, M., Sarti, E. and Pawlowski, Z. (1993) Potential eradicability of taeniasis and cysticercosis. *Bulletin of Pan American Health Organisation* **27**, 397–403.

Schantz, P.M., Sarti, E., Plancarte, A., Wilson, M., Criales, J.L., Roberts, J. and Flisser, A. (1994). Community based epidemiological investigations of cysticercosis due to *Taenia solium*: comparison of serological screening tests and clinical findings in two populations in Mexico. *Clinical Infectious Diseases* **18**, 879–885.

Schantz, P.M., Chai, J., Craig, P.S., Eckert, J., Jenkins, D.J., Macpherson, C.N.L. and Thakur, A. (1995). Epidemiology and control of hydatid disease. In: *Echinococcus and Hydatid Disease* (R.C.A. Thompson and A.J. Lymbery, eds), pp. 233–332. Wallingford: CAB International.

Schenone, H., Villarroel, F., Rojas, A. and Ramirez, R. (1982). Epidemiology of human cysticercosis in Latin America. In: *Cysticercosis: Present State of Knowledge and Perspectives* (A. Flisser, K. Willms, J.P. Laclette, C. Larralde, C. Ridaura and F. Beltran, eds), pp. 25–38. New York: Academic Press.

Schiller, E. (1955). Studies on the helminth fauna of Alaska XXVI. Some observations on the cold-resistance of eggs of *Echinococcus sibiricensis* Raush and Schiller, 1954. *Journal of Parasitology* **41**, 578–582.

Schwabe, C.W. (1964). *Veterinary Medicine and Human Health*. Baltimore: Williams and Wilkins, pp. 211 and 395.

Schwabe, C.W. (1991). Helminth zoonoses in African perspective. In: *Parasitic Helminths and Zoonoses in Africa* (C.N.L. Macpherson and P.S. Craig, eds), pp. 1–24. London: Unwin Hyman.

Shambesh, M.K., Craig, P.S., Wen, H., Rogan, M.T. and Paolillo, E. (1996a). IgG1 and IgG4 serum antibody responses in asymptomatic and advanced cystic echinococcosis patients. *Parasite Immunology*. (In press).

Shambesh, M.K., Craig, P.S., Rogan, M.T., Gusbi, A.M., Macpherson, C.N.L. and Echtuish, E.F. (1996b). A mass ultrasound and serological study to investigate the prevalence of human cystic echinococcosis in Libya. (Submitted).

Shandera, W.X., White, A.C., Chen, J.C., Diaz, P. and Armstrong, R. (1994). Neurocysticercosis in Houston, Texas. *Medicine* **73**, 37–52.

Singh, B.P. and Dhar, D.N. (1988). *Echinococcus granulosus* in animals in northern India. *Veterinary Parasitology* **28**, 261–266.

Smyth, J.D. (1979). Importance of larval cestodiasis in man in developing countries. In: "Health Policies in Developing Countries": *Royal Society of Medicine. International Congress and Symposium Series No. 24*, pp. 89–95. London: Academic Press and Royal Society of Medicine.

Sorvillo, F.J., Waterman, S.H., Richards, F.O. and Schantz, P.M. (1992). Cysticercosis surveillance: locally acquired and travel-related infections and detection of intestinal tapeworm carriers in Los Angeles County. *American Journal of Tropical Medicine and Hygiene* **47**, 365–371.

Stallbaumer, M.F., Clarkson, M.J., Bailey, J.W. and Pritchard, J.E. (1986). The epidemiology of hydatid disease in England and Wales. *Journal of Hygiene (Cambridge)* **96**, 121–127.

Stehr-Green, J.K., Stehr-Green, P.A., Schantz, P.M., Wilson, J.F. and Lanier, A. (1988). Risk factors for infection with *Echinococcus multilocularis* in Alaska. *American Journal of Tropical Medicine and Hygiene* **38**, 380–385.

Storandt, S.T. and Kazacos, K.R. (1993). *Echinococcus multilocularis* identified in Indiania, Ohio, and East Central Illinois. *Journal of Parasitology* **79**, 301–305.

Suzuki, K., Sato, N. and Uchino, J. (1993). Epidemiology. In: *Alveolar Echinococcosis of the Liver* (J. Uchino and N. Sato, eds), pp. 1–9. Sapporo: Hokkaido University Medical School.

Takahashi, K., Yagi, K., Uragnuchi, K. and Kondo, N. (1989). Infection of larval *Echinococcus multilocularis* in red-backed vole, *Clethrionymus rufocanus bedfordiai*, captured around fox dens. *Report of the Hokkaido Institute of Public Health* **39**, 5–8.

Tang, C. (1988). Investigation of *Echinococcus multilocularis* in Huluenbeir grassland, Inner Mongolia. *Zoologica Acta Sinica* **34**, 172–179.

Thompson, R.C.A. and Lymbery, A.J. (1988). The nature, extent and significance of variation within the genus *Echinococcus*. *Advances in Parasitology* **27**, 210–258.

Thompson, R.C.A. and Smyth, J.D. (1975). Equine hydatidosis: a review of the current status in Great Britain and the results of an epidemiological survey. *Veterinary Parasitology* **1**, 107–127.

Thompson, R.C.A., Lymbery, A.J. and Constantine, C.C. (1995). Variation in *Echinococcus*: Towards a taxonomic revision of the genus. *Advances in Parasitology* **35**, 145–176.

Torgerson, P.R., Pilkington, J., Gulland, F.M.D. and Gemmell, M.A. (1995). Further evidence for the long distance dispersal of taeniid eggs. *International Journal for Parasitology* **25**, 265–267.

Tsang, V.C.W. and Wilson, M. (1995). *Taenia solium* cysticercosis: an under recognised but serious public health problem. *Parasitology Today* **11**, 124–126.

Tsang, V.C.W., Brand, A.J. and Boyer, A.E. (1989). An enzyme imunoelectrotransfer blot assay and glycoprotein antigens for diagnosing human *Taenia solium* cysticercosis. *Journal of Infectious Diseases* **159**, 50–59.

Uchino, J., Suzuki, K., Sato, W., Choji, K. and Irie, G. (1991). Mass screening for alveolar hydatid disease of the liver. *Archivos de la Hidatidosis* **30**, 835–838.

Vazquez, V. and Sotelo, J. (1992). The course of seizures after treatment of cerebral cysticercosis. *New England Journal of Medicine* **327**, 696–701.

Veit, P., Bilger, B., Schad, V., Schafer, J., Frank, W. and Lucius, R. (1995). Influence of environmental factors on the infectivity of *Echinococcus multilocularis* eggs. *Parasitology* **110**, 79–86.

Verastegui, M., Moro, P., Guevara, A., Rodriguez, T., Miranda, E. and Gilman, R.H. (1992). Enzyme-linked immunoelectrotransfer blot test for diagnosis of human hydatid disease. *Journal of Clinical Microbiology* **30**, 1557–1561.

Verster, A. (1967). Redescription of *Taenia solium* Linnaeus, 1758 and *Taenia saginata* Goeze, 1782. *Zeitschrift fur Parasitenkunde* **29**, 313–328

Vicary, F.R., Cusick, G., Shirley, M. and Blackwell, R.J. (1977). Ultrasound and abdominal hydatid disease. *Transactions of the Royal Society of Tropical Medicine and Hygiene* **71**, 29–31.

Viljoen, N.F. (1937). Cysticercosis in swine and bovines, with special reference to South African conditions. *Ondestepoort Journal of Veterinary Research* **9**, 337–570.

Vogel, H. (1955). Uber den Entwicklungszykulus und die Artzugehorigkeit des europäischen Alveolarechinococcus. *Deutsche Medizinsche Wochenschrift* **80**, 931–932.

Vogel, M., Muller, N., Gottstein, B., Flury, B., Eckert, J. and Seebeck, T. (1991). *Echinococcus multilocularis* characterisation of a DNA probe. *Acta Tropica* **48**, 109–116.

Von Sinner, W.N. (1991). New diagnostic signs in hydatid disease; radiography, ultrasound, CT and MRI correlated to pathology. *European Journal of Radiology* **12**, 150–159.

Von Sinner, W.N. (1992) Hydatidosis mimicking metastases. *Annals of Saudi Medicine* **12**, 316–320.

Von Sinner, W.N. (1993). Radiographic, CT and MRI spectrum of hydatid disease of the chest: a pictorial essay. *European Radiology* **3**, 62–70.

Vuitton, D.A. (1990). Alveolar echinococcosis of the liver: a parasitic disease in search of a treatment. *Hepatology* **12**, 617–618.

Wachira, T.M. (1993). Host influence on the rate of maturation of *Echinococcus granulosus* in dogs in Kenya. *Annals of Tropical Medicine and Parasitology* **87**, 607–609.

Wachira, T.M. and Zehyle, E. (1993). Hydatid Control Programme. In: *Turkana Rural Health Programme Annual Report*. AMREF, Nairobi, Kenya, pp. 25–31.

Wachira, T.M., Macpherson, C.N.L. and Gathuma, J.M. (1990). Hydatid disease in the Turkana District of Kenya VII. Analysis of the infection pressure on definitive and intermediate hosts of *E. granulosus*. *Annals of Tropical Medicine and Parasitology* **84**, 361–368.

Wachira, T.M., Macpherson, C.N.L. and Gathuma, J.M. (1991). Release and survival of *Echinococcus* eggs in different environments in Turkana and their possible impact on the incidence of hydatidosis in man and livestock. *Journal of Helminthology* **65**, 55–61.

Wachira, T.M., Bowles, J., Zeyhle, E. and McManus, D.P. (1993). Molecular examination of the sympatry and distributions of sheep and camel strains of *Echinococcus granulosus* in Kenya. *American Journal of Tropical Medicine and Hygiene*, **48**, 473–479.

Walters, T.M.H. (1978). Hydatid disease in Wales: I Epidemiology. *Veterinary Record* **102**, 257–259.

Walters, T.M.H. (1984). Hydatid control scheme in South Powys, Wales, U.K. *Annals of Tropical Medicine and Parasitology* **78**, 183–187.

Walters, T.M.H. and Clarkson, M.J. (1980). The prevalence of *Echinococcus granulosus* in farm dogs in mid-Wales. *Veterinary Parasitology* **7**, 185–190.

Wang, C.Y., Zhang, H.H. and Ge, L.Y. (1992). A monoclonal antibody based ELISA for detecting circulating antigen in CSF of patients with neurocysticercosis. *Hybridoma* **11**, 825–827.

Wang, H. and Ding, Z. (1989). Investigation of *Echinococcus multilocularis* adults in wolves in China. *Endemic Diseases Bulletin* **4**, 8–11.

Watson-Jones, D.L. and Macpherson, C.N.L. (1988). Hydatid disease in Turkana district of Kenya. VI. Man–dog contact and its role in the transmission and control of hydatidosis amongst the Turkana. *Annals of Tropical Medicine and Parasitology* **82**, 343–356.

Wen, H., New, R.R.C. and Craig, P.S. (1993). Diagnosis and treatment of human hydatidosis. *British Journal of Pharmacology* **35**, 565–574.

Wen, H., Craig, P.S., Ito, A., Vuitton, D.A., Bresson-Hadni, S., Allan, J.C., Rogan, M.T., Paolillo, E. and Shambesh, M. (1995). Immunoblot evaluation of IgG and IgG subclass antibody responses for immunodiagnosis of human alveolar echinococcosis. *Annals of Tropical Medicine and Parasitology* **89**. 485–495.

WHO (1996). Guidelines for treatment of cystic and alveolar echinococcosis in humans. *Bulletin of the World Health Organisation*, in press.

Wilson, J.F. and Rausch, R.L. (1980). Alveolar hydatid disease. A review of clinical features of 33 indigenous cases of *Echinococcus multicocularis* infection in Alaskan Eskimos. *American Journal of Tropical Medicine and Hygiene* **29**, 1340–1355.

Wilson, J.F., Diddams, A.C. and Rausch, R.L. (1968). Cystic hydatid disease in Alaska. A review of 101 autochthonous cases of *Echinococcus granulosus* infection. *American Review of Respiratory Disease* **98**, 1–12.

Wilson, J.F., Rausch, R.L., McMahon, B.J. and Schantz, P.M. (1992). Parasiticidal effect of chemotherapy in alveolar hydatid disease. Review of experience with mebendazole and albendazole in Alaskan Eskimos. *Clinical Infectious Diseases* **15**, 234–249.

Wilson, M., Bryan, R.T., Fried, J.A., Ware, D.A., Schantz, P.M., Pilcher, J.B.

Tsang, V.C.W. (1991). Clinical evaluation of cysticercosis enzyme linked immunoelectrotransfer blot in patients with neurocysticercosis. *Journal of Infectious Diseases* **164**, 1007–1009.

Wray, J.R. (1958). Note on human hydatid disease in Kenya. *East African Medical Journal* **35**, 37–39.

Wyn-Jones, G. and Clarkson, M.J. (1984). Radiologic detection of ovine hydatidosis. *Veterinary Radiology* **25**, 182–186.

Yamashita, J. (1973). *Echinococcus* and echinococcosis. *Progress of Medical Parasitology in Japan* **5**, 65–123. Tokyo: Meguro Parasitological Museum.

Yao, P. (1965). First clinical report of alveolar hydatid disease in China. *Chinese Journal of Surgery* **13**, 461.

Yoshino, K. (1934). On the evacuation of eggs from detached gravid proglottids of *Taenia solium* and on the structure of its eggs. *Taiwan Igakkai Zasshi* **33**, 47–58.

Zittouna, M.M., Boubaker, S., Dellagi, K., Safta, B.Z., Hadja, H., Robbana, M. and Rachid, B.M.S. (1985). L'échinococcose alvéolaire en Tunisie. *Bulletin Société Pathologique Exotique* **78**, 723–728.

Human Strongyloidiasis

David I. Grove

*Department of Clinical Microbiology and Infectious Diseases,
The Queen Elizabeth Hospital, Adelaide, South Australia*

1. Introduction 252
2. History 252
3. Species of *Strongyloides* 255
4. Morphology of *Strongyloides stercoralis* 255
    4.1. Parasitic female adult worm 256
    4.2. Free-living adult female worm 256
    4.3. Free-living adult male worm 257
    4.4. Eggs 258
    4.5. First-stage larva 259
    4.6. Second-, third- and fourth-stage free-living larvae 259
    4.7. Pre-infective second-stage larva 259
    4.8. Third-stage infective larva 259
5. Other Taxonomic Parameters 260
    5.1. Isoenzyme electrophoresis 260
    5.2. Genetic analysis 260
6. Cryopreservation 260
7. Life Cycle of *Strongyloides stercoralis* 261
    7.1. Free-living cycle 261
    7.2. Parasitic cycle 264
8. The Host–Parasite Relationship 267
    8.1. Eradication of infection 267
    8.2. Chronic infection 268
    8.3. Hyperinfection and disseminated infection 270
    8.4. Animal models 275
9. Clinical Features 276
    9.1. Chronic uncomplicated strongyloidiasis 276
    9.2. Severe complicated strongyloidiasis 278
10. Diagnosis 280
    10.1. Morphological identification of *Strongyloides* infecting humans 280

ADVANCES IN PARASITOLOGY VOL 38
ISBN 0–12–031738–9

    10.2. Immunodiagnosis 285
    10.3. Imaging 288
    10.4. Endoscopy 290
  11. Treatment 290
    11.1. Benzimidazoles 290
    11.2. Ivermectin 292
    11.3. Miscellanea 293
    11.4. Approaches to management 293
  12. Prevention and Control 295
  13. *Strongyloides fuelleborni* 296
  14. Future Directions 297
  References 297

1. INTRODUCTION

Strongyloides stercoralis remains an enigma. In a number of ways, it differs from other worms that infect humans. There are both free-living and parasitic life cycles with reproduction in the latter occurring by parthenogenesis. Replication within infected humans permits infection to persist for decades, usually with host and parasite reaching amicable agreement of mutual toleration. Should host defences become impaired, however, then the balance is disturbed, worms multiply enormously and disseminate throughout the body, and a severe illness supervenes. Like some other helminthiases, diagnosis of infection may sometimes be difficult and treatment may be problematic.

Much of our knowledge of this infection and the disease that it causes has been reviewed by Grove (1989). The purpose of this paper is to briefly summarize knowledge to that point then highlight advances that have been made over the past few years.

2. HISTORY

In July 1876, Louis Normand, physician to the Naval Hospital in Toulon, France, found a novel minute worm about 0.25 mm in length in the faeces of troops who had been repatriated from Cochin-China (Vietnam) with diarrhoea (see Grove (1989) or Grove (1990a) for all references prior to 1930). The parasite was then named *Anguillula stercoralis* by his colleague, Bavay, to reflect its shape (Latin: "anguillula" = "eel" and "stercus" = "dung"). A few months later at an autopsy of another soldier who had died from Cochin-China diarrhoea, he found another worm, about 2 mm long. He thought this was a distinct species, as did Bavay who named it *Anguillula intestinalis*. To complicate matters even further, Bavay then found a new type of larva in cultured stools which he believed was the larval form of *A. intestinalis*. These worms were in fact rhabditiform larvae, parasitic adults and infective third-stage larvae, respectively, of the one worm we now know as *Strongyloides stercoralis*. The correct relationship between these stages was then worked out by Grassi and Parona in Italy in 1878. To accommodate these findings, Grassi in 1879 erected a new genus which he called *Strongyloides* (Greek στρογγύλος, (STRONGYLOS) meaning "round" and εἶδος (EIDOS) meaning "similar") and named the parasite *Strongyloides intestinalis*. Perroncito in 1881 then cultivated free-living adult worms (which he called *Pseudorhabditis stercoralis*) from larvae (now called rhabditiform) identical with Normand's original *A. stercoralis*. Two years later, Leuckart realized that all these worms were but different phases in the life cycle of a single parasite. Finally, in 1902 Stiles and Hassall pointed out that the parasite should in fact be called *Strongyloides stercoralis*.

At first, it seemed as though the most likely mode of transmission was by ingestion of larvae in food and water. However, in 1902, van Durme showed that *Strongyloides* larvae could penetrate intact skin. Subsequently, following his work with hookworm larvae, Looss in 1904 by self-experimentation showed that patent infections followed filariform larvae penetrating the intact skin. Subsequently, there have been a number of other experimental infections of humans with various species of *Strongyloides* (reviewed in Freedman, 1991).

The path which these larvae took after skin penetration became a matter of continuing debate. Fülleborn showed in infected dogs that the majority of worms migrated to the lungs via the bloodstream, ascended the airways to the pharynx, then were swallowed and developed in the intestines. Others, however, believed that larvae migrated through the lymphatics or directly through the connective tissues. Askanazy (1900) found larvae in the deeper layers of the intestinal wall, Gage (1911) found filariform larvae in the sputum, then Fülleborn (1926) described patients with urticaria on the buttocks and suggested that autoinfection may occur by larvae moulting in the perianal folds and then penetrating the skin. Over the next few decades, increasing numbers of patients who had massive, disseminated infections with enormous worm burdens were recognized and such events were correlated with a breakdown in host defences. Specific treatment with thiabendazole was introduced in 1962 but this drug has not proven to be a panacea.

Table 1 List of species of *Strongyloides* and their type hosts (from Speare, 1989).

Species	Host – scientific name	Host – common name
S. agoutii	*Dasyprocta agouti*	Golden rumped agouti
S. akbari	*Crocidura coerula*	Musk rat
S. amphibiophilus	*Peltophryne peltocephala*	Toad
S. ardeae	*Nyctanassa violacea*	Yellow-crowned night heron
S. avium	*Gallus gallus*	Domestic fowl
S. bufonis	*Bufo melanostictus*	Malayan toad
S. carinii	*Leptodactylus gracilis*	Frog
S. cebus	*Cebus capucinus*	Capucin monkey
S. chapini	*Hydrochoerus hydrochaeris*	Capybara (rodent)
S. cruzi	*Hemidactylus mabouia*	Skink
S. cubaensis	*Butorides virescens maculatus*	Cuban green heron
S. darevskyi	*Lacerta saxicola*	Skink
S. dasypodis	*Dasypodis novemcinctus*	Armadillo
S. elephantis	*Elephas indicus*	Indian elephant
S. erschowi	*Nyctereutes procyonoides usuriensis*	Raccoon, dog
S. felis	*Felis catus*	Domestic cat
S. fuelleborni	*Pan troglodytes*	Chimpanzee
S. fuelleborni	*Papio cyanocephalus*	Yellow baboon
S. gulae	*Natrix cyclopyon cyclopyon*	Green water snake
S. herodiae	*Ardea herodias herodias*	Great blue heron
S. lutrae	*Lutra canadensis*	North American otter
S. martis	*Martes zibellina*	Sable
S. martis	*Martela ermina*	Stoat
S. minimum	*Anas bahamensis*	Duck
S. mirzai	*Zamensis mucosus*	Rat snake
S. mustelorum	*Mustela ermina*	Stoat
S. myopotami	*Myocaster coypus*	Coypu rat
S. nasua	*Nasua narica panamensis*	Coatimundi
S. ophidiae	*Drymobius bifossatus*	Snake
S. oswaldoi	*Gallus gallus*	Domestic fowl
S. papillosus	*Ovis aries*	Domestic sheep
S. pavonis	*Pavo muticus*	Green peafowl
S. pereirai	*Elosia rustica*	Frog
S. physali	*Bufo valiceps*	Wiegman's toad
S. planiceps	*Felis catus*	Domestic cat
S. procyonis	*Procyon lotor*	Raccoon
S. putorii	*Mustela putorius*	Polecat
S. quiscali	*Quiscalus niger caribaeus*	Bird
S. ransomi	*Sus scrofa*	Domestic pig
S. ratti	*Rattus norvegicus*	Brown rat
S. ratti v *ondatrae*	*Ondatra zibethicus*	Musk rat
S. robustus	*Scirius niger rufiventer*	Fox squirrel
S. rostombekowi	*Erinaceus europea*	Hedgehog
S. serpentis	*Natrix cyclopyon cyclopyon*	Green water snake
S. sigmodontis	*Sigmodon hispidus*	Cotton rat
S. spiralis	*Rana esculenta, R. lessoni*	Edible frog

Table 1 Continued.

Species	Host – scientific name	Host – common name
S. stercoralis	*Homo sapiens*	Man
S. stercoralis v *vulpi*	*Vulpes alopex*	Arctic fox
S. thylacis	*Isoodon macrouris*	Short-nosed bandicoot
S. tumefaciens	*Felis catus*	Domestic cat
S. turkmenica	*Himantopus candidus*	Stilt
S. venezuelensis	*Rattus norvegicus*	Brown rat
S. vulpis	*Vulpes vulpes*	Red fox
S. westeri	*Equus caballus*	Domestic horse

3. SPECIES OF *STRONGYLOIDES*

The genus *Strongyloides* is classified in the order Rhabditoidea, most members of which are soil-dwelling, microbiverous nematodes. Species of the genus *Strongyloides* bridge the gap between free-living saprobe and animal parasite. At least 52 valid species of *Strongyloides* have been described and are listed in Table 1 (Speare, 1989). The vast majority of these species do not infect humans. By far the most frequent human pathogen is *S. stercoralis* which is spread widely throughout the world, especially in the tropics. *S. fuelleborni* is found sporadically in Africa and perhaps elsewhere. A similar *S. fuelleborni*-like worm has been described in Papua New Guinea. In addition, patent infections in humans have been reported to have been produced experimentally with *S. procyonis* (Little, 1965) and *S. ransomi* (Kotlan and Vajda, 1934), although the latter could not be reproduced by other workers. Patent infections were not found after experimental infections of humans with *S. canis, S. cebus, S. felis, S. myopotami, S. planiceps* and *S. simiae* (reviewed in Speare, 1989). The various species of *Strongyloides* can often be differentiated by using taxonomic criteria visible by light microscopy. These criteria and the most appropriate techniques have been described in detail by Speare (1989).

4. MORPHOLOGY OF *STRONGYLOIDES STERCORALIS*

S. stercoralis has a parasitic life cycle as well as a free-living phase. The major morphological features of the various forms of the parasite will be described first then the life cycle reviewed. Key measurements are given in Table 2.

Table 2 Measurements of components of various stages of *Strongyloides stercoralis*.

	Parasitic female	Free-living female	Free-living male	Rhabditiform larva	Filariform larva
Length (μm)	2100–2700	900–1700	800–1000	180–240	490–630
Width (μm)	30–40	50–85	40–50	15	15–16
Pharynx length (μm)	50–65	125–150	110–125	90–95	220–225
Tail length (μm)	40–70	80–170	55–95	–	60–80

4.1. Parasitic Female Adult Worm

This stage is a minute, slender, almost transparent worm just over 2 mm long (Figure 1). The mouth is hexagonal in shape and surrounded by six papillae. The body is attenuated anteriorly and contains the long cylindrical oesophagus (sometimes called the pharynx); the anterior 25% of the oesophagus is muscular whereas the posterior 75% is largely glandular. The oesophagus is succeeded by the intestine, a long thin tube, one cell thick, that ends in a short rectum that opens at the anus close to the tip of the tail. The oesophagus is surrounded by the nerve ring at about one quarter of the way down its length. There are two longitudinal excretory canals running the length of the worm, one in each lateral chord, and connected by a transverse duct to each other and a single excretory cell just behind the nerve ring. The vulva is located in the mid-ventral line in the posterior third of the body. Paired uteri extend anteriorly and posteriorly from a very short vagina. The uteri contain a small number of eggs aligned in single file and occupy most of the body of the worm in this region. Each uterus opens into an oviduct which leads to thin-walled ovaries; both branches are not spiralled and the anterior branch reaches almost to the oesophagus before it reflexes back upon itself and lies parallel to the intestine.

4.2. Free-living Adult Female Worm

This small worm resembles the free-living nematodes of the genus *Rhabditis* (Figure 2). The cuticle is thin, transparent and has fine striations. The muscular oesophagus is divided into three parts — the anterior, cylindrical procorpus, the narrow isthmus, and the rounded, posterior bulb. This in turn leads to the intestine then rectum which opens at the anus near the tail. The

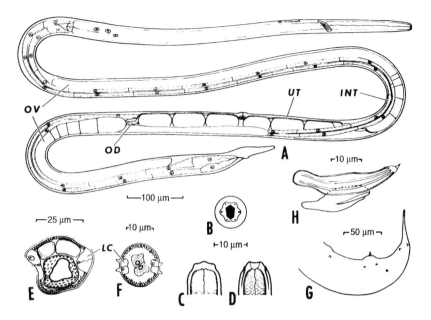

Figure 1 Major morphological features of *S. stercoralis*. A–E, Parasitic female worm. A, Whole worm in lateral view (OV, ovary; OD, oviduct; UT, uterus; INT, intestine). B, Anterior end in apical view. C, Anterior end, lateral view. D, Anterior end, dorsal view. E, Cross-section at level of ovary (LC, lateral chord). F, Filariform larva, cross-section at intestinal level. G, Free-living male lateral view of tail. H, Right spicule with underlying gubernaculum. (From Little, 1966.)

reproductive system is similar to that seen in the parasitic female except that each uterus contains numerous eggs.

4.3. Free-living Adult Male Worm

The smaller males have a sharply pointed tail that bends anteriorly and gives the worm a "J-shape" (Figure 2). The gut is similar to that seen in the female worm. The reproductive system is a simple straight tube. At the anterior end is a blind-ended testis that merges without clear demarcation first into the vas deferens then into the seminal vesicle. These organs contain spermatogonia, spermatocytes and sperms and open into the cloaca. The cloaca is surrounded by a pair of copulatory spicules that are inserted into the female during copulation. The spicules are guided in their extrusion by a chitinoid structure in the dorsal wall of the cloaca called the gubernaculum. There are six pairs of caudal papillae and a single mid-ventral precloacal papilla.

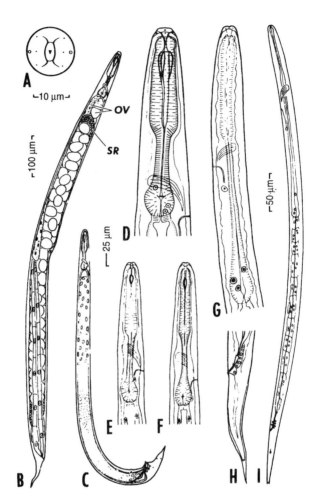

Figure 2 Major morphological features of *S. stercoralis*. A, Apical view of anterior end of free-living female worm. B, Free-living female worm, lateral view (OV, ovary; SR, seminal receptacle). C, Free-living male worm. D, Anterior end of free-living female worm. E, First-stage larva in duodenal aspirate. F, Later rhabditiform larva in faeces. G, Second-stage larva, anterior end. H, Second-stage larva, tail; cuticle is beginning to separate. I, Filariform larva. (From Little, 1966.)

4.4. Eggs

Ova laid by the free-living female worms are thin-shelled, ellipsoidal and measure about 40 × 70 μm in size. They then undergo a number of divisions and become fully embryonated. Eggs of the parasitic female are similar but hatch in the crypts of Lieberkühn and are rarely seen.

4.5. First-stage Larva

This stage is also known as a rhabditiform larva and is the form usually seen in faeces or intestinal fluid. It measures approximately 210 μm in length, is wider anteriorly and has a conical tail. The mouth opens into a simple, shallow, cup-shaped buccal capsule (Figure 2); this contrasts with the chitinoid buccal tube characteristic of rhabditiform larvae of hookworms. The oesophagus is the most prominent internal structure and occupies the anterior third of the body; it is similar to that seen in the free-living adult worm. The intestine extends through most of the rest of the body.

4.6. Second-, Third- and Fourth-stage Free-living Larvae

These stages are merely progressively larger versions of the free-living rhabditiform larva with general growth, re-organization of the head, and progressive sexual differentiation prior to the fourth moult leading to mature male and female free-living worms.

4.7. Pre-infective Second-stage Larva

In the case of second-stage larvae that are destined to become infective, marked changes occur. The oesophagus lengthens and becomes less muscular posteriorly.

4.8. Third-stage Infective Larva

This stage is also known as a filariform larva. It is the form capable of percutaneous infection and is the parasite that may be seen migrating through the tissues. The worm is long and thin with a finely striated cuticle. Lateral alae extend to the tip of the tail and give it a notched appearance under light microscopy although it may be like *S. ratti* in which scanning electron microscopy shows that there is really a circular aperture surrounded by eight projections (Zaman *et al.*, 1980). The mouth is probably closed and impervious to small particles and perhaps to liquids. The cylindrical oesophagus extends for about 40% of the length of the larva and does not have a posterior bulb, and is succeeded by a long, straight intestine. As with the parasitic female adult worm, a nerve ring encircles the oesophagus one quarter of the way down its length. A genital rudiment lies in the pseudocoelom at about the midpoint of the intestine.

It may be that there are two distinct forms of third-stage larvae. Schad *et al.* (1993) have described autoinfective larvae, i.e. larvae arising in the infected host, as being larger in diameter, shorter in length and having a more strongyliform oesophagus than the free-living infective larvae.

The amphids, two of the anterior sense organs, or sensillae, have been studied in ultrastructural detail (Ashton *et al.*, 1995). These organs are important in that they are the only sensillae open to the external environment and are likely to be chemoreceptors that accept chemical signals from the environment that initiate resumption of feeding and development. They are large, paired, goblet-shaped structures containing 13 neuronal dendrites that connect to cell bodies in the lateral ganglion posterior to the nerve ring.

5. OTHER TAXONOMIC PARAMETERS

5.1. Isoenzyme Electrophoresis

This technique is a powerful tool for addressing taxonomic problems that cannot be solved by traditional methods. Viney and Ashford (1990) studied isolates from a number of hosts (human, subhuman primates, domestic pigs and miscellaneous animals) in a variety of locations, although they were heavily weighted to Papua New Guinea. They concluded that (i) differentiation of *S. stercoralis* on morphological grounds from all other species occurring in primates is valid, (ii) *Strongyloides* cf. *fuelleborni* in humans in Papua New Guinea is very similar to *Strongyloides fuelleborni* found in Africa, and (iii) the species of *Strongyloides* in pigs in Papua New Guinea is different from the species occurring in humans.

5.2. Genetic Analysis

The entire 1766 bases of the 18S ribosomal RNA gene of *S. stercoralis* have been sequenced. The gene has a 38% G+C content and 69% homology with the same gene of *Caenorhabditis elegans*, the only other completely sequenced 18S rRNA gene (Putland *et al.*, 1993).

6. CRYOPRESERVATION

First-stage and third-stage larvae of *S. stercoralis* can both be successfully cryopreserved by first incubating them for 30–60 minutes in dimethyl sulphoxide (DMSO) as a cryoprotectant, then freezing them with liquid nitrogen in RPMI medium (Nolan *et al.*, 1988).

7. LIFE CYCLE OF *STRONGYLOIDES STERCORALIS*

The life cycle of *S. stercoralis* is complex (Figure 3) and has long been a matter of controversy or uncertainty. Development occurs within the human host (the parasitic cycle) or in the environment (the free-living cycle).

7.1. Free-living Cycle

First-stage larvae undergo further development in the soil when they are passed into the external environment. However, development may proceed by one of two routes. Indeed, both routes usually occur contemporaneously.

7.1.1. *Homogonic Development*

Some larvae may develop directly into infective third-stage larvae without the interposition of a free-living adult stage. Rhabditiform larvae feed on the microflora of faecally enriched soils. After 1–2 days, the organisms moult to become second-stage larvae, which are also feeding forms. After another 2 days, these transform into active, non-feeding infective larvae which seek the surface of soil or vegetation insofar as it is permitted by the film of moisture. These larvae die within 1–2 weeks unless they find a suitable host.

7.1.2. *Heterogonic Development*

Other larvae undergo a more complex series of changes in a process that is sometimes termed indirect development. In this process, the microbiverous worms maintain their rhabditiform structure, merely becoming larger until they attain adulthood. There are four moults (ecdyses) during which the parasites stop feeding, become inactive, generate a new cuticle and shed the old one. The fourth of these moults culminates in the appearance of free-living male and female adult worms. The adult worms mate and the female worms release eggs which hatch into rhabditiform larvae which then moult twice to become infective larvae; none of these eggs develop into a second free-living cycle (Yamada *et al.*, 1991).

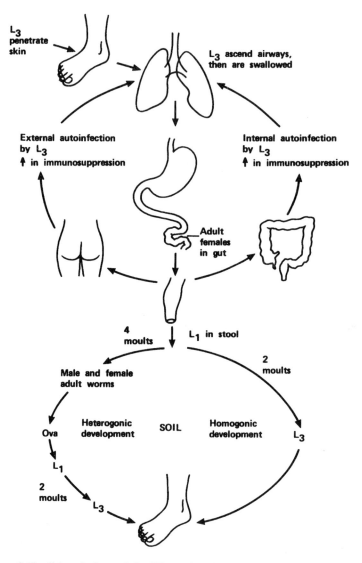

Figure 3 Traditional view of the life cycle of *S. stercoralis*. L_1 = rhabditiform larva; L_3 = infective larva; ↑ = increased. (From Grove, 1990b.)

Hammond and Robinson (1994) have studied the cytogenetics of *S. stercoralis* in preparations of free-living female and male worms. In female worms, cells at the blind apical end (vegetative zone) of the ovaries were small with large nuclei containing much DNA but the nuclei became progressively smaller as the cells moved towards the vulval end of the gonad. In the germinative zone, elements probably representing bivalents in the late prophase of meiosis were visible. Oocytes became larger as they progressed down the oviduct and the chromosomes condensed. Eventually, two sets of three chromosomes were seen with the chromosomes containing two chromatids, indicating anaphase I of meiosis. Subsequently, a sperm nucleus was seen within the cytoplasm and the egg chromosomes were believed to be in anaphase II. Embryogenesis occurred as the zygote moved down the uterus; the diploid complement of the embryos was six chromosomes. Between 5 and 45 embryos were seen in inseminated females, with the oldest embryos being fully developed and motile within the eggshell. In the testis of male worms, two, three or five dot-like chromosomes were seen within spermatocytes; this may reflect a diploid chromosome number in male germ-line tissue of five. It was concluded that reproduction appears to be by meiotic parthenogenesis (automictic thelytoky) in which eggs are activated by sperm (pseudogamy).

7.1.3. *Determinants of Mode of Development*

The factors that determine whether rhabditiform larvae of *S. stercoralis* develop directly or indirectly are poorly understood and some reliance needs to be placed upon what is known for other species of *Strongyloides*. Possible factors include genetically determined strain differences, the status of the host, and the environment of the free-living phase.

The potential for genetic control of development is perhaps best seen in the related worm, *S. ratti*, in which strains have been selected that predictably develop either homogonically or heterogonically (Graham, 1938; Wertheim and Lengy, 1964). Studies of different geographical isolates in coproculture have shown that tropical strains develop predominantly in an indirect manner at all temperatures whereas temperate isolates tended to develop directly, particularly at lower temperatures (Berezhnaia *et al.*, 1991).

Host factors are somewhat more nebulous. In studies of the swine parasite, *S. ransomi*, it was found that larvae in cultures that contained "baby pig substrate", which is said to be associated with a favourable internal environment, favoured direct development (Moncol and Triantaphyllou, 1978). The roles of several environmental factors influencing *S. stercoralis* have been studied by Shiwaku *et al.* (1988). Neither temperature nor degree of dilution of faeces had any significant effect on the number of male adult worms that developed. On the other hand, the numbers of female adult worms were maximal at 20–30°C and the numbers of filariform larvae were greatest at temperatures >30°C. Further, the numbers of female adult worms fell while filariform larvae increased as the faeces were diluted progressively. These results may indicate that free-living males are fixed as male in the egg stage but that potential female eggs develop into female adult worms or filariform larvae according to environmental circumstances.

7.2. Parasitic Cycle

The question of what happens to the parasite within the human host has been even more vexatious than uncertainty over external development. It has been known since the beginning of this century that *Strongyloides* filariform larvae can penetrate intact skin and this has generally been held to be the usual mode of infection. However, there is no reason to think that patent infections would not develop if infective larvae were ingested; that this can indeed happen was shown by Wilms as long ago as 1897. Unlike hookworms, *S. stercoralis* infective larvae do not retain the loose cuticle of the previous stage as a protective sheath. Nevertheless, *S. ratti* sheds a diaphanous surface coat during skin penetration (Grove *et al.*, 1987a) and this may also be true for *S. stercoralis*. It is now known that infective larvae secrete a metalloprotease that may facilitate penetration of the skin and migration through the tissues. This protease has elastase activity and catalyses the degradation of a model of dermal extracellular matrix. It has a molecular weight of 40 kDa and is immunogenic (Brindley *et al.*, 1995). Furthermore, skin invasion by larvae is prevented by metalloprotease inhibitors, thus emphasizing the importance of this enzyme as a virulence factor of *S. stercoralis* (McKerrow *et al.*, 1990).

The major questions, however, relate to the route or routes which infective larvae take once they have entered the host. The first major studies were done in 1914 in tracheotomized dogs by Fülleborn (1914) who concluded that the majority of larvae passed via the bloodstream to the lungs, ascended the respiratory tree, were swallowed, then arrived in the small bowel where they completed their development. Several years later, Yoshida (1920) studied oral infections and suggested that larvae may migrate directly through connective tissues.

More recently, Schad and his colleagues developed a technique for radiolabelling the soma of *S. stercoralis* infective larvae with [75]Se (Aikens and Schad, 1989) then used these parasites to examine the migratory route in dogs (Schad *et al.*, 1989). Direct sampling of larvae traversing the trachea found insufficient numbers to account for those reaching the trachea. Further, compartmental analysis of various organ compartments was consistent with the simpler idea that larvae may reach the duodenum by migration via other viscera or tissues. This technique was then extended to use whole body gamma camera scintigraphy of 10-day-old pups. Radioactivity spread radially and diffusely from the injection site then concentrated in the abdomen 2–7 days after infection. These data led the authors to conclude that the majority of larvae moved to the gut by means other than the respiratory route (Mansfield *et al.*, 1995). However, this technique is not really sensitive enough to exclude pulmonary migration if the larvae only spend a short time in the lungs. It seems likely that larvae may migrate via several routes, including the traditionally accepted route through the lungs.

Once the infective larvae reach the small intestine, they moult twice to become female adult worms. These worms, at least in dogs, are found in tunnels between the enterocytes and on the luminal side of the basement lamina and which have openings into the lumen of the intestine (Grove *et al.*, 1987b). Except for Kreis (1932) and Faust (1933) working together in New Orleans, no workers have ever found male parasitic adult worms and their existence is not generally accepted. The female worms are believed to produce eggs by parthenogenesis but we remain in ignorance of the precise mechanisms. Ova deposited in the intestinal lumen rapidly hatch, traverse the length of the bowel and are passed in the stools.

7.2.1. Autoinfection

The characteristic that separates *S. stercoralis* from almost all other worms that infect humans is its capacity to replicate within the host. Whereas one hookworm larva becomes one hookworm adult, or one *Ascaris* egg becomes one adult ascarid, and these worms eventually die and the infection subsides, *S. stercoralis* has the ability to replicate *in vivo* with the result that infection persists for many years; the current record appears to be 65 years (Leighton and MacSween, 1990).

How this happens still remains largely an enigma. In 1911, Gage reported a case of heavy protracted infection and postulated that autoinfection might occur by two routes, internal and external. It seems likely that a very small proportion of rhabditiform larvae in the bowel lumen develop into second- or third-stage larvae then penetrate the bowel mucosa and undergo another migratory cycle to return to the small bowel; this is internal autoinfection but our understanding of this process is surmise rather than fact. The second postulate is that other larvae, probably after moulting in the crevices of the perianal region, penetrate the perianal skin then retrace their predecesors' steps; this process is called external autoinfection.

On the basis of studies of infected dogs, Schad *et al.* (1993) described what they term an "autoinfective burst". They have confirmed that in chronic infections, the majority of intestinal larvae do not attain infectivity but are passed in the faeces as pre-infective rhabditiform larvae. However, in primary infections of immunologically naive animals, larval development proceeds without host hindrance and many larvae attain infectivity. There is a spectrum of rate of development of larvae and those that become infective may be the ones that develop at the most rapid rate. Thus for a brief period, there is hyperinfection during which the intestinal adult worm population increases sharply until gut level resistance supervenes and larval development is retarded. These findings need to be confirmed and the mechanisms ascertained.

The worm burden and rate of egg production in chronically infected individuals is uncertain. Rhabditiform larvae in the stools are generally sparse with most patients passing 100–2000 parasites per day, being equivalent to approximately 0–20 larvae per gram of faeces. In dogs infected with 1000 filariform larvae, 50–300 adult worms can usually be recovered from the intestines and 0–20 larvae per gram are seen in the faeces. Each adult worm has about a dozen eggs but the rate at which they are released is unknown. Using the foregoing figures, it has been suggested that each adult worm may produce about 10 larvae per day (Genta, 1992). Using a cycling time (the time required to pass through one complete life cycle *in vivo*) of 12 days, an average egg production of 10 eggs per day, and a worm mortality rate of 10% per year in an immunocompetent host, Genta (1992) calculated that 0.003% of rhabditiform larvae needed to transform into infective larvae (the moulting rate) to maintain a balanced population of worms. Further, he estimated that if the moulting rate increased to 3%, 100 adult worms would become 1379 after 10 cycles and for a rate of 30%, there would be 358000 adult worms after six cycles, and if the rate were 100% (which there clearly is not as many rhabditiform larvae are seen in the stools), there would be 1.5 million adult worms after four cycles. By using a spreadsheet and the same basic parameters I have also reached an estimated moulting rate of 0.003% to maintain a balance. However, I have been unable to reproduce Genta's figures for moulting rates of 3% and 30%. For example, I calculate that with a moulting rate of 3%, there would be 2 941 128 worms after 10 cycles; I think he may have applied the moulting rate once each cycle rather than to each day's larval output. Genta's estimates of cycling time and fecundity seem quite reasonable but the assessment for longevity may be a gross overestimate. For example, if the average life span of an adult worm were two weeks as in *Trichinella spiralis* in mice (Grove and Civil, 1978), the moulting rate required to maintain balance would be 0.72%. We do not know what the longevity of *S. stercoralis* adult worms is. Furthermore, the Genta model assumes that all filariform larvae are able to complete their cycle and maturation. It is conceivable (and it is the standard explanation) that many larvae may be destroyed by the host defences during migration. Table 3 indicates the moulting rates needed to maintain a balanced popula-

Table 3 Rates of moulting of rhabditiform larvae to filariform larvae required within the gut to maintain a balanced infection after five years, assuming various life-spans for adult *Strongyloides* in the bowel and assuming that 100%, 10% or 1% of migrating larvae are able to complete the cycle of development.

Life-span of adult worms	Per cent moulting rate required		
	100% survival of migrating worms	10% survival of migrating worms	1% survival of migrating worms
2 weeks	0.72	7.2	72
4 weeks	0.59	5.9	59
8 weeks	0.21	2.1	21
12 weeks	0.098	0.98	9.8
26 weeks	0.044	0.44	4.4
1 year	0.021	0.21	2.1
2 years	0.0099	0.099	0.99
5 years	0.0003	0.003	0.03

tion with a wide range of life-spans and with 100%, 10% or 1% of migrating worms being able to complete the cycle and mature.

8. THE HOST–PARASITE RELATIONSHIP

The interaction between *S. stercoralis* and its human host is a puzzle. Given the intrinsic capacity of *S. stercoralis* to replicate, there appear to be three possible outcomes in infected individuals, depending on the adequacy of the host's immune system and the parasite's ability to evade those responses.

8.1. Eradication of Infection

Although different biotypes of *S. stercoralis* with varying capacities to infect animals have been described (Galliard, 1967) as has dissipation of the ability of one strain to disseminate after passage through dogs (Genta, 1989b), it seems to me more likely that eradication of infection generally depends on the effectiveness of the infected person's immune response. There is no solid evidence to support this view but anecdotal evidence points in this direction. For example, tens of thousands of British and Australian soldiers were imprisoned under appalling living conditions for several years as prisoners-of-war on the Burma–Thailand railway. Upon release, almost all of them had hookworm infection (personal observations), yet 35 years later, only 25–30% of them had chronic strongyloidiasis (Gill and Bell, 1979; Grove, 1980). Environmental conditions were such that all of them should have been exposed to *Strongyloides* yet three-quarters of them seem to have dealt with the infection. In the same way, strongyloidiasis is significantly less common than hookworm infection among inhabitants of tropical countries, yet the manner and conditions of transmission of the two parasites are very similar. The mechanisms by which some subjects can eradicate infection are unknown but presumably reflect either innate or acquired immune processes that may well be determined genetically.

8.2. Chronic Infection

On the other hand, some people mount a partially effective immune response. They cannot eradicate the infection but they do contain the intensity of infection. Our understanding of the immune processes controlling *S. stercoralis* is still rudimentary. The relative roles of humoral and cell-mediated immunity are ill-defined and we are woefully ignorant of the precise mechanisms involved.

8.2.1. *Immunology*

Some insights have arisen from attempts to improve the serodiagnosis of infection. A number of proteins on the surface or in the excretory/secretory products of *S. stercoralis* infective larvae have been radiolabelled; some of these were immunogenic and precipitated human serum antibodies or stimulated IgE-mediated release of histamine from basophils of infected humans (Brindley *et al.*, 1988). Sato *et al.* (1990) found 33–39 bands on sodium dodecylsulphate-polyacrylamide gel electrophoresis (SDS-PAGE) analysis of *S. stercoralis* whole body larval extract. On subsequent immunoblot analysis, there was considerable variability among sera from infected patients with the number of reactive bands varying between one and more than 18. The most prominent bands were 26, 41, 66 and 99 kDa in size. In another study, three proteins of *S. stercoralis* filariform larvae of 28, 31 and 41 kDa in size were recognized on Western blots by serum IgG antibodies from infected patients (Conway *et al.*, 1993b; Lindo *et al.*, 1994). Further studies showed that these antigens were soluble in both aqueous and detergent extracts, did not share epitopes between them, were not found in *S. cebus* or *S. ratti*, and reacted with antibodies predominantly in the IgG_1 and IgG_4 but not IgG_2 and IgG_3 fractions of human serum (Conway *et al.*, 1994). Similarly, Genta and Lillibridge (1989) had earlier noted that IgG_4 antibodies were most prominent in human strongyloidiasis.

It is believed that three forms of infective larvae exist — free-living, host-adapted and host-restricted autoinfective (Brigandi *et al.*, 1994). If mice are infected with *S. stercoralis* larvae, the host-adapted but not host-restricted forms are found. Such mice generate IgG_1, IgA, IgM and IgE antibodies to larval antigens. Only some of these antigens are common to free-living, host-adapted and host-restricted larvae and immunity in mice exposed to free-living larvae does not cross-protect against host-restricted autoinfective forms introduced into mice (Brigandi *et al.*, 1994)

In an animal model in which Balb/cBYJ mice were immunized with infective *S. stercoralis* larvae, 97% of parasites in a challenge infection were killed within 24 hours and this effect was dependent on the presence of eosinophils (Rotman *et al.*, 1994). When *S. stercoralis* infective larvae were placed in diffusion chambers constructed with membranes containing pores varying from 0.1 to 2.0 μm in size and were then implanted in previously immunized mice, larvae were only killed if there was contact between leucocytes and parasites (Abraham *et al.*, 1995). Studies *in vitro* have shown that living *S. stercoralis* filariform larvae activate the complement system by both classical and alternate pathways and that activated complement, in the absence of specific antibody, facilitates adhesion of peripheral blood mononuclear and polymorphonuclear cells to larvae. Complement components C1q, C3, C4, C8 and properdin can be found on the larval surface and larvae coated with white cells lose their motility (de Messias *et al.*, 1994). The complement system in association with effector cells may therefore play an important non-specific role in the first line of defence against migrating parasites.

Lymphocyte activity was assessed in 64 patients with human strongyloidiasis by Sato and Shiroma (1989b) in Okinawa. They found a significant increase of $CD4^+$ and $OKIa1^+$ cells, a relative decrease in $CD8^+$ cells, increased spontaneous mitogenesis and interleukin 2 production, but a lowered lymphoproliferative response to mitogens. Unfortunately, more than half the patients with strongyloidiasis in Okinawa are concurrently infected with HTLV-1 (see below) and in this paper infection with this retrovirus was not delineated. These findings may therefore reflect the retroviral infection rather than strongyloidiasis. However, Sato and Shiroma (1989a) did show that lymphoproliferative responses to mitogens were decreased and unstimulated lymphoproliferative activity was increased in patients with strongyloidiasis who did not have HTLV-I infection compared with controls, but they still did not dissect out the effects on lymphocyte populations as assessed by cell markers. More detailed investigations are needed, particularly as other authors have also reported increased levels of interleukin 2 in strongyloidiasis (Josimovic-Alasevic *et al.*, 1988).

Even more mysterious are the means by which larvae that are circulating in the tissues and bloodstream are able to evade the cellular and humoral defence mechanisms in these patients. Although subtle increases in mast cell numbers occur in the intestine of *Erythrocebus patas* monkeys infected with *S. stercoralis* (Barrett *et al.*, 1988), parasites in the human small intestinal mucosa elicit little in the way of an inflammatory response thus implying that in some way they are able to hide from or suppress immune mechanisms. In patients with chronic infections, a balance is reached with neither the parasite nor the patient gaining the upper hand.

The points at which defence mechanisms could be operative in controlling worm numbers include:

1. shortened life-span of adult worms in the bowel;
2. reduced fecundity of adult worms in the bowel;
3. impaired transformation of rhabditiform into infective larve *in vivo*
4. shortened survival of infective larvae in the bowel lumen;
5. destruction of filariform larvae migrating through the tissues;
6. failure of development of third-stage larvae into parasitic adult female worms.

8.2.2. *Small Bowel Function*

Two papers have addressed the effect of *S. stercoralis* infection on small bowel function and small bowel bacterial overgrowth. Duodenal juice and faeces from 54 Russians with strongyloidiasis, 37 of whom lived in a temperate area and 17 of whom were from the tropics were studied. Enterokinase activity was reduced in 60.7% and 80% of patients from these two areas, respectively. Similarly in faeces, there was an 87.8% and 71.4% reduction in alkaline phosphatase activity (Tikhomirova and Lysakova, 1989). Reduced enzyme activity could play a part in malabsorption in strongyloidiasis but it is not clear from the abstract how well these studies were controlled. In any event, normalization did not occur 1–6 months after treatment.

Small bowel bacterial growth, which also has the potential for contributing to malabsorption, was studied in 12 patients and 11 control subjects in Brazil. A catheter was placed 20 cm into the jejunum, cholecystokinin was injected intravenously, then intestinal juice was aspirated and bacterial numbers quantified. Significantly increased numbers of both aerobic and anaerobic organisms were found in patients with strongyloidiasis (Sipahi *et al.*, 1991).

8.3. Hyperinfection and Disseminated Infection

Under some circumstances, defences are disturbed and the balance is tipped in favour of the parasite. Replication of worms exceeds destruction of offspring and worm numbers increase enormously and are detectable in a variety of tissues. Some authors have termed this process "hyperinfection" although the term was originally used by Faust (1930) to denote autoinfection. When large numbers of worms are present, they are often easier to find in the tissues and this has variously been labelled as "disseminated", "massive" or "overwhelming" infection, although the term "disseminated" has been restricted by some authorities to infections in which adult worms are found in ectopic sites. A simpler approach, recognizing that there is a spectrum of severity of infection, and that it is often impossible to quantify it precisely, is simply to categorize disease as "uncomplicated strongyloidiasis" or as "severe, complicated strongyloidiasis" (Grove, 1989).

Although severe infections were described sporadically in the earlier parts of this century, it is only in the last 20 years that the condition has become increasingly recognized and its association with immunosuppression, especially impairment of cell-mediated immunity, appreciated. Underlying illnesses in patients with severe, complicated strongyloidiasis include lymphoma, acute and chronic leukaemia, carcinoma, chronic glomerulonephritis, nephrotic syndrome and renal transplantation, chronic lung disease, systemic lupus erythematosus, idiopathic thrombocytopenic purpura, polymyositis, hypercalcaemia, eye diseases, skin diseases including burns and leprosy, chronic alcoholism, malnutrition and retroviral infections. In terms of reported cases, by far the most common underlying condition has been in patients who have undergone renal transplantation (DeVault et al., 1990).

8.3.1. Histopathology

A detailed autopsy examination of seven patients with severe, complicated strongyloidiasis has been reported recently (Haque et al., 1994). Six patients had received corticosteroid therapy (two asthma, two lymphoma, one renal transplant, one adenocarcinoma of the lung) whereas the seventh was not given immunosuppressive therapy but had AIDS. High worm burdens were noted in those who had not received anthelmintic treatment or had been treated for only a short period. In the intestines, parasites were seen at all levels but the highest burdens were found in the proximal jejunum. Mucosal penetration by filariform larvae was most prominent in the distal small bowel and proximal large intestine. Most of these larvae were seen in intestinal lymphatics and were also highly concentrated in the mesenteric and retroperitoneal lymph nodes. There was minimal inflammation in the lamina propria but granulomas were sometimes seen in the deeper layers of the intestine. Larvae were not seen in the spleen and very few were found in the liver, adrenals, lower urinary tract and heart. Large numbers of filariform larvae were seen in the lungs; some of these were not surrounded by an inflammatory reaction but others were associated with neutrophils (in areas of bronchopneumonia) or small granulomas. On the basis of these observations, the authors postulated that during autoinfection, larvae traverse the intestinal lymphatics to the thoracic duct, then pass via the bloodstream to the lungs, penetrate the alveoli and ascend the airways.

8.3.2. *Role of Immunosuppression*

Although impaired immunity has generally been supposed to be the basis for severe, complicated infections (Igra-Siegman *et al.*, 1981; Genta, 1986; Grove, 1989), Genta (1992) has argued against immunity being important on the grounds that:

1. there have been few reports of disseminated infection in protein-calorie malnutrition (a major cause of impaired immunity);
2. disseminated infection is not common in lepromatous leprosy unless patients had been given corticosteroids for treatment of lepra reactions;
3. disseminated infection has been uncommon in renal transplantation since the introduction of cyclosporin;
4. disseminated infection is not prominent in AIDS or infection with HTLV.

It needs to be said, however, that there may be reasons for these apparent contradictions. Severe, complicated strongyloidiasis in the context of protein-calorie malnutrition is likely to occur in areas where medical services are poor and reporting is non-existent, cyclosporin given for renal transplantation may well have an anti-*Strongyloides* effect (Schad, 1986) and such infections do occur in retroviral infections, but may well be grossly under-reported in endemic areas for both HIV plus *S. stercoralis* for the reasons alluded to above.

8.3.3. *Corticosteroids*

Genta (1992) has advanced an alternative hypothesis to explain the onset of dissemination. He has proposed that administration of corticosteroids and their subsequent metabolism may result in increased production of ecdysteroid-like molecules. Ecdysteroids are moulting hormones that control moulting in insects and perhaps in helminths. Increased quantities of these substances may upregulate the moulting rate leading to an increased worm burden and disseminated infection.

This is an interesting theory worthy of investigation. At the moment, it remains speculative at best. In his support for this thesis, Genta (1992) has mentioned the difficulty of inducing disseminated infection in experimental dogs with corticosteroids alone unless azathioprine is added; this also was my experience (Grove *et al.*, 1983). I would have thought, however, that this was an argument in favour of immunity rather than an ecdysteroid-like effect of corticosteroids. Furthermore, in the 180 cases reported in the literature up to 1988, 53 apparently did not receive corticosteroids (Grove, 1989). Indeed, Neva (1993) remarked that there have been many cases of severe strongyloidiasis in the Caribbean who were not given corticosteroids and speculated that hyperinfection was likely to be due to co-infection with HTLV-I (see below) which had suppressed immunity. Of course, these two theories are not mutually exclusive and it is possible that both immunity and ecdysteroids are important.

8.3.4. *Human Immunodeficiency Virus Infection*

When the acquired immune deficiency syndrome (AIDS) was first related to infection with human immunodeficiency virus (HIV) and was associated with greatly reduced numbers of $CD4^+$ lymphocytes, it was anticipated that disseminated strongyloidiasis would be a feature of AIDS as a consequence of impaired cell-mediated immunity. Despite the fact that HIV and *S. stercoralis* co-exist in many countries, particularly in Africa, there have been very few reported cases of severe, complicated strongyloidiasis in AIDS. So much so, in fact, that disseminated strongyloidiasis was withdrawn as an opportunistic infection in the Center for Disease Control's criteria for the definition of AIDS (Anonymous, 1987) and has been cited as a "missing infection" in AIDS (Lucas, 1990). As already indicated, this has been used as an argument to advance the hypothesis that disseminated strongyloidiasis is not due to impaired immunity (Genta, 1992). Furthermore, intestinal strongyloidiasis does not appear to be any more common in HIV-infected individuals than in individuals without the latter infection (Conlon *et al.*, 1990; Dias *et al.*, 1992).

It is not true to say that there have been no cases of disseminated strongyloidiasis in AIDS. At least 24 such patients have been described, the vast majority of whom did not receive corticosteroid therapy (Maayan *et al.*, 1987; Vieyra-Herrera *et al.*, 1988; Armignacco *et al.*, 1989; Schainberg *et al.*, 1989; Dutcher *et al.*, 1990; Glezerov and Masci, 1990; Kramer *et al.*, 1990; Stey *et al.*, 1990; Gompels *et al.*, 1991; Harcourt-Webster *et al.*, 1991; Batista *et al.*, 1992; Couprie *et al.*, 1993; Lessnau *et al.*, 1993; Makris *et al.*, 1993; Morgello *et al.*, 1993; Torres *et al.*, 1993; Celedon *et al.*, 1994; Jain *et al.*, 1994; Takayanagui *et al.*, 1995). Nevertheless, the reported frequency of systemic strongyloidiasis in AIDS is much less than might have been predicted. Perhaps some of this is apparent rather than real given the nature of health services in areas in which both infections are common.

Even if it is accepted that the incidence of disseminated strongyloidiasis is relatively low in AIDS, it does not necessarily follow that impaired immunity is not important in the genesis of complicated strongyloidiasis. Rather, one can only say that the component of the immune system that is depressed in AIDS is not particularly important in the array of defences against *S. stercoralis*. Such partitioning of immune mechanisms with differential effects on different classes of pathogens is well recognized: encapsulated bacterial infections in splenectomy; tuberculosis, *Nocardia* and viral infections in impaired cell-mediated immunity; giardiasis and bacterial infections in hypogammaglobulinaemia (poor humoral immunity); and bacterial and fungal infections in neutropenia.

8.3.5. *Human T cell Lymphotropic Virus Infection*

This virus is a retrovirus that has been shown to cause adult T cell leukaemia and tropical spastic paraparesis (also known as HTLV-I-associated myelopathy). It is endemic in a number of areas but is especially prevalent in parts of Japan and the Caribbean. Nakada *et al.* (1987) reported that in Japan the incidence of antibodies to this virus was increased in patients with chronic strongyloidiasis. This association has been confirmed in Japan by some investigators (Hanada *et al.*, 1989; Sato *et al.*, 1994) but denied by Arakaki *et al.* (1992a,b) who used agar-plate detection of parasites rather than the presence of antibody for the diagnosis of strongyloidiasis. Neither Neva *et al.* (1989) nor Robinson *et al.* (1994) could find an association between seropositivity for *Strongyloides* and seropositivity for HTLV-I, although the latter authors did observe that faecal larvae were found more frequently in patients with concurrent HTLV-I infection. Many cases have now been described from the Caribbean, South America, Japan and Africa who have had both T-cell leukaemia or lymphoma and strongyloidiasis, with many of the latter being severe, complicated infections (Rio *et al.*, 1990; Phelps *et al.*, 1991; Kudeken *et al.*, 1992; Newton *et al.*, 1992; Patey *et al.*, 1992; Blank *et al.*, 1993; Plumelle *et al.*, 1993; Adachi *et al.*, 1994; D'Incan *et al.*, 1994). It seems likely that, at the least, HTLV-I infection predisposes to more severe infections with *S. stercoralis* (Brosset *et al.*, 1991), perhaps via depression of cell-mediated immunity or IgE responses (Matsumoto *et al.*, 1990; Newton *et al.*, 1992; Robinson *et al.*, 1994). In addition, it has been suggested that strongyloidiasis may be a co-factor that promotes evolution of disease due to HTLV-I (Nakada *et al.*, 1987; Yamaguchi *et al.*, 1987). Finally, patients with concurrent HTLV infection seem less responsive to anthelmintic therapy (Sato *et al.*, 1994).

8.4. Animal Models

The inability of *S. stercoralis* to develop in most small animals has been a major impediment to our understanding of the pathogenesis of infection with this parasite. A number of non-human hosts have now been described.

8.4.1. *Dogs*

Dogs have long been known to be susceptible to infection and have been used intermittently as experimental models for 60 years. Interest was revived in the 1980s (Grove and Northern, 1982; Grove *et al.*, 1983; Schad *et al.*, 1984). Recently, infection in IgA-deficient dogs has been studied; no differences in either the severity or course of infection were found (Mansfield and Schad, 1992a).

8.4.2. *Cats*

These animals have varying susceptibility to *S. stercoralis*. Such infections have not been studied in any depth.

8.4.3. *Ferrets*

Ferrets (*Mustela putorius furo*) do not appear to be susceptible to infection under normal circumstances. However, after treatment with methylprednisolone, patent infections developed but dissemination of infection did not occur (Davidson, 1988).

8.4.4. *Gerbils*

It has been shown recently that gerbils (*Meriones unguiculatus*) can be infected with *S. stercoralis* (Nolan *et al.*, 1993). Rhabditiform larvae and adult worms appear to have been recovered from male gerbils for at least 131 days after infection. On the other hand, in the case of female gerbils, intestinal worms were not seen after day 70. The authors interpreted the findings in male gerbils as indicating a life span of adult worms of greater than 131 days as they did not find any circulating autoinfective larvae. This may be true but does not agree well with the observations in female gerbils. Furthermore, low-grade autoinfection may well have been occurring at a level below the sensitivity of the necropsy in finding migrating larvae. Administration of methylprednisolone resulted in greatly increased numbers of worms. Finally, rare trans-mammary transmission by lactating females to their pups was demonstrated.

8.4.5. *Subhuman Primates*

Gibbons (*Hylobates lar*) that had been splenectomized developed overwhelming infections (De Paoli, 1974). Infections in the monkey, *Erythrocebus patas*, sometimes result in disseminated infections (Harper *et al.*, 1984).

9. CLINICAL FEATURES

9.1. Chronic Uncomplicated Strongyloidiasis

Many patients chronically infected with *S. stercoralis* are asymptomatic. Others have a variety of symptoms, mostly referable to the skin or gastrointestinal system. The frequencies of various symptoms in patients with chronic strongyloidiasis but not other intestinal parasitic symptoms and in two types of control subjects are shown in Table 4.

Table 4 Frequencies of symptoms in an unselected series of patients with chronic strongyloidiasis expressed as a percentage and compared with two control groups (adapted from Grove, 1980). All subjects were veterans who had been prisoners-of-war in World War II.

	Prisoners in southeast Asia		Prisoners in Europe
	Strongyloidiasis proven ($n = 44$)	Strongyloidiasis not proven ($n = 114$)	($n = 45$)
Abdominal pain	57	41	30
Heartburn	50	33	33
Indigestion	73	46**	44
Reflux	39	39	40
Anorexia	25	19	20
Nausea ± vomiting	23	20	23
Diarrhoea	45	23**	16*
Constipation	2	17*	16
Pruritus ani	59	43*	40*
Weight loss	23	11*	5*
Urticaria	66	3***	0***
Larva currens	30	0***	0**
Chest pain	9	16	30
Cough	34	29	30
Dyspnoea	32	45	51
Malaise/weakness	43	51	47
Nervousness	43	54	30

Compared with men with proven strongyloidiasis, * = $P<0.05$, ** = $P<0.005$, *** = $P<0.0005$ (χ^2 test)

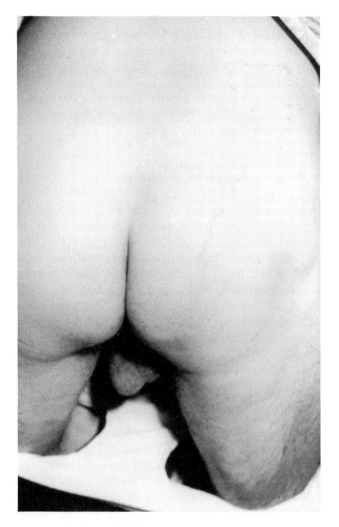

Figure 4 Larva currens: the pathognomonic sign of *Strongyloides* infection.

Two types of skin lesions occur. Larva currens (Figure 4) is pathognomonic; an urticarial rash migrates in a serpiginous fashion at the rate of several centimetres per hour for up to 1–2 days. This pattern may be seen more frequently in patients infected with Asian compared with European strains of the parasite (Genta *et al.*, 1988). More common than larva currens is a non-specific urticarial rash in which crops of stationary wheals lasting 1–2 days, appear, particularly around the waist and on the buttocks.

Gastrointestinal symptoms include indigestion, cramping lower abdominal pains, intermittent or persistent diarrhoea, pruritus ani, and sometimes weight loss. These patients may be misdiagnosed as having irritable bowel syndrome. Miscellaneous symptoms include cough and arthralgias, although it is often difficult to be certain these symptoms are due to *S. stercoralis* infection. Co-existent asthma is not more common in strongyloidiasis (Leeman and Cabrera, 1995) and does not improve after treatment for strongyloidiasis (Wehner *et al.*, 1994).

9.2. Severe Complicated Strongyloidiasis

Just as there is a range of symptoms in chronic, uncomplicated strongyloidiasis, so there is a spectrum of severity in the more significant and complicated forms of disease. Severe disease has had various labels applied to it, including disseminated strongyloidiasis, overwhelming strongyloidiasis, hyperinfection and massive strongyloidiasis. These heavy infections have protean manifestations depending on the intensity of infection and the organs involved, as well as upon the presence or absence of secondary bacterial infection. The major targets of infection are the bowel, lungs and central nervous system. The mortality in patients with massive infections is high, reflecting not only the parasitic infection, but also the underlying condition which predisposes to dissemination.

9.2.1. *Gastrointestinal Complications*

A variety of gastrointestinal syndromes have been described, including a sprue-like syndrome with steatorrhoea, protein-losing enteropathy, hypoalbuminaemia and generalized oedema (Sullivan *et al.*, 1992). Patients may complain of flatulence, nausea, abdominal distension, frequent foul-smelling stools, and weight loss or puffiness of the face and ankles. Some patients have features of intestinal obstruction with fever, tachycardia, hypotension and abdominal distension, tenderness and absent bowel sounds indicative of paralytic ileus (Bannon *et al.*, 1995). Laparotomy reveals dilated and thickened loops of jejunum without evidence of mechanical obstruction; this has been described as pseudo-obstruction. Other presentations include necrotizing jejunitis, arteriomesenteric occlusion (Lee and Terry, 1989), small bowel infarction (Kennedy *et al.*, 1989), papillary stenosis with biliary obstruction (Delarocque-Astagneau *et al.*, 1994), aphthoid ulceration of the colon (Stoopack and Raufman, 1991), massive upper (Bhatt *et al.*, 1990) and lower gastrointestinal haemorrhage and anorectitis (Bili *et al.*, 1991).

Achlorhydria (often brought about by treatment with histamine–2 blockers or proton pump inhibitors) may facilitate gastric strongyloidiasis (Wurtz *et al.*, 1994). In some patients, strongyloidiasis may be part of a polymicrobial infection. For example, in a series of children with strongyloidiasis who died in a refugee camp in Thailand while suffering from bloody diarrhoea and abdominal pain, concurrent infections with hookworm or possibly *Clostridium perfringens* type C may have played a role (Boyajian, 1992; Coninx, 1993).

9.2.2. *Pulmonary Complications*

Patients generally complain of an irritative or productive cough and shortness of breath, often associated with wheezing. Examination usually reveals scattered crackles and wheezes. In extreme cases, respiratory failure develops and may require intubation and ventilation. Indeed, one case has been described in which fatal adult respiratory distress syndrome appeared to develop after successful therapy of the parasitic infection (Thompson and Berger, 1991). It has been claimed that *S. stercoralis* causes granulomatous lung disease leading to fibrosis (Lin *et al.*, 1995), but it is possible that *Strongyloides* infection was coincidental in a patient with interstitial lung disease.

9.2.3. *Neurological Complications*

Invasion of the central nervous system by migrating larvae is often accompanied by secondary bacterial infection. This may result in meningitis or brain abscess causing any combination of fever, headache, nausea, vomiting, neck stiffness, or convulsions or coma. Lumbar puncture may reveal evidence of bacterial meningitis with increased neutrophils and protein concentration but a reduced glucose level in the cerebrospinal fluid.

9.2.4. *Other Presentations*

Urinary tract infections are common. Septicaemia due to enteric organisms including coliforms, anaerobes and *Pseudomonas aeruginosa* may result in shock. Pelvic inflammatory disease has been ascribed to *S. stercoralis* (Young *et al.*, 1989), but since the patient had concomitant gonococcal infection, this conclusion must be doubted. Reactive arthritis combined with uveitis was attributed to strongyloidiasis since anthelmintic treatment resulted in prompt improvement (Patey *et al.*, 1990), but this may have been a misdiagnosis of Reiter's syndrome with coincidental *Strongyloides* infection. Invasion of the skin by larvae may cause petechial or purpuric lesions (Gordon *et al.*, 1994; Chaudhary *et al.*, 1994).

9.2.5. *Recent Trends*

In the past six years, over 50 case reports of severe, complicated strongyloidiasis have appeared. These have added little to our understanding of the syndrome although they have heightened individual practitioners' awareness of the condition. Interestingly, only three of these reports were of patients who developed severe disease in the context of renal transplantation, despite the fact that this was the most common underlying condition in cases reported up till 1988. Whether this indicates a changing pattern is uncertain. Changes in management of transplanted patients including less use of corticosteroids and more use of cylosporin (with its possible anti-*Strongyloides* activity) may have reduced the frequency of this complication. On the other hand, practitioners may not be reporting the condition any more or authors may have difficulty in getting their papers accepted as the material presented is no longer novel. Apart from patients with HIV or HTLV-I infections, most of the patients reported received corticosteroids; underlying conditions included polyarteritis, temporal arteritis, systemic lupus erythematosus, rheumatoid arthritis, lymphoma, multiple myeloma, carcinoma, asthma, chronic obstructive airways disease, interstitial pulmonary fibrosis, uveitis, Bell's palsy, diabetes mellitus and cirrhosis.

10. DIAGNOSIS

The classic triad of diarrhoea, abdominal pain and urticaria is suggestive of a diagnosis of strongyloidiasis and clues are given by the discovery of an eosinophilia or suggestive radiological findings and supported by serological tests (Figure 5). These pointers are of less value in patients living in endemic areas where polyparasitism is rife (de Messias *et al.*, 1987). In all patients, the key is demonstration of the parasite.

10.1. Morphological Identification of *Strongyloides* Infecting Humans

Parasites are usually found in faeces but sometimes they are seen in other body fluids or tissue samples. Identification is most accurate when adult as well as larval stages are available for examination but this is rarely possible with clinical specimens.

The parasitological diagnosis is usually made after examination of faeces. Most clinical laboratories will identify *S. strongyloides* simply upon observation of rhabditiform larvae in faeces and will usually be right. It has been suggested that rhabditiform larvae of *S. stercoralis* lash

Suspected because of:

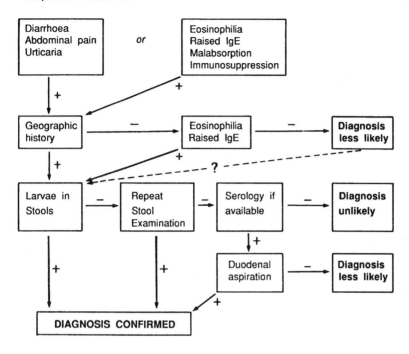

Figure 5 Flow chart for the diagnosis of strongyloidiasis. (From Grove, 1989.)

out like a whip whereas those of hookworm glide sinuously like a snake (Gustavo dos Santos, 1993). However, if there is any doubt as to the speciation of *Strongyloides* or differentiation from rhabditiform larvae of hookworm that may have developed in old stools, then culture *in vitro* and inspection of free-living adult worms is necessary. It needs to be remembered that the intestinal worm burden is often small and worms may not be found in a cursory examination of a small quantity of faeces. Repeated examinations of stool specimens improve the chances of finding parasites (Grove, 1980; Nielsen and Mojon, 1987). Various techniques have been used to demonstrate the worms including direct smear of faeces in saline, Baermann concentration, formalin–ether concentration (Ritchie), commercial concentrator kits, test-tube culture (Harada-Mori), culture on wash glasses and nutrient agar plate culture. There are conflicting reports on the relative yields of the various methods of examining faeces (Grove, 1980; Arakaki *et al.*, 1990; Perry *et al.*, 1990; Assefa *et al.*, 1991; Koga *et al.*, 1991; Koga *et al.*, 1992; Mahdi *et al.*, 1993; de Kaminsky, 1993; Sukhavat *et al.*, 1994); the balance of opinion probably favours the agar plate culture method but this is perhaps more expensive and complex.

Table 5 Key for the identification of species of *Strongyloides* in humans (adapted from Speare, 1989)

Freshly voided faeces

1. Thin-shelled eggs (50–60 μm × 30–40 μm) typical of *Strongyloides*, containing morula or larva — *S. fuelleborni*

   Larva typical of *Strongyloides* — 2

2. Rhabditiform larva — *S. stercoralis*

   Filariform larvae — *S. stercoralis*

Free-living female adult cultured *in vitro*

3. Vaginal axis rotated posteriorly — 4

   Vaginal axis not rotated; no post-vulval narrowing — *S. stercoralis*

4. Degree of rotation 90–105° — *S. stercoralis or S. fuelleborni*

   Degree of rotation >105°; post-vulval narrowing — *S. fuelleborni*

Free-living male adult cultured *in vitro*

5. Tip of spicule pointed — 6

   Tip of spicule not pointed — other

6. Ventral membrane of spicule straight — 7

   Ventral membrane of spicule convex or concave — other

7. Anterior adanal papilla dorsal to the line between the subventral preanal and posterior adanal papillae — *S. fuelleborni*

   Anterior adanal papilla on the line between the subventral preanal and posterior adanal papillae — *S. stercoralis*

Parasitic female

8. Ovary directly recurrent — 9

   Ovary spiral — 10

9. Stoma hexagonal on apical view — *S. stercoralis*

   Stoma not hexagonal on apical view — other

10. Stoma modified X shape on apical view — 11

    Stoma not modified X shape on apical view — other

11. Tail bluntly rounded — *S. fuelleborni*

    Tail narrowly tapered or pointed — other

Parasite in lung or sputum

12. Parasitic female (confirm identity as above) — *S. stercoralis*

Table 5 Continued.

immature stages only	13
13. Eggs (60–75 μm × 30–40 μm)	*S. stercoralis*
Larvae	14
14. Rhabditiform larvae	*S. stercoralis*
Filariform larvae	15
15. Larva < 500 μm	*S. stercoralis*
Larva > 500 μm	*S. fuelleborni* or *S. stercoralis*
Parasites in tissues	
16. Larvae	17
Adults	19
17. Rhabditiform larvae	*S. stercoralis*
Filariform larvae	18
18. Filariform larvae in lung only	*S. stercoralis* or *S. fuelleborni*
Filariform larvae in other tissues, especially gut wall and mesenteric lymph nodes	*S. stercoralis*
19. Parasitic female with spiral ovaries	*S. stercoralis*
Spiral ovaries not seen	*S. stercoralis or S. fuelleborni*

Techniques that are claimed to facilitate culturing clean larvae from faeces have been described by Ndalahwa and Wambo (1989) and van Swinderin *et al.* (1994). Likewise, there is uncertainty concerning the relative merits of looking at faeces versus duodenal fluid (Grove, 1980; Goka *et al.*, 1990). Parasites found in duodenal fluid of patients with strongyloidiasis are usually rhabditiform larvae. On the other hand, filariform larvae are the usual forms noted in sputum, bronchial lavage washings (Williams *et al.*, 1988), blood (Onuigbo and Ibeachum, 1991) or other body fluids although rhabditiform larvae have been seen in vaginal swabs (Murty *et al.*, 1994). Sometimes parasites are seen only in biopsy or autopsy specimens; in these circumstances, there is less confidence in the specificity of identification.

A key for the identification of larvae in faeces, free-living adults cultured from faeces, parasitic female worms isolated from the small bowel, larvae in lungs or sputum, and parasites elsewhere in the tissues is presented in Table 5. This key assumes that only *S. stercoralis* and *S. fuelleborni* will be

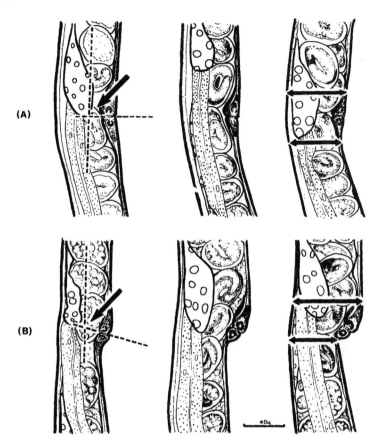

Figure 6 Perivulval regions of free-living female worms showing range of variation. Transverse arrows indicate sites for determining body diameter before and after vulva. Oblique arrows indicate angle measured between longitudinal axis of anterior body and line of vagina. (A) *S. stercoralis*. (B) *S. fuelleborni*. (From Speare, 1989.)

found in humans and cannot differentiate between *S. fuelleborni* and *S. fuelleborni kellyi* of Papua New Guinea.

Morphological features useful for identification include the angle of vaginal rotation and presence or absence of post-vulval narrowing in adult free-living females (Figure 6); the natures of the tip and ventral membrane of the spicules (Figure 7) and the locations of the peri-anal papillae (Figure 8) in adult free-living males; and the presence or absence of spiralling of the ovaries (Figure 9) and shapes of the stoma (Figure 10) and tail (Figure 11) of parasitic female adult worms.

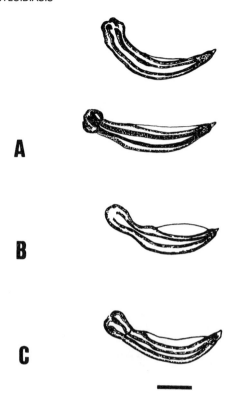

Figure 7 Spicules of the free-living male worms of *Strongyloides* showing the shape of the ventral membrane. A, straight, upper, *S. westeri*, lower, *S. stercoralis*. B, convex, *S. cebus*. C, concave, *S. papillosus*. All the tips are pointed. (From Speare, 1989.)

10.2. Immunodiagnosis

Serological diagnosis using an enzyme-linked immunosorbent assay is now available in many centres. Sensitivity and specificity may both be improved by pre-incubation of serum with *Onchocerca gutturosa* antigens (Conway *et al.*, 1993a; Lindo *et al.*, 1994). A gelatine particle test which is simple to perform, does not require specialized equipment, and has comparable efficacy has been described by Sato *et al.* (1991).

There has been considerable confusion about the applicability of sero-diagnosis when applied to individuals and to populations. Genta (1988) claimed that the ELISA was 88% sensitive, 99% specific, and had positive and negative predictive values of 97% and 95%, respectively, but was then roundly criticized for using incorrect statistics, and also for failure to take

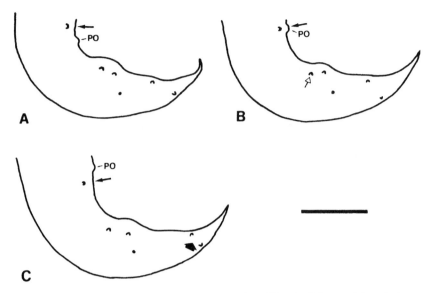

Figure 8 Positions of the subventral pre-anal papillae (solid arrow) in relation to the pre-anal organ of free-living male worms of *Strongyloides*. A, anterior, *S. serpentis*. B, level with, *S. papillosus*. C, posterior, *S. stercoralis*. PO = pre-anal organ. Broad arrow indicates subventral and subdorsal post-anal papillae in close proximity. Scale bar = 30 µm. (From Speare, 1989.)

into account the prevalence in the population (Lo and Kajioka, 1989). In another study of Indochinese refugees in Canada, these figures were calculated as 95%, 29%, 30% and 95%, respectively (Gyorkos *et al.*, 1990).

In fact, the problem may not be completely soluble in strongyloidiasis. The parameters used to calculate these four measures are shown in Table 6. The difficulty in calculating them in strongyloidiasis relates to our inability to properly define the benchmarks. Demonstration of a parasite is proof-positive of infection. On the other hand, failure to find a parasite does not exclude infection as larvae are often sparse. Thus, sensitivity of serodiagnosis can be well-delineated as the number of positive serological tests in parasitologically proven patients can be easily determined. Specificity can be estimated if the serodiagnostic test is applied to a population in which strongyloidiasis in not endemic as it can be safely assumed that they are really uninfected, but this cannot necessarily be extrapolated to the specificity in an endemic area as concomitant infections with other parasites may cause false-positive cross-reactions. However, it may not be possible to calculate positive or negative predictive values with any degree of certainty as it is not possible to truly define the numbers of infected and non-infected individuals in a population. Furthermore, these values will

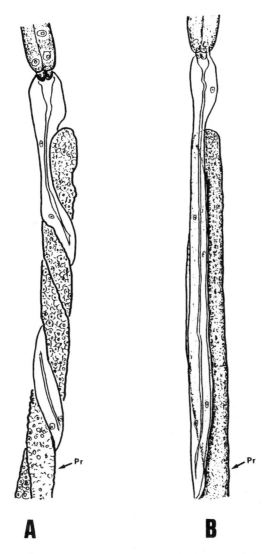

A   **B**

Figure 9 Ovary of parasitic female worms. A, spiral, *S. ransomi*. B, directly recurrent, *S. stercoralis*. (From Speare, 1989.)

change according to the prevalence of infection in the population; as the prevalence falls, the positive predictive value drops dramatically (Lo and Kajioka, 1989; Altman, 1991). Despite all these reservations, Joseph *et al.* (1995) have described a complex mathematical approach which they believe may be useful in this situation.

Finally, demonstration of antibodies, even when correct, does not

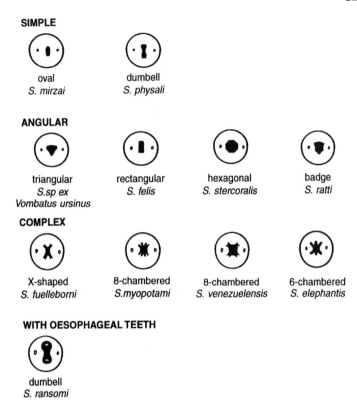

Figure 10 Stomal shapes of parasitic female adult *Strongyloides*. *En face* view. (From Speare, 1989.)

differentiate between past and present infection and it is often difficult with many patients to know whether or not low-level autoinfection is continuing. The major value in serology is the provision of a screening test which, if positive, can stimulate further searches for the parasite.

10.3. Imaging

Most patients have normal chest radiographs although the occasional patient has infiltrates consistent with Loeffler's syndrome. In patients with massive strongyloidiasis, there may be extensive, diffuse pulmonary opacities (Woodring *et al.*, 1994). In patients with severe intestinal infections, loops of dilated small bowel may be seen on a plain abdominal radiograph. Most patients have normal barium meal examinations but in heavy infections, duodenal dilatation or duodenal stricture with ulceration

BLUNTLY ROUNDED

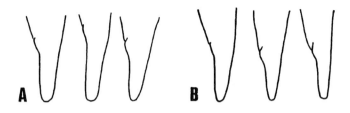

NARROWLY TAPERED

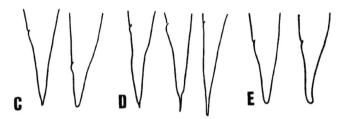

Figure 11 Lateral views of the tails of parasitic female adult *Strongyloides*. A, *S. papillosus*; B, *S. fuelleborni*; C, *S. ransomi*; D, *S. felis*; E, *S. stercoralis*. (From Speare, 1989.)

Table 6 Statistical methods for assessment of the validity of serological tests for the diagnosis of strongyloidiasis.

sensitivity:	a/a+c
specificity:	d/b+d
positive predictive value:	a/a+b
negative predictive value:	d/c+d

Serological test	Proven strongyloidiasis	
	Positive	Negative
Positive	a	b
Negative	c	d

have been described; the pipestem appearance may reverse after successful treatment (Medina *et al.*, 1992). Similarly, barium enema examination is usually normal although rarely localized or generalized colitis has been seen. It has been claimed in one patient that *S. stercoralis* caused focal hepatic lesions (Rawat and Simons, 1993) but the lesions shown may have been either amoebic or pyogenic in origin.

10.4. Endoscopy

The duodenum looks normal in the vast majority of infected patients with uncomplicated disease although there have been occasional reports of gastritis or duodenitis. In patients with severe disease, flattening of the duodenal folds, mucosal oedema and ulcerations may be seen (Chen *et al.*, 1994). Sigmoidoscopy and colonoscopy are generally normal but again there have been occasional reports of ulceration. In each of these unusual cases, it is difficult to be certain that the abnormalities observed are actually due to *Strongyloides* infection.

11. TREATMENT

The aim of treatment in most worm infections is to reduce the number of worms to the point where the infection is unlikely to cause disease. This simple approach is not applicable in strongyloidiasis as any parasites remaining after administration of an anthelmintic have the capacity to replicate and build up the worm burden again. Since the parasites are often difficult to detect in the first place, the problem is compounded frequently by uncertainty in determining whether or not they have all been eradicated. *S. stercoralis* is relatively resistant to anthelmintics and most attention has focused on benzimidazole agents and ivermectin.

11.1. Benzimidazoles

Drugs belonging to this class of anthelmintics appear to act by binding to tubulin and disrupting the assembly of microtubules, and by altering transmembrane proton discharge.

11.1.1. *Thiabendazole*

This drug has been the traditional agent used to treat strongyloidiasis. It is unreliable in its efficacy and significantly toxic. It is has been suggested in one study that improved cure rates are achieved with recurrent courses of treatment but the side-effect rate was high and nearly half the patients dropped out of the trial (Oyakawa *et al.*, 1991). Antibody levels often change little during the first year after treatment (Grove, 1982b; Kobayashi *et al.*, 1994) but tend to fall after 3–4 years (Pelletier *et al.*, 1988). Consequently, this test is usually of little benefit in assessing the effectiveness of treatment. A major problem in the treatment of patients with disseminated strongyloidiasis, many of whom have vomiting or intestinal obstruction, has been the unavailability of a preparation of thiabendazole suitable for intravenous administration. In one patient, rectal administration of thiabendazole suspension (1.5 g in 15 ml) as a retention enema twice daily for 14 days was successful (Boken et al., 1993).

11.1.2. Mebendazole

This drug has had a disappointing record when given for several days. This was confirmed by Beus (1989) who compared mebendazole given for 5 days with thiabendazole administered for 2 days; efficacy after three months was 44% and 96%, respectively. However, Mravak et al. (1983) reported that mebendazole given daily for 3 weeks was completely effective. Various regimens of mebendazole have been studied by Shikiya and colleagues and support this suggestion. In their latest report (Shikiya et al., 1992) which summarized findings in 245 patients, the efficacy of long-term or repeated administration of mebendazole was confirmed when reassessed after 8 months to 2 years; however, many patients had abnormal liver function tests (Table 7). Animal studies (Grove, 1982a) have shown that mebendazole is active against intestinal adult worms but not against migrating larvae. It is likely that treatment for several weeks is effective by eliminating adult worms initially in the gut and then removing those that develop over the next two or three weeks from autoinfecting larvae that migrate from the tissues back to the bowel.

Table 7 Studies of various anthelmintic regimens utilising mebendazole by Shikiya et al. (1992).

Regimen	Number	Eradication rate	Liver abnormality rate
Mebendazole 100 mg bd for 28 days	16	94%	71%
Thiabendazole 500 mg tds then mebendazole 100 mg daily for 9 days then repeated once	16	100%	50%
Mebendazole 100 bd for 5 days then repeated at weeks 1, 3 and 4	31	87%	51%
Mebendazole 100 bd for 5 days then repeated at weeks 1 and 3	7	100%	31%
Mebendazole 100 bd for 4 days then repeated at weeks 1, 3 and 4	48	96%	58%
Mebendazole 100 bd for 4 days then repeated after 1 week	13	69%	25%

11.1.3. *Albendazole*

Albendazole is the benzimidazole agent most recently introduced into widespread clinical practice. In a previous review (Grove, 1989), it was indicated that reported studies have given conflicting results as to its efficacy in strongyloidiasis. The confusion continues with current titles of papers ranging from "albendazole is effective treatment for chronic strongyloidiasis" to "the weak performance of albendazole in the treatment of strongyloidiasis". Cure rates after short-term treatment for several days vary between 50% and 85% (Mojon and Nielsen, 1987; Chanthavanich *et al.*, 1989; Niimura *et al.*, 1992; Gryschek *et al.*, 1992; Archibald *et al.*, 1993). Since albendazole has a greater larvicidal activity than mebendazole in animals (Grove *et al.*, 1988; Grove and Northern, 1988), it is likely that prolonged courses of treatment with albendazole as described above for mebendazole will be effective.

11.2. Ivermectin

This drug is structurally related to the macrolide antibiotics and has exceptional potency against a number of nematodes and arthropods. It has been introduced for the treatment of strongyloidiasis in the past few years. A number of series have now been reported indicating cure rates

Table 8 Recent studies of the efficacy of ivermectin in uncomplicated strongyloidiasis.

Study	Regimen	Number	Time till evaluation	Cure rate (%)
1	Ivermectin 50–200 µg kg^{-1} once or twice on consecutive days	101	1 month	67–100
2	Ivermectin 6 mg twice 2 weeks apart	125	?	86
3	(i) ivermectin 200 µg kg^{-1} once	15	6 months	100
	(ii) ivermectin 200 µg kg^{-1} twice on consecutive days	17		100
	(iii) thiabendazole 50 mg kg^{-1} bd for 3 days	17		94
4	(i) ivermectin 150–200 µg kg^{-1} once	24	3 months	83
	(ii) albendazole 400 mg daily for 3 days	29		38

1, Naquira *et al.* (1989); 2, Shikiya *et al.* (1994); 3, Gann *et al.* (1994); 4, Datry *et al.* (1994).

varying between 67% and 100% (Table 8). It must be remembered that evaluations performed relatively shortly after administration of anthelmintic are likely to overestimate efficacy as the parasite may require a longer time to build up worm numbers to a detectable level. In direct comparisons, ivermectin has been shown to be more effective and better tolerated than thiabendazole (Gann *et al.*, 1994) and more effective than albendazole (Datry *et al.*, 1994). Patients who fail to respond to the first dose, often respond to a second. Likewise, two patients with AIDS who were given a single dose of ivermectin failed to respond whereas seven patients with severe HIV infection who were treated with ivermectin four times over 16 days were cured of the parasitic infection. These observations are perhaps explained by the demonstration in dogs with corticosteroid-enhanced infections that ivermectin eradicated adult worms and almost all larvae from the gut but was not reliable in removing infective larvae from other organs (Mansfield and Schad, 1992b). It is likely that further studies will show, like mebendazole, that repeated courses or a long course of treatment are required to eradicate infection.

11.3. Miscellanea

11.3.1. *Medamine*

Several papers have appeared in the Russian literature describing the anti-*Strongyloides* activity of medamine (or medamin). I have been unable to find out anything about the pharmacology of medamine. In one study of 166 patients, 66% of patients were cured after a single course of treatment and all patients were said to be cured after four courses (Kuliev *et al.*, 1993).

11.3.2. *Plant Extracts*

In vitro activity against *S. stercoralis* filariform larvae has been reported for five Jamaican plants that were selected for investigation because of folklore and demonstrated insecticidal activity: mimosa (*Mimosa pudica*), love weed (*Cuscuta americana*), vervine (*Stachytarpheta jamaicensis*), chicken weed (*Salvia serotina*) and breadfruit (*Artocarpus altilis*) (Robinson *et al.*, 1990).

11.4. Approaches to Management

Strongyloidiasis is often not easy to treat. A single course of treatment cannot always be relied on to eradicate infection. Failed responses may be

Table 9 Approaches to anthelmintic therapy in patients with strongyloidiasis

---

Initial regimens for uncomplicated strongyloidiasis

- thiabendazole 25 mg kg^{-1} twice daily for three days
- albendazole 400 mg twice daily for three days
- ivermectin 200 μg kg^{-1} mg once

Regimens for eradication of *S. stercoralis* in patients with uncomplicated strongyloidiasis and who fail to respond to the initial regimen or for the attempted eradication of *S. stercoralis* in patients with disseminated strongyloidiasis who are immunosuppressed

- mebendazole 100 mg daily for 3 weeks
- albendazole 400 mg daily for 3 weeks
- ivermectin 200 μg kg^{-1} once weekly for 4 weeks

Regimens for containment of strongyloidiasis in immunosuppressed patients in whom infection cannot be eradicated

- thiabendazole 25 mg kg^{-1} twice a day for one day each month
- albendazole 400 mg for one day each month
- ivermection 200 μg kg^{-1} for one day each month

---

devastating in patients with impaired defences and massive, overwhelming infections. We have probably placed too much faith in the efficacy of a single short course of therapy. Several studies now suggest that a prolonged course of primary treatment is likely to be much more reliable. A suggested approach for treatment of uncomplicated strongyloidiasis and severe infections is detailed in Table 9. Thiabendazole is the traditional agent and is widely available but is probably the least satisfactory of the various agents available. Mebendazole is of little value unless given for an extended period. Albendazole is now available in many countries and may be the preferred option. Ivermectin has perhaps the greatest potential but is not widely available.

Patients with disseminated strongyloidiasis and intestinal obstruction are a difficult problem. Such patients usually need suction and drainage as well as intravenous fluids. Bacterial superinfections require treatment with an appropriate antibiotic. No benzimidazole preparations suitable for intravenous use are available but thiabendazole administered rectally may be successful. If patients are *in extremis*, consideration could be given to attempting to use a parenteral formulation of ivermectin, considering the use of cyclosporin (3 mg kg^{-1} daily) since this has been effective in experimental animals, and finally, stopping immunosuppressive therapy. Despite intensive measures, many patients with severe, complicated infections still die.

12. PREVENTION AND CONTROL

No one really knows how many people are infected with *S. stercoralis* around the globe but it is probably of the order of 100 million. Infection is spread widely throughout the tropics with prevalences in various faecal surveys generally varying from <1% to 15% of samples tested, although some isolated surveys have given much higher figures (Genta, 1989a). Since examination of faeces is insensitive for the diagnosis of strongyloidiasis, these figures probably underestimate the true prevalence of infection. It can be said with a reasonable degree of confidence, however, that wherever hookworm infections are common, *S. stercoralis* is likely to coexist since both worms are skin-penetrating, soil-transmitted parasites endemic in areas where sanitation is poor. Infection is much less common in temperate countries but sporadic cases occasionally occur among the general population. Furthermore, infection may be endemic in closed communities such as institutions for the intellectually retarded where sanitation may not be ideal (Proctor *et al.*, 1987, Braun *et al.*, 1988), in selected populations groups such as returned military personnel (Grove, 1980; Genta *et al.*, 1987) and refugees from countries with endemic strongyloidiasis (Catanzaro and Moser, 1982; Sampson and Grove, 1987).

In endemic areas, age-prevalence studies show that infection is acquired progressively during childhood, then prevalence rates generally remain more or less constant for the rest of adult life (Whitworth *et al.*, 1991; Ashford *et al.*, 1992; Hall *et al.*, 1994).

The factors determining transmission vary from location to location but are usually associated with poverty and insanitary conditions, in particular poor disposal systems for human waste and inadequate water supplies to permit elementary hygiene, as well as lack of education in sanitary practices. For example, in one recent study from an urban slum community in Bangladesh, factors associated with increased prevalence of strongyloidiasis were use of a community rather than private latrine, living in a house with an earthen rather than cement floor, Bihari ethnicity and being aged 7–10 years (Hall *et al.*, 1994). Not surprisingly, infection is often concentrated in certain households (Lindo *et al.*, 1995). There is usually little difference in infection rates between the sexes although some studies have shown an increased frequency in males (Hall *et al.*, 1994).

Control of strongyloidiasis is clearly dependent upon improving economic circumstances with installation of adequate human waste disposal systems and reliable water supplies. This approach is far more likely to be successful than control measures directed against the parasite itself, especially as the available anthelmintics cannot always be relied upon, especially if given in inadequate dosages and for too short a period.

Confirmation of this concept is provided by observations in Okinawa, Japan, which have shown that most infected inhabitants were aged over 40 years and the peak prevalence was in those aged 50 years and more (Arakaki et al., 1992b). In other words, infection is likely to disappear from a community with improving socio-economic status as the infected population ages and dies. Mass chemotherapy is not an appropriate method for community control of strongyloidiasis but it has been suggested that the efficacy of targeted chemotherapy directed at groups particularly at risk of infection is worthy of investigation (Conway et al., 1995).

One problem in areas where water is in short supply, is the desire to recycle water from sewage systems and use it for cultivation, on gardens, and on grassed areas. Despite treatment of water in processing plants, large numbers of Strongyloides larvae may be present (Bolbol, 1992). Since these worms can penetrate the skin, such water may provide a significant risk if it is used on recreational areas where children or adults playing games or sports may come in contact with the ground.

13. *STRONGYLOIDES FUELLEBORNI*

The only species of Strongyloides other than S. stercoralis that infects humans to any significant extent is S. fuelleborni. The morphological features that differentiate adult worms of S. fuelleborni from those of S. stercoralis have been described earlier. In contrast to S. stercoralis in which larvae are passed in the stools, S. fuelleborni embryos are passed as eggs, and autoinfection probably does not occur, although this is controversial (Ashford and Barnish, 1989a). As with S. stercoralis, there may be both direct and indirect development of free-living stages and infection is acquired by percutaneous penetration by infective larvae.

S. fuelleborni infects both humans and some primates. Almost all cases of human infection with this parasite have been found in Africa and Papua New Guinea. Very little has been written in recent years about the African infection other than to note its prevalence among bushmen children in Southwest Africa (Evans et al., 1991).

In contrast, a number of papers have appeared concerning the form prevalent in Papua New Guinea. The taxonomic position of this parasite has been reviewed and it has been recently described as a subspecies of S. fuelleborni as S. fuelleborni kellyi (Viney et al., 1991). In contrast to the usual pattern with S. stercoralis, in some Papua New Guinean communities, the peak age prevalence is in young children (Barnish and Ashford, 1989a, b). Rainfall, altitude, slope of the ground and population densities did not determine distribution of the parasite but infection was uncommon in limestone areas (Barnish and Ashford, 1990). Likewise, host genetic factors did not influence variations in parasite loads (Smith *et al.*, 1991). Heavy infections in young children cause gross abdominal distension, hypoproteinaemia and peripheral oedema. Thiabendazole is an effective treatment, but not surprisingly in endemic situations, reinfection occurs (Ashford and Barnish, 1989b).

14. FUTURE DIRECTIONS

There is much that remains to be learned about *Strongyloides stercoralis*. The biology of the parasite, in particular its genetics and the mode of replication of the female parasitic worm, need to be further elucidated. We still know very little about the interactions between the parasite and host. Are genetic factors important in determining resistance? What are the mechanisms by which the parasite evades the host's defences? What arms of the immune response or other defence mechanisms contain infection and are lost in disseminated infections? Do corticosteroids act as ecdysteroids and promote multiplication of *S. stercoralis*? We need better methods of diagnosing occult strongyloidiasis and differentiating between past and current infections — would a polymerase chain reaction-based test of faeces be helpful? Finally, we need better anthelmintics and must optimize the regimens for those drugs that are currently available.

REFERENCES

Abraham, D., Rotman, H.L., Haberstroh, H.F., Yutanawiboonchai, W., Brigandi, R.A., Leon, O. *et al.* (1995). *Strongyloides stercoralis*: protective immunity to third-stage larvae in BALB/cByJ mice. *Experimental Parasitology* **80**, 297–307.

Adachi, N., Ohta, K., Negoro, N., Kurihara, N. and Takeda, T. (1994). [A case of adult T cell leukemia complicated with strongyloidiasis and amplification of *Pneumocystis carinii* DNA in bronchoalveolar lavage fluid.] *Nippon Kyobu Shikkan Gakki Zasshi* **32**, 348–352.

Aikens, L.M. and Schad, G.A. (1989). Radiolabelling of infective third-stage larvae of *Strongyloides stercoralis* by feeding [^{75}Se]selenomethionine-labelled *Escherichia coli* to first- and second-stage larvae. *Journal of Parasitology* **75**, 735–739.

Altman D.G. (1991). *Practical Statistics for Medical Research*. London: Chapman and Hall, 611 pp.

Anonymous (1987). Revision of the CDC surveillance case definition for acquired immunodeficiency syndrome. *Morbidity and Mortality Weekly Report* **36**(1S), 3S–15S.

Arakaki, T., Iwanaga, M., Kinjo, F., Saito, A., Asato, R. and Ikeshiro, T. (1990). Efficacy of agar-plate culture in detection of *Strongyloides stercoralis* infection. *Journal of Parasitology* **76**, 428–430.

Arakaki, T., Asato, R., Ikeshiro, T., Sakiyama, K. and Iwanaga, M. (1992a). Is the prevalence of HTLV-1 infection higher in *Strongyloides* carriers than in non-carriers? *Tropical Medicine and Parasitology* **43**, 199–200.

Arakaki, T., Iwanaga, M., Asato, R. and Ikeshiro, T. (1992b). Age-related prevalence of *Strongyloides stercoralis* infection in Okinawa, Japan. *Tropical and Geographical Medicine* **44**, 299–303.

Archibald, L.K., Beeching, N.J., Gill, G.V., Bailey, J.W. and Bell, D.R. (1993). Albendazole is effective treatment for chronic strongyloidiasis. *Quarterly Journal of Medicine* **86**, 191–195.

Armignacco, O., Capecchi, A., De Mori, P. and Grillo, L.R. (1989). *Strongyloides stercoralis* hyperinfection and the acquired immunodeficiency syndrome. *American Journal of Medicine* **86**, 258.

Ashford, R.W. and Barnish, G. (1989a). *Strongyloides fuelleborni* and similar parasites in animals and man. In: *Strongyloidiasis: a Major Roundworm Infection of Man* (D.I. Grove ed.), pp. 271–286. London: Taylor & Francis.

Ashford, R.W. and Barnish, G. (1989b). *Strongyloides* and hookworm in Papua New Guinea: longitudinal studies on treated and untreated children. *Annals of Tropical Medicine and Parasitology* **83**, 583–589.

Ashford, R.W., Craig, P.S. and Oppenheimer, S.J. (1992). Polyparasitism on the Kenya coast. 1. Prevalence, and association between parasitic infections. *Annals of Tropical Medicine and Parasitology* **86**, 671–679.

Ashton, F.T., Bhopale, V.M., Fine, A.E. and Schad, G.A. (1995). Sensory neuroanatomy of a skin-penetrating nematode parasite: *Strongyloides stercoralis*. I. Amphidial neurons. *Journal of Comparative Neurology* **357**, 281–295.

Assefa, T., Wodemichael, T. and Seyoum, T. (1991). Evaluation of the modified Baermann's method in the laboratory diagnosis of *Strongyloides stercoralis*. *Ethiopian Medical Journal* **29**, 193–198.

Bannon, J.P., Fater, M. and Solit, R. (1995). Intestinal ileus secondary to *Strongyloides stercoralis* infection: case report and review of the literature. *American Surgeon* **61**, 377–380.

Barnish, G. and Ashford, R.W. (1989a). *Strongyloides* cf *fuelleborni* and hookworm in Papua New Guinea: patterns of infection within the community. *Transactions of the Royal Society of Tropical Medicine and Hygiene* **83**, 684–688.

Barnish, G. and Ashford, R.W. (1989b). *Strongyloides* cf *fuellborni* in Papua New Guinea: epidemiology in an isolated community, and results of an intervention study. *Annals of Tropical Medicine and Parasitology* **83**, 499–506.

Barnish, G. and Ashford, R.W. (1990). *Strongyloides* cf. *fuelleborni* and other intestinal helminths in Papua New Guinea: distribution according to environmental factors. *Parassitologia* **32**, 245–263.

Barrett, K.E., Neva, F.A., Gam, A.A., Cicmanec, J., London, W.T., Phillips, J.M. and Metcalfe, D.D. (1988). The immune response to nematode parasites: modulation of mast cell numbers and function during *Strongyloides stercoralis* infections in nonhuman primates. *American Journal of Tropical Medicine and Hygiene* **38**, 574–581.

Batista, N., Davila, M.F., Gijon, H. and Perez, M.A. (1992). Estrongiloidiasis en un pacienta con el sindrome de immunodeficencia adquirida. *Enfermedades Infecciosas y Microbiologia Clinica (Barcelona)* **10**, 431–432.

Berezhnaia, V.G., Prokhorov, A.F. and Semiashkina, L.R. (1991). [The characteristics of the development of different geographical strains of *Strongyloides stercoralis* in a fecal culture.] *Medtsinskaia Parazitologiia i Parazitarnyē Bolezni (Moskva)* **2**, 26–28.

Beus, A. (1989). Poredbeno ispitivanje tiabendazola i mebendazola u strongiloidozi. *Lijec-Vjesn* **111**, 98–101.

Bhatt, B.D., Cappell, M.S., Smilow, P.C. and Das, K.M. (1990). Recurrent massive upper gastrointestinal hemorrhage due to *Strongyloides stercoralis* infection. *American Journal of Gastroenterology* **85**, 1034–1036.

Bili. H., Foll, Y., Morand, C. and Abgrall, J. (1991). Anorectite revelactrice d'une anguillulose. *La Presse Médicale* **20**, 1457.

Blank, A., Yamaguchi, K., Blank, M., Zaninovic, V., Sonoda, S. and Takatsuki, K. (1993). Six Colombian patients with adult T-cell leukemia/lymphoma. *Leukemia Lymphoma* **9**, 407–412.

Boken, D.J., Leoni, P.A. and Preheim, L.C. (1993). Treatment of *Strongyloides stercoralis* hyperinfection syndrome with thiabendazole administered per rectum. *Clinical Infectious Diseases* **16**, 123–126.

Bolbol, A.S. (1992) Risk of contamination of human and agricultural environment with parasites through reuse of treated municipal wastewater in Riyadh, Saudi Arabia. *Journal of Hygiene, Epidemiology, Microbiology and Immunology* **36**, 330–337.

Boyajian, T. (1992). Strongyloidiasis on the Thai-Cambodian border. *Transactions of the Royal Society of Tropical Medicine and Hygiene* **86**, 661–662.

Braun, T.I., Fekete, T. and Lynch, A. (1988). Strongyloidiasis in an institution for mentally retarded adults. *Archives of Internal Medicine* **148**, 634–636.

Brigandi, R.A., Rotman, H.L., Yutanawiboonchai, W., Nolan, T.J., Schad, G.A. and Abraham, D. (1994). Control of hyperinfective strongyloidiasis: comparative recognition of infective and autoinfective larval antigens by immunized mice. *American Journal of Tropical Medicine and Hygiene* Abstracts, pp. 233–234.

Brindley, P.J., Gam, A.A., Pearce, E.J., Poindexter, R.W. and Neva, F.A. (1988). Antigens from the surface and excretions/secretions of the filariform larva of *Strongyloides stercoralis*. *Molecular and Biochemical Parasitology* **28**, 171–180.

Brindley, P.J., Gam, A.A., McKerrow, J.H. and Neva, F.A. (1995). Ss40: The zinc endopeptidase secreted by infective larvae of *Strongyloides stercoralis*. *Experimental Parasitology* **80**, 1–7.

Brosset, C., Hovette, P., Raphenon, G., Debonne, J.M., Dano, P. and Laroche, R. (1991). HTLV1 et coinfections. *La Médecine Tropicale* **51**, 399–406.

Catanzaro, A. and Moser, R.J. (1982). Health status of refugees from Vietnam, Laos and Cambodia. *Journal of the American Medical Association* **247**, 1303–1308.

Celedon, J.C., Mathur-Wagh, U., Fox, J., Garcia, R. and Wiest, P.M. (1994). Systemic strongyloidiasis in patients infected with the human immunodeficiency virus. A report of 3 cases and review of the literature. *Medicine* **73**, 256–263.

Chanthavanich, P., Nontasut, P., Prarinyanuparp, V. and Sa-Nguankiat, S. (1989). Repeated doses of albendazole against strongyloidiasis in Thai children. *Southeast Asian Journal of Tropical Medicine and Public Health* **20**, 221–226.

Chaudhary, K., Smith, R.J., Himelright, I.M. and Baddour, L.M. (1994). Case report: purpura in disseminated strongyloidiasis. *American Journal of the Medical Sciences* **308**, 186–191.

Chen, J.J., Lee, C.M. and Changchan, C.S. (1994). Duodenal *Strongyloides stercoralis* infection. *Endoscopy* **26**, 272.

Coninx, R. (1993). Strongyloidiasis on the Thai-Cambodian border. *Transactions of the Royal Society of Tropical Medicine and Hygiene* **87**, 350.

Conlon, C. P., Pinching A.J., Perera, C.U., Moody, A., Luo, N.P. and Lucas, S.B. (1990). HIV-related enteropathy in Zambia: a clinical, microbiological, and histological study. *American Journal of Tropical Medicine and Hygiene* **42**, 83–88.

Conway, D.J., Atkins, N.S., Lillywhite, J.E., Bailey, J.W., Robinson, R.D., Lindo, J.F., Bundy, D.A.P. and Bianco, A.E. (1993a). Immunodiagnosis of *Strongyloides stercoralis* infection: a method for increasing the specificity of the indirect ELISA. *Transactions of the Royal Society of Tropical Medicine and Hygiene* **87**, 173–176.

Conway, D.J., Bailey, J.W., Lindo, J.F., Robinson, R.D., Bundy, D.A.P. and Bianco, A.E. (1993b). Serum IgG reactivity with 41-, 31-, and 28-kDa larval proteins of *Strongyloides stercoralis* in individuals with strongyloidiasis. *Journal of Infectious Diseases* **168**, 784–787.

Conway, D.J., Lindo, J.F., Robinson, R.D., Bundy, D.A.P. and Bianco, A.E. (1994). *Strongyloides stercoralis*: characterization of immunodiagnostic larval antigens. *Experimental Parasitology* **79**, 99–105.

Conway, D.J., Lindo, J.F., Robinson, R.D. and Bundy, D.A.P. (1995). Towards effective control of *Strongyloides stercoralis*. *Parasitology Today* **11**, 420–427.

Couprie, R., Maslo, C., Buchaud, O., Matheron, S., Saimot, A.G. and Coulaud, J.P. (1993). Anguillulose disséminée au cours de l'infection par le VIH. Une nouvelle observation. *La Presse Médicale* **22**, 968.

Datry, A., Hilmarsdottir, I., Mayorga-Sagastume, R., Lyagoubi, M., Gaxotte, P., Biligui, S. *et al.* (1994). Treatment of *Strongyloides stercoralis* infection with ivermectin compared with albendazole: results of an open study of 60 cases. *Transactions of the Royal Society of Tropical Medicine and Hygiene* **88**, 344–345.

Davidson, R.A. (1988). *Strongyloides stercoralis* infection in the ferret. *Journal of Parasitology* **74**, 177–179.

de Kaminsky, R.G. (1993). Evaluation of three methods for laboratory diagnosis of *Strongyloides stercoralis* infection. *Journal of Parasitology* **79**, 277–280.

Delarocque-Astagneau, E., Hadengue, A., Degott, C., Vilgrain, V., Erlinger, S. and Benhamou, J.P. (1994). Biliary obstruction resulting from *Strongyloides stercoralis* infection. Report of a case. *Gut* **35**, 705–706.

de Messias, I.T., Telles, F.Q., Boaretti, A.C., Sliva, S., Guimarres, L.M. and Genta, R.M. (1987). Clinical, immunological and epidemiological aspects of strongyloidiasis in an endemic area of Brazil. *Allergologia et Immunopathologia (Madrid)* **15**, 37–41.

de Messias, I.J.T., Genta, R.M. and Mohren, W.D. (1994). Adherence of monocytes and polymorphonuclear cells to infective larvae of *Strongyloides stercoralis* after complement activation. *Journal of Parasitology* **80**, 267–274.

De Paoli, A. (1974). Strongyloidiasis: a comparative study in the gibbon and man. PhD thesis, George Washington University.

DeVault, G.A., King, J.W., Rohr, M.S., Landreneau, M.D., Brown S.T. III and McDonald, J.C. (1990). Opportunistic infections with *Strongyloides stercoralis* in renal transplantation. *Reviews of Infectious Diseases* **12**, 653–671.

Dias, R.M., Mangini, A.C., Torres, D.M., Vellosa, S.A., da-Silva, M.I., da-Silva, R.M. *et al.* (1992). Ocorrencia de *Strongyloides stercoralis* em pacientes portadores da sindrom de imunodeficiencia adquirida. *Revista dos Instituto Medecina Tropical de São Paulo* **34**, 15–17.

D'Incan, M., Combemale, P., Verrier, B., Garin, D., Audoly, G., Brunot, J. *et al.* (1994). Transient adult T-cell leukemia/lymphoma picture during varicella infection in an HTLV-1 carrier. *Leukemia* **8**, 682–687.

Dutcher, J.P., Marcus, S.L., Tanowitz, H.B., Wittner, M., Fuks, J.Z. and Wiernik, P.H. (1990). Disseminated strongyloidiasis with central nervous system involvement diagnosed antemortem in a patient with acquired immunodeficiency syndrome and Burkitt's lymphoma. *Cancer* **66**, 2417–2420.

Evans, A.C., Markus, M.B., Joubert J.J. and Gunders, A.E. (1991). Bushman children infected with the nematode *Strongyloides füelleborni*. *South African Medical Journal* **80**, 410–411.

Faust, E.C. (1930). The Panama strains of human *Strongyloides*. *Proceedings of the Society for Experimental Biology and Medicine* **17**, 1343–1348.

Faust, E.C. (1933). Experimental studies on human and primate species of *Strongyloides* in the experimental host. *American Journal of Hygiene* **18**, 114–132.

Freedman, D.O. (1991). Experimental infection of human subjects with *Strongyloides* species. *Reviews of Infectious Diseases* **13**, 1221–1226.

Fülleborn, F. (1914). Untersuchungen über den Infektionsweg bei *Strongyloides* und *Ankylostomum* und die Biologie dieser Parasiten. *Archiv für Schiffs-und Tropen-Hygien* **18**, 26–80.

Gage, J.G. (1911). A case of *Strongyloides intestinalis* with larvae in the sputum. *Archives of Internal Medicine* **7**, 55–59.

Galliard, H. (1967). Pathogenesis of *Strongyloides*. *Helminthological Abstracts* **36**, 247–260.

Gann, P.H., Neva, F.A. and Gam, A.A. (1994). A randomized trial of single- and two-dose ivermectin versus thiabendazole for treatment of strongyloidiasis. *Journal of Infectious Diseases* **169**, 1076–1079.

Genta, R.M. (1986). *Strongyloides stercoralis*: immunobiological considerations of an unusual worm. *Parasitology Today* **2**, 241–246.

Genta, R.M. (1988). Predictive value of an enzyme-linked immunosorbent assay (ELISA) for the serodiagnosis of strongyloidiasis. *American Journal of Clinical Pathology* **89**, 391–394.

Genta, R.M. (1989a). Global prevalence of strongyloidiasis: critical review with epidemiologic insights into the prevention of disseminated disease. *Reviews of Infectious Diseases* **11**, 755–767.

Genta, R.M. (1989b). *Strongyloides stercoralis*: loss of ability to disseminate after repeated passage in laboratory beagles. *Transactions of the Royal Society of Tropical Medicine and Hygiene* **83**, 539–541.

Genta, R.M. (1992). Dysregulation of strongyloidiasis: a new hypothesis. *Clinical Microbiology Reviews* **5**, 345–355.

Genta, R.M. and Lillibridge, J.P. (1989). Prominence of IgG4 antibodies in the human responses to *Strongyloides stercoralis* infection. *Journal of Infectious Diseases* **160**, 692–699.

Genta, R.M., Weesner, R., Douce, R.W., Huitger-O'Connor, T. and Walzer, P.D. (1987). Strongyloidiasis in US veterans of the Vietnam and other wars. *Journal of the American Medical Association* **358**, 29–52.

Genta, R.M., Gatti, S., Linke, M.J., Cenvini, C. and Scaglia, M. (1988). Endemic strongyloidiasis in northern Italy: clinical and immunological aspects. *Quarterly Journal of Medicine* **247**, 679–690.

Gill, G.V. and Bell, D.R. (1979). *Strongyloides stercoralis* infection in former Far East prisoners-of-war. *British Medical Journal* **2**, 572–574.

Glezerov, V. and Masci, J.R. (1990). Disseminated strongyloidiasis and other selected unusual infections in patients with the acquired immunodeficiency syndrome. *Progress in AIDS Pathology* **2**, 137–142.

Goka, A.K.J., Rolston, D.D.K., Matham, V.I. and Farthing, M.J.G. (1990). Diagnosis of *Strongyloides* and hookworm infections: comparison of faecal and duodenal fluid microscopy. *Transactions of the Royal Society of Tropical Medicine and Hygiene* **84**, 829–831.

Gompels, M.M., Todd, J., Peters, B.S., Main, J. and Pinching, A.J. (1991). Disseminated strongyloidiasis in AIDS: uncommon but important. *AIDS* **5**, 329–332.

Gordon, S.M., Gal, A.A., Solomon, A.R. and Bryan, J.A. (1994). Disseminated strongyloidiasis with cutaneous manifestations in an immunocompromised host. *Journal of the American Academy of Dermatology* **31**, 255–259.

Graham, G.L. (1938). Studies on *Strongyloides*. II. Homogonic and heterogonic progeny of the single homogonically derived parasite. *American Journal of Hygiene* **24**, 71–87.

Grove, D.I. (1980). Strongyloidiasis in Allied ex-prisoners of war. *British Medical Journal* **180**, 598–601.

Grove, D.I. (1982a). *Strongyloides ratti* and *S. stercoralis*: the effects of thiabendazole, mebendazole and cambendazole in infected mice. *American Journal of Tropical Medicine and Hygiene* **31**, 469–476.

Grove, D.I. (1982b). Treatment of strongyloidiasis with thiabendazole: an analysis of toxicity and effectiveness. *Transactions of the Royal Society of Tropical Medicine and Hygiene* **76**, 114–118.

Grove, D.I. (ed.) (1989). *Strongyloidiasis: a Major Roundworm Infection in Man*, London: Taylor & Francis.

Grove, D.I. (1990a). *A History of Human Helminthology*, Wallingford: CAB International.

Grove, D.I. (1990b). Strongyloidiasis. In: *Tropical and Geographical Medicine*, (K.S. Warren and A.A.F. Mahmoud eds), pp. 393–399. New York: McGraw-Hill Information Services Company.

Grove, D.I. and Civil, R.H. (1978). *Trichinella spiralis*: effects on the host–parasite relationship in mice of BCG attenuated *Mycobacterium bovis*. *Experimental Parasitology* **44**, 181–189.

Grove, D.I. and Northern, C. (1982). Infection and immunity in dogs infected with a human strain of *Strongyloides stercoralis*. *Transactions of the Royal Society of Tropical Medicine and Hygiene* **76**, 833–838.

Grove, D.I. and Northern, C. (1988). The effects of thiabendazole, mebendazole and cambendazole in normal and immunosuppressed dogs infected with a human strain of *Strongyloides stercoralis*. *Transactions of the Royal Society of Tropical Medicine and Hygiene* **82**, 146–149.

Grove, D.I., Heenan, P.J. and Northern, C. (1983). Persistent and disseminated infections with *Strongyloides stercoralis* in immunosuppressed dogs. *International Journal for Parasitology* **13**, 483–490.

Grove, D.I., Warton, A., Northern, C. and Papadimitriou, J.M. (1987a). Electron microscopical studies of *Strongyloides ratti* infective larvae: loss of the surface coat during skin penetration. *Journal of Parasitology* **73**, 1030–1034.

Grove, D.I., Warton, A., Yu, L.L., Northern, C. and Papadimitriou, J.M. (1987b). Light and electron microscopical studies of the location of *Strongyloides stercoralis* in the jejunum of the immunosuppressed dog. *International Journal for Parasitology* **73**, 1030–1034.

Grove, D.I., Lumsden, J. and Northern, C. (1988). Efficacy of albendazole against *Strongyloides ratti* and *S. stercoralis in vitro*, in mice, and in normal and immunosuppressed dogs. *Journal of Antimicrobial Chemotherapy* **21**, 75–84.

Gryschek, R.C., Amato-Nevo, V., Matsubara, L. and Campos, R. (1992). Fraco desempenho do albendazol no tratamento da estrongiloidiase. *Revista da Sociedade Brasileira de Medicina Tropical (Rio de Janeiro)* **25**, 205–206.

Gustavo dos Santos, J. (1993). Movement of the rhabditiform larva of *Strongyloides stercoralis*. *Lancet* **342**, 1310.

Gyorkos, T.W., Genta, R.M., Viens, P. and MacLean, J.D. (1990). Seroepidemiology of *Strongyloides* infection in the southeast Asian refugee population in Canada. *American Journal of Epidemiology* **132**, 257–264.

Hall, A., Conway, D.J., Anwar, K.S. and Rahman, M.L. (1994). *Strongyloides stercoralis* in an urban slum community in Bangladesh: factors independently associated with infection. *Transactions of the Royal Society of Tropical Medicine and Hygiene* **8**, 527–530.

Hammond, M.P. and Robinson, R.D. (1994) Chromosome complement, gametogenesis, and development of *Strongyloides stercoralis*. *Journal of Parasitology* **80**, 689–695.

Hanada, S., Uematsu, T., Iwahashi, M., Nomura, K., Utsunomiya, A., Kodama, M. *et al.* (1989). The prevalence of human T-cell Leukemia Virus Type 1 infection in patients with hematologic and nonhematologic diseases in an adult T-cell leukemia-endemic area of Japan. *Cancer* **64**, 1290–1295.

Haque, A.K., Schnadig, V, Rubin, S.A. and Smith, J.H. (1994). Pathogenesis of human strongyloidiasis: autopsy and quantitative parasitological analysis. *Modern Pathology* **7**, 276–288.

Harcourt-Webster, J.M., Scaravilli, F. and Darwish, A.H. (1991). *Strongyloides stercoralis* hyperinfection in an HIV positive patient. *Journal of Clinical Pathology* **44**, 346–348.

Harper, J.S., Genta, R.M., Gam, A., London., W.T. and Neva, F.A. (1984). Experimental disseminated strongyloidiasis in *Erythrocebus patas*. I. Pathology. *American Journal of Tropical Medicine and Hygiene* **33**, 431–443.

Igra-Siegman, U., Kapila, R., Sen, P., Zaminski, Z.C. and Louria, D.B. (1981). Syndrome of hyperinfection with *Strongyloides stercoralis*. *Reviews of Infectious Diseases* **3**, 397–407.

Jain, A.K., Agarwal, S.K. and El-Sadr, W. (1994). *Streptococcus bovis* bacteremia and meningitis associated with *Strongyloides stercoralis* colitis in a patient infected with human immunodeficiency virus. *Clinical Infectious Diseases* **18**, 253–254.

Joseph, L., Gyorkos, T.W. and Coupal, L. (1995). Bayesian estimation of disease prevalence and the parameters of diagnostic tests in the absence of a gold standard. *American Journal of Epidemiology* **141**, 263–272.

Josimovic-Alasevic, O., Feldmeier, H., Zwingenberger, K., Harms, G. and Hahn, H. (1988). Interleukin 2 receptor in patients with localized and systemic parasitic diseases. *Clinical Experimental Immunology* **72**, 249–254.

Kennedy, S., Campbell, R.M., Lawrence, J.E., Nichol, G.M. and Rao, D.M. (1989). A case of severe *Strongyloides stercoralis* infection with jejunal perforation in an Australian ex-prisoner-of-war. *Medical Journal of Australia* **150**, 92–93.

Kobayashi, J., Sato, Y., Toma, H., Masahiro, T. and Shiroma, Y. (1994). Application of enzyme immunoassay for postchemotherapy evaluation of human strongyloidiasis. *Diagnostic Microbiology and Infectious Diseases* **18**, 19–23.

Koga, K., Kasuya, S., Khamboonruang, C., Sukhavat, K., Ieda, M., Takatsuka, N. *et al.* (1991). A modified agar plate method for detection of *Strongyloides stercoralis*. *American Journal of Tropical Medicine and Hygiene* **45**, 518–521.

Koga, K., Kasuya, S. and Ohtomo, H. (1992). How effective is the agar plate method for *Strongyloides stercoralis*? *Journal of Parasitology* **78**, 155–156.

Kotlan, S. and Vajda, T. (1934). Strongyloides-tanulmanyok. *Allatorvosi Lapok*, **57**, 198–205.

Kramer, M.R., Gregg, P.A., Goldstein, M., Llamas, R. and Krieger, B.P. (1990). Disseminated strongyloidiasis in AIDS and non-AIDS immunocompromised hosts: diagnosis by sputum and bronchoalveolar lavage. *Southern Medical Journal* **83**, 1226–1229.

Kreis, H.A. (1932). Studies on the genus *Strongyloides* (Nematoda). *American Journal of Hygiene* **16**, 450–491.

Kudeken, N., Kitsukawa, K., Hokama, A., Saito, A., Kiyuna, M. and Toda, T. (1992). A case of smoldering adult T-cell leukemia associated with duodenal dilatation due to strongyloidiasis. *Nippon Shokakibyo Gakkai Zasshi* **89**, 2623–2628.

Kuliev, N.D., Iarotskii, L.S., Mekhraliev, M.B. and Abdullaev, A.D. (1993). [The efficacy of treating strongyloidiasis patients with medamine.] *Medtsinskaia Parazitologiia i Parazitarnyē Bolezni (Moskva)* **1**, 47–48.

Lee, M.G. and Terry, S.I. (1989). Arteriomesenteric duodenal occlusion associated with strongyloidiasis. *Journal of Tropical Medicine and Hygiene* **92**, 41–45.

Leeman, B.J. and Cabrera, M.D. (1995). No association between *Strongyloides* infestation and asthma. *Journal of Asthma* **32**, 57–62.

Leighton, P.M. and MacSween, H.M. (1990). *Strongyloides stercoralis:* the cause of an urticarial-like eruption of 65 years' duration. *Archives of Internal Medicine* **150**, 1747–1748.

Lessnau, K., Can, S. and Talavera, W. (1993). Disseminated *Strongyloides stercoralis* in human immunodeficiency virus-infected patients. Treatment failure and a review of the literature. *Chest* **104**, 119–122.

Lin, A.L., Kessimian, N. and Benditt, J.O. (1995). Restrictive pulmonary diseases due to interlobular septal fibrosis associated with disseminated infection by *Strongyloides stercoralis*. *American Journal of Respiratory and Critical Care Medicine* **151**, 205–209.

Lindo, J.F., Conway, D.J., Atkins, N.S., Bianco, A.E., Robinson, R.D. and Bundy, D.A.P. (1994). Prospective evaluation of enzyme-linked immunosorbent assay and immunoblot methods for the diagnosis of endemic *Strongyloides stercoralis* infection. *American Journal of Tropical Medicine and Hygiene* **51**, 175–179.

Lindo, J.F., Robinson, R.D., Terry, S.I., Vogel, P., Gam, A.A., Neva, F.A. and Bundy, D.A.P. (1995). Age-prevalence and household clustering of *Strongyloides stercoralis* infection in Jamaica. *Parasitology* **110**, 97–102.

Little, M.D. (1965). Dermatitis in a human volunteer infected with *Strongyloides* of nutria and raccoon. *American Journal of Tropical Medicine and Hygiene* **14**, 1007–1009.

Little, M.D. (1966). Comparative morphology of six species of *Strongyloides* (Nematoda) and redefinition of the genus. *Journal of Parasitology* **52**, 69–84.

Lo, C.Y. and Kajioka, R. (1989). Predictive values. *American Journal of Clinical Pathology* **91**, 505–506.

Lucas, S.B. (1990). Missing infections in AIDS. *Transactions of The Royal Society of Tropical Medicine and Hygiene* **84**, Supplement 1, 34–38.

Maayan S., Wormser, G.P., Widerhorn, J., Sy, E.R., Kim, Y.H. and Ernst, J.A. (1987). *Strongyloides stercoralis* hyperinfection in a patient with the acquired immunodeficiency syndrome. *American Journal of Medicine* **83**, 945–948.

Mahdi, N.K., Setrak, S.K. and Shiwaish, S.M. (1993). Diagnostic methods for intestinal parasites in southern Iraq with reference to *Strongyloides stercoralis*. *Southeast Asian Journal of Tropical Medicine and Public Health* **24**, 685–691.

Makris, A.N., Sher, S., Bertoli, C. and Latour, M.G. (1993). Pulmonary strongyloidiasis: An unusual opportunistic pneumonia in a patient with AIDS. *American Journal of Roentgenology* **161**, 545–547.

Mansfield, L.S. and Schad, G.A. (1992a). *Strongyloides stercoralis* infection in IgA-deficient dogs. *American Journal of Tropical Medicine and Hygiene* **47**, 830–836.

Mansfield, L.S. and Schad, G.A. (1992b). Ivermectin treatment of naturally acquired and experimentally induced *Strongyloides stercoralis* infections in dogs. *Journal of the American Veterinary Medical Association* **201**, 726–730.

Mansfield, L.S., Alavi, A., Wortman, J.A. and Schad, G.A. (1995). Gamma camera scintigraphy for direct vizualization of larval migration in *Strongyloides stercoralis*-infected dogs. *American Journal of Tropical Medicine and Hygiene* **52**, 236–240.

Matsumoto, T., Miike, T., Mizoguchi, K., Yamaguchi, K., Takatsuki, K, Hosoda, M. *et al.* (1990). Decreased serum levels of IgE and IgE-binding factors in individuals infected with HTLV-1. *Clinical and Experimental Immunology* **81**, 207–211.

McKerrow, J.H., Brindley, P., Brown, M. Gam, A.A., Staunton, C. and Neva, F.A. (1990). *Strongyloides stercoralis*: identification of a protease that facilitates penetration of skin by the infective larvae. *Experimental Parasitology* **70**, 134–143.

Medina L.S., Heiken, J.P. and Gold, R.P. (1992). Pipestem appearance of small bowel in strongyloidiasis is not pathognomonic of fibrosis and irreversibility. *American Journal of Roentgenology* **159**, 543–544.

Mojon, M. and Nielsen, P.B. (1987). Treatment of *Strongyloides stercoralis* with albendazole. A cure rate of 86 per cent. *Zentralblatt für Backteriologie, Microbiologie und Hygiene* **263**, 619–624.

Moncol, D.J. and Triantaphyllou, A.C. (1978). *Strongyloides ransomi*: factors influencing the *in vitro* development of the free-living generation. *Journal of Parasitology* **64**, 220–225.

Morgello, S., Soifer, F.M., Lin, C.S. and Wolfe, D.E. (1993). Central nervous system *Strongyloides stercoralis* in acquired immunodeficiency syndrome: a report of two cases and review of the literature. *Acta Neuropatholologica* **86**, Supplement, 285–288.

Mravak, S., Schopp, W. and Bienzle, U. (1983). Treament of strongyloidiasis with mebendazole. *Acta Tropica* **40**, 93–94.

Murty, D.A., Luthra, U.K., Sehgal, K. and Sodhani, P. (1994). Cytologic detection of *Strongyloides stercoralis* in a routine cervicovaginal smear. A case report. *Acta Cytologica* **38**, 223–225.

Nakada, K., Yamaguchi, K., Furugen, S., Nakasone, T., Nakasone, S., Oshiro, Y. *et al.* (1987). Monoclonal integration of HTLV-1 proviral DNA in patients with strongyloidiasis. *International Journal of Cancer* **40**, 145–148.

Naquira, C., Jimenez, G., Guerra, J.G., Bernal, R., Nalin, D.R., Neu, D. and Aziz,

M. (1989). Ivermectin for human strongyloidiasis and other intestinal helminths. *American Journal of Tropical Medicine and Hygiene* **40**, 304–309.

Ndalahwa, J.B. and Lwambo, N.J. (1989). An improved kit for culturing clean nematode larvae from faeces. *East African Medical Journal* **66**, 203–207.

Neva, F. A. (1993). *Strongyloides* hyperinfection in patients coinfected with HTLV-1 and *Strongyloides stercoralis*. *American Journal of Medicine* **94**, 447–449.

Neva, F.A., Murphy, E.L., Gam, A., Hanchard, B., Figueroa, J.P., and Blattner, W.A. (1989). Antibodies to *Strongyloides stercoralis* in healthy Jamaican carriers of HTLV-I. *New England Journal of Medicine* **320**, 252–253.

Newton, R.C., Limpuangthip, P., Greenberg, S., Gam, A. and Neva, F.A. (1992). *Strongyloides stercoralis* hyperinfection in a carrier of HTLV-1 virus with evidence of selective immunosuppression. *American Journal of Medicine* **92**, 202–208.

Nielsen, P.B. and Mojon, M. (1987). Improved diagnosis of *Strongyloides stercoralis* by seven consecutive stool specimens. *Zentralblatt für Bakteriologie, Mikrobiologie und Hygiene* **263**, 616–618.

Niimura, S., Hirata, T., Zaha, O., Nakamura, H., Kouchi, A., Uehara, T., Uechi, H., Ohshiro, Y. *et al.* (1992). [Clinical study of albendazole therapy for strongyloidiasis.] *Kansenshogaku Zasshi* **66**, 1231–1235.

Nolan, T.J., Aikens, L.M. and Schad, G.A. (1988). Cryopreservation of first-stage and infective third-stage larvae of *Strongyloides stercoralis*. *Journal of Parasitology* **74**, 387–391.

Nolan, T.J., Megyeri, Z., Bhopale, V.M. and Schad, G.A. (1993). *Strongyloides stercoralis*: the first rodent model for uncomplicated and hyperinfective strongyloidiasis, the Mongolian gerbil (*Meriones unguiculatus*). *Journal of Infectious Diseases* **168**, 1479–1484.

Onuigbo, M.A.C. and Ibeachum, G.I. (1991). *Strongyloides stercoralis* larvae in peripheral blood. *Transactions of the Royal Society of Tropical Medicine and Hygiene* **85**, 97.

Oyakawa, T., Kuniyohshi, T., Arakaki, T., Higashionna, A., Shikiya, K. and Sakugawa, H. (1991). [New trial with thiabendazole for treatment of human strongyloidiasis.] *Kansenshogaku Zasshi* **65**, 304–310.

Patey, O., Bouhali, R., Breuil, J., Chapuis, L., Courillon-Mallet, A. and Lafaix, C. (1990). Arthritis associated with *Strongyloides stercoralis*. *Scandinavian Journal of Infectious Diseases* **22**, 233–236.

Patey, O., Gessain, A., Breuil, J., Courillon-Mallet, A., Daniel, M.T., Miclea, J.M., *et al.* (1992). Seven years of recurrent severe strongyloidiasis in an HTLV-1-infected man who developed adult T-cell leukaemia. *AIDS* **6**, 575–579.

Pelletier, L.L. Jr, Baker, C.B., Gam, A.A., Nutman, T.B. and Neva, F.A. (1988). Diagnosis and evaluation of treatment of chronic strongyloidiasis in ex-prisoners of war. *Journal of Infectious Diseases* **157**, 573–576.

Perry, J.L., Matthews, J.S. and Miller, G.R. (1990). Parasite detection efficiencies of five stool concentration systems. *Journal of Clinical Microbiology* **28**, 1094–1097.

Phelps, K.R., Ginsberg, S.S., Cunningham, A.W., Tschachler, E. and Dosik, H. (1991). Case report: adult T-cell leukemia/lymphoma associated with recurrent *Strongyloides* hyperinfection. *American Journal of Medical Science* **302**, 224–228.

Plumelle, Y., Pascaline, N., Nguyen, D., Panelatti, G., Jouannelle, A., Jouault, H. and Imbert, M. (1993). Adult T-cell leukemia-lymphoma: a clinico-pathologic study of twenty-six patients from Martinique. *Hematology and Pathology* **7**, 251–262.

Proctor, E.M., Muth, H.A.V., Proudfoot, D.L., Allen, A.B., Fish, R., Isaac-Renton, J. and Black, W.A. (1987). Endemic institutional strongyloidiasis in British Columbia. *Canadian Medical Association Journal* **136**, 1173–1176.

Putland, R.A., Thomas, S.M., Grove, D.I. and Johnson, A.M. (1993). Analysis of the 18S ribosomal RNA gene of *Strongyloides stercoralis*. *International Journal for Parasitology* **23**, 149–151.

Rawat, B. and Simons, M.E. (1993). *Strongyloides stercoralis* hyperinfestation; another cause of focal hepatic lesions. *Clinical Imaging* **17**, 274–275.

Rio, B., Louvet, C., Gessain, A., Dormont, D., Gisselbrecht, C., Martoia, R. *et al.* (1990). Leucemies à cellules T de l'adulte et adenopathies non malignés associées au virus HTLV. A propos de 17 malades originaires des Caraibes et d'Afrique. *La Presse Médicale* **19**, 746–751.

Robinson, R.D., Williams, L.A.D., Lindo, J.F., Terry, S.I. and Mansingh, A. (1990). Inactivation of *Strongyloides stercoralis* filariform larvae in vitro by six Jamaican plant extracts and three commercial anthelmintics. *West Indies Medical Journal* **39**, 213–217.

Robinson, R.D., Lindo, J.F., Neva, F.A., Gam, A.A., Vogel, P., Terry, S.I. and Cooper, E.S. (1994). Immunoepidemiologic studies of *Stronglyoides stercoralis* and human T lymphotropic virus type I infections in Jamaica. *Journal of Infectious Diseases* **169**, 692–696.

Rotman, H.L., Yutanawiboonchai, W, Brigandi, R.A., Leon, O., Nolan, T.J., Schad, G.A. and Abraham, D. (1994). *Strongyloides stercoralis*: eosinophil-dependent immune-mediated killing of infective L3 in Balb/cBYJ mice. *American Journal of Tropical Medicine and Hygiene* Abstracts, 234.

Sampson, I.A. and Grove, D.I. (1987). Strongyloidiasis is endemic in another Australian population group: Indochinese immigrants. *Medical Journal of Australia* **146**, 580–582.

Sato, Y. and Shiroma, Y. (1989a). Concurrent infections with *Strongyloides* and T-cell leukemia virus and their possible effect on immune reponses of host. *Clinical Immunology and Immunopathology* **52**, 214–224.

Sato, Y. and Shiroma, Y. (1989b). Peripheral lymphocyte subsets and their responsiveness in human strongyloidiasis. *Clinical Immunology and Immunopathology* **53**, 430–438.

Sato, Y., Inoue, F., Matsuyama, R. and Shiroma, Y. (1990). Immunoblot analysis of antibodies in human strongyloidiasis. *Transactions of the Royal Society of Tropical Medicine and Hygiene* **84**, 403–406.

Sato, Y., Toma, H., Kiyuna, S. and Shiroma, Y. (1991). Gelatin particle indirect agglutination test for mass examination for strongyloidiasis. *Transactions of the Royal Society of Tropical Medicine and Hygiene* **85**, 515–518.

Sato, Y., Shiroma, Y., Kiyuna, S., Toma, H. and Kobayashi, J. (1994). Reduced efficacy of chemotherapy might accumulate concurrent HTLV-1 infection among strongyloidiasis patients in Okinawa, Japan. *Transactions of the Royal Society of Tropical Medicine and Hygiene* **88**, 59.

Satoh, M., Tsukidate, S., Fujita, K. and Yamamoto, K. (1991). Strongyloidiasis influences the elevation of adult T-cell leukemia-associated antigen antibody titer. *International Archives of Allergy and Applied Immunology* **96**, 95–96.

Schad, G.A. (1986). Cyclosporine may eliminate the threat of overwhelming strongyloidiasis in immunosuppressed patients. *Journal of Infectious Diseases* **153**, 1768.

Schad, G. A., Hellman, M.E. and Muncey, D.W. (1984). *Strongyloides stercoralis*: hyperinfection in immunosuppressed dogs. *Experimental Parasitology* **57**, 287–296.

Schad, G.A., Aikens, L.M. and Smith, G. (1989). *Strongyloides stercoralis*: is there a canonical migratory route through the host? *Journal of Parasitology* **75**, 740–749.

Schad, G.A., Smith, G., Megyeri, Z., Bhopale, V.M., Niamatali, S. and Maze, R. (1993). *Strongyloides stercoralis*: an initial autoinfective burst amplifies primary infection. *American Journal of Tropical Medicine and Hygiene* **48**, 716–725.

Schainberg, L., Scheinberg, M.A. and Mota, R.C. (1989). Recovery of *Strongyloides stercoralis* by bronchoalveolar lavage in a patient with acquired immunodeficiency syndrome in a heterosexual population in Zaire. *American Journal of Medicine* **87**, 486.

Shikiya, K., Zaha, O., Niimura, S., Ikema, M., Nakamura, H., Nakayoshi, T. *et al.* (1992). [Long term eradication rate of mebendazole therapy for strongyloidiasis.] *Kansenshogaku Zasshi* **66**, 354–359.

Shikiya, K., Zaha, O., Niimura, S., Uehara, T., Ohshiro, J., Kinjo, F. *et al.* (1994). [Clinical study of ivermectin against 125 strongyloidiasis patients.] *Kansenshogaku Zasshi* **68**, 13–20.

Shiwaku, K., Chigusa, Y., Kadosaka, T. and Kaneko, K. (1988). Factors influencing development of free-living generations of *Strongyloides stercoralis*. *Parasitology* **97**, 129–138.

Sipahi, A.M., Damião, A.O.M.C., Simionato, C.S., Bonini, N., Santos, M.A.A., de Moraes-Filho, J.P.P. *et al.* (1991). Small bowel bacterial overgrowth in strongyloidiasis. *Digestion* **49**, 120–124.

Smith, T., Bhatia, K., Barnish, G. and Ashford, R.W. (1991). Host genetic factors do not account for variation in parasite loads in *Strongyloides fuelleborni kellyi*. *Annals of Tropical Medicine and Parasitology* **85**, 533–537.

Speare, R. (1989). Identification of species of *Strongyloides*. In: *Strongyloidiasis: a Major Roundworm Infection of Man* (D.I. Grove, ed.), pp. 11–83. London: Taylor & Francis.

Stey, C., Jost, J. and Luthy, R. (1990). Extraintestine Strongyloidiasis bei erworbenem Immunmangelsyndrom. *Deutsches medizinische Wochenschrift* **115**, 1716–1719.

Stoopack, P.M. and Raufman, J.P. (1991). Aphthoid ulceration of the colon in strongyloidiasis. *American Journal of Gastroenterology* **86**, 639–642.

Sukhavat, K., Morakote, N., Chaiwong, P. and Piangjai, S. (1994). Comparative efficacy of four methods for the detection of *Strongyloides stercoralis* in human stool specimens. *Annals of Tropical Medicine and Parasitology* **88**, 95–96.

Sullivan, P.B., Lunn, P.G., Northrop-Clewes, C.A. and Farthing, M.J.G. (1992). Parasitic infection of the gut and protein-losing enteropathy. *Journal of Paediatric Gastroenterology and Nutrition* **15**, 404–407.

Takayanagui, O.M., Lofrano, M.M., Araújo, M.B.M. and Chimelli, L. (1995). Detection of *Strongyloides stercoralis* in the cerebrospinal fluid of a patient with acquired immunodeficiency syndrome. *Neurology* **45**, 193–194.

Thompson, J.R. and Berger, R. (1991). Fatal adult respiratory distress syndrome following successful treatment of pulmonary strongyloidiasis. *Chest* **99**, 772–774.

Tikhomirova, E.P. and Lysakova, L.A. (1989). [Intestinal enzyme activity in patients with strongyloidiasis.] *Medtsinskaia Parazitologiia i Parazitarnyē Bolezni (Moskva)* **5**, 33–36.

Torres, J.R., Isturiz, R., Murillo, J., Guzman, M. and Contreras, R. (1993). Efficacy of ivermectin in the treatment of strongyloidiasis complicating AIDS. *Clinical Infectious Diseases* **17**, 900–902.

van Swinderin, B., Steinbert, T.H. and Weil, G.J. (1994). *Strongyloides stercoralis*: the movie. *American Journal of Tropical Medicine and Hygiene* **51**, Abstracts, p. 121.

Vieyra-Herrera, G., Becerril-Carmona, G., Padua-Gabriel, A., Jessurun, J. and Alonso-de-Ruiz, P. (1988). *Strongyloides stercoralis* hyperinfection in a patient with the acquired immunodeficiency syndrome. *Acta Cytologica* **32**, 277–278.

Viney, M.E. and Ashford, R.W. (1990). The use of isoenzyme electrophoresis in the taxonomy of *Strongyloides*. *Annals of Tropical Medicine and Parasitology* **84**, 35–47.

Viney, M.E., Ashford, R.W. and Barnish, G. (1991). A taxonomic study of *Strongyloides grassi*, 1879 (Nematoda) with special reference to *Strongyloides fuelleborni* von Linstow, 1905 in man in Papua New Guinea and the description of a new subspecies. *Systematic Parasitology* **18**, 95–109.

Wehner, J.H., Kirsch, C.M., Kagawa, F.T., Jensen, W.A., Campagna, A.C. and Wilson, M. (1994). The prevalence and response to therapy of *Strongyloides stercoralis* in patients with asthma from endemic areas. *Chest* **106**, 762–766.

Wertheim, G. and Lengy, J. (1965). Growth and development of *Strongyloides ratti* Sandground, 1925, in the albino rat. *Journal of Parasitology* **51**, 636–639.

Whitworth, J.A.G., Morgan, D., Maude, G.H., McNicholas, A.M. and Taylor, D.W. (1991). A field study of the effect of ivermectin on intestinal helminths in man. *Transactions of the Royal Society of Tropical Medicine and Hygiene* **85**, 232–234.

Williams, J., Nunley, D., Dralle, W., Berk, S.L. and Verghese, A. (1988). Diagnosis of pulmonary strongyloidiasis by bronchoalveolar lavage. *Chest* **94**, 643–644.

Woodring, J.H., Halfhill, H. II and Reed, J.C. (1994). Pulmonary strongyloidiasis: clinical and imaging features. *American Journal of Roentgenology* **162**, 537–542.

Wurtz, R., Mirot, M., Fronda, G., Peters, C. and Kocka, F. (1994). Short report: gastric infection by *Strongyloides stercoralis*. *American Journal of Tropical Medicine and Hygiene* **51**, 339–340.

Yamada, M., Matsuda, S., Nakazawa, M. and Arizono, N. (1991). Species-specific differences in heterogonic development of serially transferred free-living generations of *Strongyloides planiceps* and *Strongyloides stercoralis*. *Journal of Parasitology* **77**, 592–594.

Yamaguchi, K., Matutes, E., Catovski, D., Galton, D.A.G., Nakada, K. and Takatsuki, K. (1987). *Strongyloides stercoralis* as a candidate co-factor for HTLV–1-induced leukaemogenesis. *Lancet* **ii**, 94–95.

Yoshida, S. (1920). A new course for migrating *Ancylostoma* and *Strongyloides* larvae after oral infection. *Journal of Parasitology* **7**, 46–48.

Young, R.L., Zund, G., Mason, B.A. and Faro, S. (1989). Pelvic inflammatory disease complicated by massive helminthic hyperinfection. *Obstetrics and Gynecology* **74**, 484–486.

Zaman, V., Dawkins, H.J.S. and Grove, D.I. (1980). Scanning electron microscopy of the penetration of newborn mouse skin by *Strongyloides ratti* and *Ancylostoma caninum* larvae. *Southeast Asian Journal of Tropical Medicine and Public Health* **112**, 212–219.

The Biology of the Intestinal Trematode *Echinostoma caproni*

Bernard Fried[1] and Jane E. Huffman[2]

*[1]Department of Biology, Lafayette College, Easton, PA 18042, USA and
[2]Department of Biological Sciences, East Stroudsburg University, East
Stroudsburg, PA 18301, USA*

1. Introduction to the Biology of *Echinostoma caproni* 312
2. Maintenance of the Cycle in the Laboratory 315
3. Adults 319
4. Eggs and Miracidia 326
5. Sporocysts and Rediae 330
6. Cercariae and Metacercariae 333
7. Infectivity of *E. caproni* in Intermediate Hosts 337
8. Echinostomes, Schistosomes and Snails 340
9. Infectivity, Growth and Development of *E. caproni* in Definitive Hosts 341
10. Concurrent Studies Using *E. caproni* and Other Helminths 343
11. Effects of *E. caproni* Infection on Pregnancy in Mice 344
12. Excystation, Implantation and *In Vitro* and *In Ovo* Cultivation of
    E. caproni 345
13. Pathobiochemical Effects of Larval and Adult *E. caproni* on Their Hosts . . . 347
14. Gross, Histopathological and Clinical Effects of Adult *E. caproni* on
    Their Hosts 350
15. Immunobiology of *E. caproni* in Definitive Hosts 354
16. Electrophoretic and Polymerase Chain Reaction Studies on *E. caproni* 357
17. Summary and Concluding Remarks 359
References 360

ADVANCES IN PARASITOLOGY VOL 38
ISBN 0–12–031738–9

1. INTRODUCTION TO THE BIOLOGY OF *ECHINOSTOMA CAPRONI*

Huffman and Fried (1990) reviewed the salient literature on the biology of various species of *Echinostoma*, but were concerned mainly with species of 37-collar-spined echinostomes in the *E. revolutum* group of Kanev (1985). Kanev's work considered the European–Asian form, *E. revolutum*, an echinostome of aquatic birds with first intermediate hosts in the genus *Lymnaea* (see Kanev, 1994), the African species *E. caproni* that lives mainly in rodent and avian definitive hosts and in planorbid snails of the genus *Biomphalaria* and *Bulinus* as first intermediate hosts, and a North American species, *E. trivolvis*, that lives in aquatic birds and mammals and uses planorbid snails in the genus *Helisoma* as first intermediate hosts (see Kanev *et al.*, 1995). Kanev (1985) considered *E. paraensei* a synonym of *E. caproni* based mainly on the fact that this echinostome also uses species of *Biomphalaria* as first intermediate hosts. Recent isoenzymatic studies by Sloss *et al.* (1995) and DNA sequencing studies by Morgan and Blair (1995) suggest that *E. paraensei* is a distinct species and their views are accepted in this review. Information on *E. paraensei* only as it relates to *E. caproni* is covered in this review.

Since the review by Huffman and Fried (1990) we have noted about 50 new papers on various aspects of the biology of *E. caproni* and realize that this species is a useful one for research in parasitology. The species is convenient for studies on trematodes since it cycles between the medically important *Biomphalaria glabrata* snail and laboratory mice. Therefore, all stages of the life cycle are available for experimentation. It should be remembered that relatively few digeneans are maintained in the laboratory and much of what we know about these trematodes is based on the blood fluke *Schistosoma mansoni* and the liver fluke *Fasciola hepatica*. The presence of an intestinal digenean that is readily available and easy to manipulate in the laboratory makes this echinostome a convenient one for research. The purpose of this review is to examine salient features on the biology, life history, infectivity, immunology, pathology, physiology, cultivation, biochemistry and molecular biology of this organism. The point of reference of this review is Huffman and Fried (1990) and *E. caproni* studies covered therein will only be mentioned briefly herein.

Because this organism has been referred to by different names, mainly *E. revolutum*, *E. liei* and *E. togoensis*, Table 1 presents a list of studies of *E. caproni* in which the other names were used. Moreover, several laboratories have begun work on *E. caproni* and either maintain its life cycle or obtain metacercarial cysts from one of us (BF). Table 2 lists some of the investigators and their research topics.

Table 1 Studies on *Echinostoma caproni* in which the organism was referred to as either *Echinostoma revolutum, E. liei* or *E. togoensis*

Authors	Subject of the study
Echinostoma revolutum	
Barus *et al.* (1974)[a]	Concurrent infections with *Echinoparyphium*
Bindseil and Christensen (1984)[a]	Histopathological effects in mice
Bindseil *et al.* (1989)	Concurrent infections in mice
Christensen (1980)[a]	Miracidial labelling and host finding
Christensen *et al.* (1981a)[a]	Resistance in mice
Christensen *et al.* (1981b)[a]	Resistance in mice
Christensen *et al.* (1984)[a]	Resistance in mice
Christensen *et al.* (1985)[a]	Resistance in mice
Christensen *et al.* (1986)[a]	Resistance in mice
Christensen *et al.* (1988)[a]	Resistance in mice
Simonsen and Andersen (1986)[a]	Antibody dynamics on worm surface
Sirag *et al.* (1980)[a]	Resistance in mice
Echinostoma liei	
Donovick and Fried (1988)	Scanning electron microscopy of adults
Fried and Emili (1988)[a]	*In vitro* excystation of metacercariae
Fried *et al.* (1988)[a]	Development of adults in chicks
Hosier *et al.* (1988)[a]	Resistance in mice
Huffman *et al.* (1988)[a]	Concurrent infections in hamsters
Jeyarasasingham *et al.* (1972)[a]	Life cycle studies
Jourdane *et al.* (1990)	Antagonism in snails with *Schistosoma mansoni*
Kuris (1980a, b)[a]	Effects of miracidial infection on snails
Kuris and Warren (1980)[a]	Cercarial penetration and encystment
Mohamed (1992)	Ultrastructure of intramolluscan stages
Moore *et al.* (1989)	Effects of adults on enterochromaffin cell proliferation in mice
Mounkassa and Jourdane (1990)	Snail leucocytic response to larvae
Riddell *et al.* (1991)	Neuropeptides in adults
Ross *et al.* (1989)	Studies on enzymes and pigments in adults
Schaefer *et al.* (1977)	Glucose metabolism in adults
Sullivan (1985)	Juvenile snails as second intermediate hosts
Thorndyke and Whitfield (1987)[a]	Neuropeptides in adults
Thorndyke *et al.* (1988)	Effects of adults on gastrointestinal hormones of the host
Voltz *et al.* (1988)[a]	Isoenzyme analysis of adults
Yousif and Haroun (1986)	Intramolluscan development in *Biomphalaria alexandrina*
Yousif *et al.* (1989)	Effects of miracidia on *B. alexandrina*
Echinostoma togoensis	
Jourdane and Kulo (1981)[a]	Life cycle studies
Jourdane and Kulo (1982)	Biological control studies
Voltz *et al.* (1985)	Chemotaxonomy studies
Voltz *et al.* (1986)	Chemotaxonomy studies

[a] Cited previously in Huffman and Fried (1990).

Table 2 Laboratories using *Echinostoma caproni* for various studies.

Laboratory	Study
David Blair Zoology Department, James Cook University, Townsville, Queensland 4811, Australia	DNA analysis of adults
Niels Ø. Christensen[a] Danish Bilharziasis Laboratory, Jaegersborg Alle ID, DK, Charlottenlund, Denmark	Population regulation of adults in rodent hosts
Douglas D. Colwell[a] Livestock Sciences Section, Agriculture Canada Research Station, PO Box 3000, Main Lethbridge, Alberta, Canada T1J4B1	Opioid peptides as mediators of the infection in rodents
Bernard Fried[a] Department of Biology, Lafayette College, Easton, PA 18042, USA	Intramolluscan studies; studies on the physiology and biochemistry of larval and adult stages
Takahiro Fujino Department of Parasitology, Faculty of Medicine, Kyushu University, Fukuoka 812, Japan	Expulsion phenomena in echinostomes
David W. Halton School of Biology and Biochemistry, The Queen's University, Belfast BTF 1NN, UK	Neuropeptides in larvae and adults; neuro-endocrinology in infected mice
Peter Hotez[a] MacArthur Molecular Parasitology Center, Yale University 700 LEPH, PO Box 3333, New Haven, CT 06510, USA	Proteases in larvae and adults
Donald Hosier Department of Biology, Moravian College Bethlehem, PA 18018, USA	Immunobiology of adults in mice
Jane E. Huffman Department of Biology, East Stroudsburg University, East Stroudsburg, PA 18304, USA	Biology of the adult in the golden hamster
Paul Nollen[a] Department of Biological Sciences, Western Illinois University, Macomb, IL 61455, USA	Biology and behaviour of larval and adult stages; reproductive physiology of the adults

[a] Life cycle is maintained in the laboratory; the other workers obtain metacercarial cysts from one of us (BF) to infect rodents.

2. MAINTENANCE OF THE CYCLE IN THE LABORATORY

The life cycle of *E. caproni* can be maintained in the laboratory between *Biomphalaria glabrata* snails and laboratory mice. A schematic of the life cycle of *E. caproni* is shown in Figure 1 and second intermediate hosts and definitive hosts of *E. caproni* are shown in Table 3.

The following information is based on its routine maintenance by one of us (BF) beginning with the encysted metacercarial stage. Encysted meta-cercariae can be maintained for about 6 months in Locke's 1: 1 solution at 4 ± 1°C (see Fried and Emili, 1988 for details). Cysts of this species are shipped by mail to workers who use this echinostome for teaching or research purposes. A list of some of these workers is provided in Table 2.

Instructions for maintaining the life cycle beginning with the encysted metacercarial stage are given below. The appearance of a viable cyst is shown in Figure 2. A discussion of the characteristics of viable versus non-viable cysts of echinostomes together with photomicrographs is given in Huffman and Fried (1990).

To obtain adult parasites, 6–8-week-old female ICR mice are each infected with 25–50 cysts of *E. caproni* by stomach tube. Instructions for handling and infecting small rodents with parasites are given by Voge (1970). Male mice and strains other than ICR also may be used, but

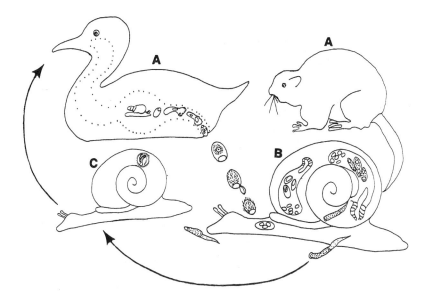

Figure 1 Generalized life cycle of *Echinostoma caproni*. A, definitive hosts; B, first intermediate host; C, second intermediate host.

Table 3 First and second intermediate hosts and definitive hosts of *Echinostoma caproni.*

First intermediate hosts (Gastropods)	Second intermediate hosts (Gastropods and Amphibians)	Definitive hosts (Birds and Mammals)
	Gastropods	
Biomphalaria alexandrina (N, E)	*Biomphalaria alexandrina* (N, E)	**Experimental** mice, hamsters, rats,
Biomphalaria camerunensis	*Biomphalaria glabrata*	domestic chicks,
Biomphalaria glabrata	*Biomphalaria pfeifferi*	pigeons, finches
Biomphalaria pfeifferi (N, E)	*Biomphalaria straminea*	(*Lochuru striata*)
	Numerous species of ***Bulinus*** **as follows:**	**Natural**
	B. permembranaceus;	Domestic ducklings,
	B. liratus; *B. forskalii*;	*Rattus rattus*,
	B. tropicus;	The Egyptian giant
	B. reticulatus; *B. truncatus*	shrew, *Crocidura olivieri*; *Falco newtoni*
	Other snails *Helisoma duryi* *Lymnaea natalensis* *Physa acuta* *Planorbarius corneus*	
	Amphibians Kidney of tadpoles, *Rana esculenta* and in *Ptychadaena mascareniensis*	

Compiled from Richard (1964), Jeyarasasingam *et al.* (1972), Richard and Brygoo (1978) and Christensen *et al.* (1980). N = natural infection; E = experimental infection. No designation after the intermediate host indicates that the infection is an experimental one. Numerous strains of *Biomphalaria glabrata* have been used as experimental intermediate hosts with varying degrees of success.

considerable success has been obtained with ICR female mice (see Hosier and Fried, 1991; Kaufman and Fried, 1994). The mice are lightly anaesthetized with diethyl ether and killed by cervical flexure about 2 to 3 weeks post-infection. Eggs can easily be recognized in wet mounts by examining the faeces of infected hosts. The amber colour, operculum and lack of embryonic development in the eggs should be noted. About 50% worm yield can be expected at necropsy. Worms will be found mainly in the ileum and jejunum and will mainly be clustered. The infected areas of the

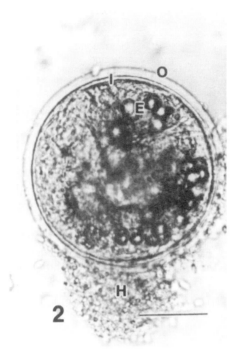

Figure 2 A viable cyst of *E. caproni* from an experimentally infected *Biomphalaria glabrata*. Scale bar = 100 μm. Note the outer cyst (O), inner cyst (I), excretory concretions (E), and snail haemocytes (H) attached to the cyst.

small intestines will be dilated. The intestines are then removed into a saline solution (we routinely use full-strength Locke's solution consisting of NaCl, 9 g l^{-1}; KCl, 0.4 g l^{-1}; $CaCl_2$, 0.2 g l^{-1}; and $NaHCO_3$, 0.2 g l^{-1}) for handling the adults, and the gut opened longitudinally. The worms can be removed with a pipette and transfered to fresh Locke's solution. The worms can be fixed for histological purposes or used live for biological, physiological, biochemical or behavioural studies. For life cycle maintenance studies, the eggs are removed with needles from the uteri of worms; expect to get about 200–400 eggs per worm. Worms can also be homogenized in a Waring blender as described in Fried and Weaver (1969) for studies on *E. trivolvis* (see Figure 3). Eggs should then be transferred to a Petri dish (about 1000 eggs) containing 30 ml of artificial spring water (see Ulmer, 1970 for the formula of such a solution). Deionized water or dechlorinated tap water can also be used for maintaining and embryonating the eggs. The eggs should be kept in the dark at 28°C for 10 days or at 22–24°C for 15 days until eggs contain fully developed miracidia (see Figure 4).

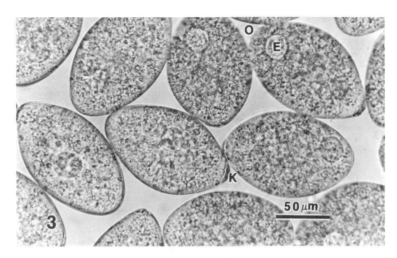

Figure 3 Eggs of *E. caproni* obtained following homogenization of gravid adults. Note the embryo (E), the operculum (O) and the knob (K).

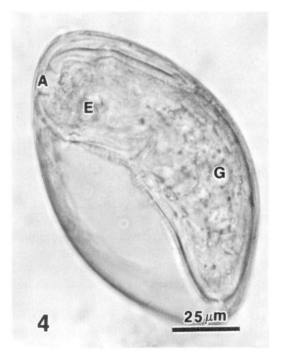

Figure 4 Egg of *E. caproni* with a fully developed miracidium. Note the apical papilla (A), eyespots (E) and germ cells (G).

Eggs with fully developed miracidia are useful for infecting 5–10 mm shell diameter *Biomphalaria glabrata*. Such eggs should be placed in glass or plastic containers in the approximate ratio of 15–20 eggs per snail. A container should hold about 20 snails in approximately 1000 ml of artificial spring water. At least one-half the snails can be expected to have patent infections (cercariae will be released into the water when snails are isolated) at 5–6 weeks post-infection when maintained at 22–24°C in aerated jars. Cercariae released from snails will infect the pericardium–kidney regions of infected and uninfected snails in the same aquarium. By 7–10 weeks post-infection each snail in an aquarium will contain hundreds of metacercarial cysts in the pericardial–kidney region. There will be relatively high death rates of snails with both redial and metacercarial infections. Cysts from the pericardial–kidney regions of infected snails should be removed and stored in vials with about 1000 cysts/15 ml of Locke's 1:1 solution at 4°C. The cysts are viable as determined by chemical excystation (see Fried and Emili, 1988) or *in vivo* infectivity studies (see Hosier and Fried, 1991; Kaufman and Fried, 1994).

3. ADULTS

Echinostoma caproni becomes an ovigerous adult in domestic chicks, mice and hamsters at 7 days post-infection (PI) (Fried *et al.*, 1988; Hosier and Fried, 1991; Yao *et al.*, 1991) and at 8 days PI in chick embryos (Chien and Fried, 1992). However, development is delayed when hosts are infected with heavy cyst inocula (>50 metacercariae per host; see Manger and Fried, 1993). A good description of 2- to 3-week-old ovigerous adults from hamsters has been provided by Jeyarasasingam *et al.* (1972). Because this echinostome can be maintained in mice and hamsters for at least 4 months, it is a good subject for studies on the effects of worm ageing in rodent hosts (Hosier and Fried, 1991; Yao *et al.*, 1991; Ursone and Fried, 1995).

Excysted metacercariae are non-progenetic and contain only genital anlage. Preovigerous adults from rodents or chicks show distinct testes by day 2 or 3, an ovary distinct from the ootype by day 4 and coiling of the uterus by day 5; the vitellaria are present by day 6 and the worms become ovigerous by day 7 or 8 (Fried *et al.*, 1988; Chien and Fried, 1992; Manger and Fried, 1993). During development from the excysted metacercaria to the ovigerous adult in rodents or chicks, the worm body area may increase some 80 times (about 0.02 mm^2 as the excysted metacercaria to 1.6 mm^2 for the 7-day-old ovigerous worm). Details of the morphogenesis of this organism from excysted metacercaria to ovigerous adult, with the exception of brief light microscopical observations (Fried *et al.*, 1988; Manger and Fried, 1993), are lacking.

A line drawing of a 14-day-old adult grown in a hamster has been given by Jeyarasasingam *et al.* (1972, their Figure 16) and is typical of the appearance of *E. caproni*. In living adults, the region anterior to the acetabulum is attenuated (Figure 5) and the length of 2–3-week-old worms is about 5 mm. Because of considerable extension and contraction of these echinostomes, measurements of live organisms are not reliable. To obtain

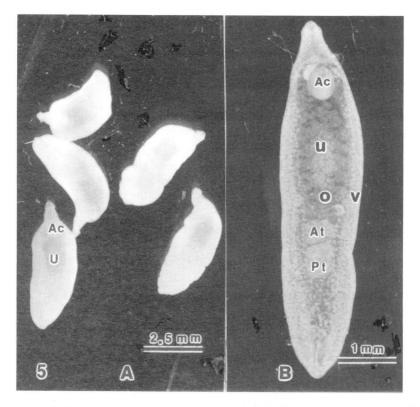

Figure 5 Live *E. caproni* adults from experimentally infected ICR mouse at 2 weeks post-infection. A, dark field at low magnification; B, dark field at higher magnification. Ac, acetabulum; At, anterior testis; O, ovary; Pt, posterior testis; U, uterus; V, vitellaria.

consistent measurements, organisms should be fixed without flattening in hot (80–90°C) alcohol–formalin–acetic acid (AFA) and then prepared as stained whole mounts.

Live *E. caproni* adults are attenuated anteriorly, and the broadest region is just posterior to the acetabulum (Figure 5). The posterior aspect of the worm tapers almost to a point (see Figure 5). *E. caproni* is more attenuated posteriorly than either *E. trivolvis* (see Figure 1 in Huffman and Fried,

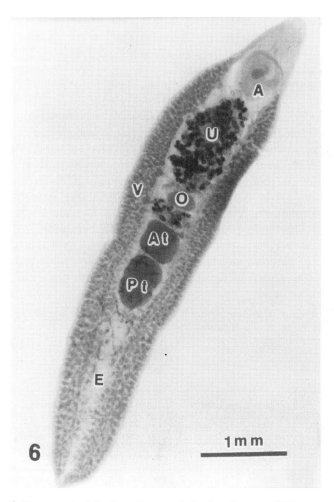

Figure 6 E. caproni adult, 2 weeks post-infection from an ICR mouse, fixed in hot alcohol–formalin–acetic acid and stained in Gower's carmine. Note the acetabulum (A), anterior testis (At), excretory bladder (E), ovary (O), posterior testis (Pt), uterus (U), and vitellaria (V).

1990) or *E. paraensei* (see Figure 15 in Lie and Basch, 1967). A characteristic of some live adults of *E. caproni* is the apparent confluence of the testes (see Figure 1 in Huffman and Fried, 1990), but fixed and stained specimens show two distinct testes (see Figure 6) and the area of the posterior testis is greater than that of the anterior testis. Other characteristics of live organisms include a red uterus, amber-coloured eggs, yellow–orange vitellaria, white gonads and seminal vesicles.

The length of live adults ranges from 2 to 8 mm with the maximum width about one-third the length (see Figure 5). Dry weights for ovigerous worms grown for 1, 2 and 3 weeks in domestic chicks or ICR mice average about 0.1, 0.5 and 0.7 mg/worm, respectively (Fried *et al.*, 1988; Hosier and Fried, 1991). As mentioned in Huffman and Fried (1990), numerous factors influence the size of this worm including the host, cyst inoculum, number of worms recovered, crowding, worm microhabitat in the small intestines, and other factors. Most trematodologists rely on body area measurements for size determinations with the body area being considered proportional to the length × maximal width of the worm (see Berntzen and Macy, 1969; Shostak *et al.*, 1993; Ursone and Fried, 1995). The body area of digeneans is usually determined from whole worms stained in carmine after fixation in hot AFA. The body area of *E. caproni* adults is about the same in domestic chicks and ICR mice (see Table 4), but is greater in worms grown in hamsters and less in worms grown in chick embryos. If relatively large worms are required for research, the hamster is an ideal host. Sexual maturation with eggs capable of producing viable miracidia that infect *Biomphalaria glabrata* can be achieved in hamsters, mice and domestic chicks. Worms grown in the allantoic cavity of the domestic chick are

Table 4 Average body area (mm^2) of *Echinostoma caproni* adults at various days post-infection in experimental hosts.

Host	Number of days post-infection[a]			
	7	14	21	28
Hamsters	2.5	10	12	13
Domestic chicks	1.5	4.7	6.2	–
Mice	1.3	4.0	5.0	7.0
Allantois	0.5	1.3[b]	–	–
Chick chorioallantois	0.4	0.7	–	–

[a] Worms were ovigerous on day 7 in all these studies, except on day 8 for worms grown in the allantois or the chick chorioallantois. Data compiled from various sources as follows: Fried *et al.* (1988); Hosier and Fried (1991); Chien and Fried (1992); Rosa-Brunet and Fried (1992); Yao *et al.* (1991).
[b] Refers to day 11 data, the oldest day in which these flukes were grown in the allantois.

capable of producing eggs that contain miracidia, but infectivity studies in *Biomphalaria glabrata* snails have not been done (Chien and Fried, 1992).

Recent studies have used scanning electron microscopy (SEM) (Fried *et al.*, 1990a; Ursone and Fried, 1995) to show topographical features of the tegument including developmental changes in the collar spines from the excysted metacercaria to the adult; changes in tegumentary spines from single-pointed in young adults to multiple-pointed in older adults; the appearance of uniciliate and aciliate papillae on the tegument; the appearance of the spinose ventral tegument compared to the aspinose dorsal tegument; the appearance of the smooth cirrus in this species confirming a previous observation by light microscopy (LM) of this structure by Jeyarasasingam *et al.* (1972).

Light microscopy (LM) observations of the collar spines, suckers, digestive, excretory and reproductive systems have been described by Huffman and Fried (1990). The appearance of ovigerous adults by light microscopy are shown in Figures 5 and 6. Selected scanning electron micrographs of adults are also shown in Figures 7 and 8.

Adults of *E. caproni* have been used in various biological studies some of which are summarized below.

Thorndyke and Whitfield (1987) did immunocytochemical studies on the tegument of *E. caproni*. Tests with well-characterized antisera to mammalian vasoactive intestinal polypeptide (VIP) showed a subpopulation of tegumental cells with material immunologically similar to VIP. Control experiments with antisera for a related peptide together with absorption of primary antisera with VIP or secretin confirmed the specificity of the reaction. Immunoreactivity could be traced from the tegumentary cells to the distal cytoplasm of the tegument. The finding of VIP-like immunoreactive material in nonneuronal and non-neurosecretory-like cells in this intestinal digenean of rodents is of interest in terms of helminth biology and host pathophysiology. Richard *et al.* (1989) also did immunocytochemical studies using antibodies against various peptides of adult *E. caproni* and noted the presence of substance P or a related peptide. Substance P-immunoreactive neurons were observed in the nervous system and consisted of two populations according to their size; labelled axons were seen in the ventral and dorsal nerve cords. The comparative study of immature 6-day-old worms and sexually mature 14-day-old worms showed an increase in substance P-immunoreactivity in the mature worms. Strong immunoreactivity for substance P was seen in the prostate cells of sexually mature worms, but no reactivity occurred in immature worms suggesting that immunoreactivity was related to maturation. Riddell *et al.* (1991) described phenylalanyl-methionyl-arginyl-phenylalamine-amide-like peptides (usually referred to as FMRF) in the nervous system and presumptive neurosecretory cells of *E. caproni* and suggested that these substances

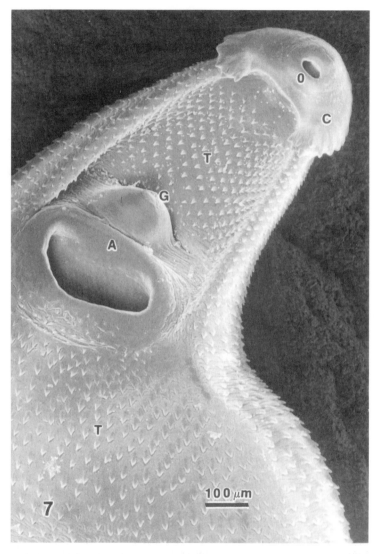

Figure 7 Scanning electron micrograph (ventral aspect) of a 15-day-old *E. caproni* adult from the golden hamster. Note the oral sucker (O), oral collar with cephalic spines (C), genital pore (G), acetabulum (A), and tegumentary spines (T). Micrograph courtesy of Sam Irwin.

function as neurotransmitters. They also described a single neuroendocrine-like cell with FMRFamide-like immunoreactivity in the cirrus pouch of the male reproductive system.

Nollen (1990) used this echinostome to study mating behaviour and

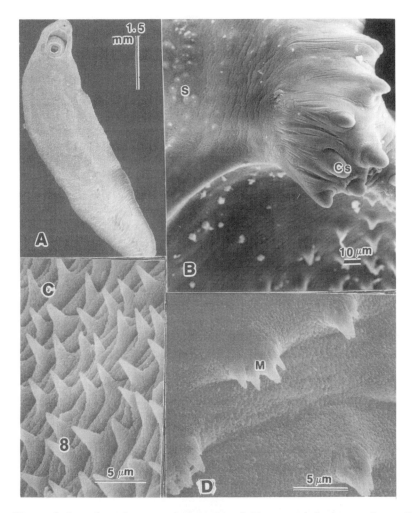

Figure 8 Scanning electron micrographs of *E. caproni* from experimentally infected mice except for B from the golden hamster. A, ventral view of a 12-week-old worm; B, collar region from a 15-day-old worm showing sensory papillae (S) and collar spines (Cs); C, ventral view of a 2–week-old worm showing tegumentary spines; D, ventral view of a 10-week-old worm showing multipointed (M) tegumentary spines. Micrographs A, C and D reproduced with the permission of the editors of *Parasitology Research* from Ursone and Fried (1995). Micrograph B courtesy of Sam Irwin.

other aspects of trematode reproductive physiology. The development and movement of reproductive cells were determined in these digeneans on autoradiograms by labelling nuclei of stem cells during exposure to [³H]thymidine and then surgically transplanting the worms to laboratory mice. The development and movement of sperms, primary oocytes, and vitelline cells were more rapid in *E. caproni* than other digeneans investigated previously. Mating behaviour was determined by labelling the sperms of a single adult *in vitro* by exposure to tritiated tyrosine and then transplanting the single worm alone or with unlabelled worms to the mice for 4 and 6 days. *E. caproni* adults self-inseminated when isolated and self- and cross-inseminated when in groups. Unrelated to the observations on reproductive physiology was the occurrence of an increased number of structural anomalies in transplanted worms with the most common anomaly being an accumulation of vitelline cells in the vitelline reservoir and ducts. The significance of structural anomalies in digenetic trematodes is not clear and this subject has been reviewed by Bakke (1988).

4. EGGS AND MIRACIDIA

The eggs of *E. caproni* are oval in shape, amber in colour, operculate and contain a distinct knob at the abopercular end (Figure 3). Mean measurements ($n = 25$) of the eggs as determined by Krejci and Fried (1994) are (in µm): length, 117; width, 75; opercular diameter, 20; abopercular diameter, 18; embryo diameter, 23. Scanning electron microscopy of the eggs revealed that the topography of the operculum was smooth and the operculum was surrounded by a ridge in the eggshell; the eggshell topography was also smooth; the abopercular knob of the *E. caproni* egg was shallow and the knob is probably a point of attachment of the forming shell with the ootype of the worm (Krejci and Fried, 1994). For additional light microscopical descriptions of the eggs see Richard and Brygoo (1978), Jeyarasasingam *et al.* (1972) and Christensen *et al.* (1980).

Eggs of this species appear in the faeces of various definitive hosts by 9–10 days (Jeyarasasingam *et al.*, 1972; Huffman and Fried, 1990). They can be obtained from the faeces by sedimentation (Pritchard and Kruse, 1982; Huffman and Fried, 1990) or from adults by homogenizing worms in a Waring blender as described for *E. trivolvis* by Fried and Weaver (1969). They can also be obtained from the uteri of gravid worms. The ovum within the egg capsule is undeveloped when laid and the developing embryo is surrounded by yolk globules. It takes 10 days at 28°C in spring water to obtain an egg with a fully developed miracidium (see Figure 4). To prevent premature hatching, eggs should be kept in the dark until they contain fully developed miracidia. Exposure of eggs with fully developed miracidia to a light stimulus results in a synchronous hatch. Methods of obtaining a synchronous hatch have been described by Behrens and Nollen (1993) who found that *E. caproni* eggs produced fully developed miracidia from hamster-source eggs in 9 days and from mouse-source eggs in 10 days under either light or dark conditions at 27°C. Incubation of egg cultures under constant light resulted in miracidial hatching from hamster-source eggs in 11 days and from mouse-source eggs in 13 days. Exposure to light was essential to trigger hatching, with incandescent light providing more consistent stimulation than fluorescent light. Eggs stored in a dark environment for 14 days required 6 days to reach maximal hatching after exposure to light. Eggs stored in the dark for 46 and 56 days hatched on the same day after exposure to light. Miracidia that hatched after 56 days of dark storage showed aberrant swimming behaviour; miracidia stored in a dark environment for 70 days or longer did not hatch when exposed to light.

E. *caproni* is suitable for laboratory infection studies in *Biomphalaria glabrata* because it is possible to get a uniform population of miracidia. Infectivity studies of *E. caproni* miracidia in *Biomphalaria* have been described by Huffman and Fried (1990). Large-scale laboratory maintenance studies on *E. caproni*, with observations on echinostome/snail host combinations have not been done. An excellent model for such studies is the work of Lewis *et al.* (1986) on *Schistosoma mansoni* with considerable details on schistosome/snail host combinations.

A recent technique developed in the Laboratory of Medical Helminthology, Yale University, New Haven, CT, by Mr Brian Jones has been used to obtain eggs of *E. caproni* for molecular biology studies. Using this technique, *E. caproni* adults, 2 to 3 weeks old, are obtained from experimentally infected mice or hamsters (see Section 3). Worms are washed rapidly in two or three changes of sterile Locke's solution, and then transfered to a RPMI–antibiotic medium (consisting of RPMI 1640; 25 mM HEPES, pH 7.2; 100 units ml^{-1} penicillin; 100 μg ml^{-1} streptomycin; 100 μg ml^{-1} gentmycin; 100 μg ml^{-1} fungizone=amphotericin B, 1X). Approximately 10 ml of medium is used per 30 worms in a 6 cm Petri dish for 20 min at 37°C. The worms are then transferred (about 30 worms/10 ml of fresh medium) to a fresh 6 cm Petri dish and incubated at 37°C with a gas phase of 95% air and 5% CO_2. Eggs released from the worms are noticeable on the bottom of the Petri dish after a few hours. Worms usually remain alive for about 3 days under the conditions described herein. Eggs are collected by swirling the Petri dish and removing the eggs that collect in the centre of the dish with a pipette.

The effects of environmental conditions on *E. caproni* eggs have been studied by Christensen *et al.* (1980) who found that freezing, desiccation and pH 5 killed the eggs immediately or after partial development; salinity of about 4 parts per thousand and temperatures greater than 33°C reduced the development capability. Likewise, temperatures of 6 and 8°C and maintenance of eggs in mouse faeces interfered with development. Increasing temperatures decreased the time of development from 40 days at 18°C to nine days at 35°C. Maintenance of unembryonated eggs for 14 weeks in faeces at 12°C and for 20 weeks in filtered pond water at 4°C allowed for subsequent normal development. Embryonated eggs maintained at 4°C showed unchanged hatchability for 3 weeks.

Light microscopical (LM) studies of *E. caproni* miracidia carried out by Richard and Brygoo (1978) and Jeyarasasingam *et al.* (1972) showed typical morphological features of echinostomatid miracidia, i.e. ciliated epidermal plates, apical papillae, paired eyespots, germinal cells (see Figure 9). The body size of live specimens was in the range 105–110 μm by 43–52 μm. Miracidia fixed in hot 2% silver nitrate showed the typical epidermal plate formula for members of the 37-collar-spined *Echinostoma*, i.e. 6–6–4–2 (row 1 of 6 nearly triangular plates, 2 dorsal, 2 ventral and 2 lateral; row 2 of 3 nearly square plates, 3 dorsal and 3 ventral; row 3 of 4 elongate plates, 1 dorsal, 1 ventral and 2 lateral; and row 4 of 2 square plates, 1 dorsal and 1 ventral). Miracidia are short lived but retain infectivity for up to 8 hours. Detailed studies on the longevity of these miracidia at different environmental temperatures have not been done. Various factors influence infectivity of miracidia to *B. glabrata* including age of snail, number of miracidia, age of miracidia, volume of water, but detailed studies on the biotic and abiotic factors that influence miracidial penetration have not been studied (see Section 7). Strain differences in susceptibility of snails in the genus *Biomphalaria* have been noted (Jeyarasasingam *et al.*, 1972; Christensen *et al.*, 1980).

Lie *et al.* (1980) showed that *B. glabrata* exposed to 37-collar-spined echinostome miracidia developed miracidial immobilizing substances in their haemolymph, i.e. substances capable of killing echinostome miracidia *in vitro*. They noted that of numerous species of echinostome-infected albino *B. glabrata* laboratory strains studied, more than 85% of the snails developed miracidial immobilizing substances (MIS) in the haemolymph, whereas less than 5% of control uninfected snails showed this capability. Snails infected with *Echinostoma lindoense* showed a strong miracidial immobilizing test (MIT) when homologous miracidia were exposed to the haemolymph and a moderate response when *E. caproni* miracidia were used. Infection with *E. paraensei* resulted in a high level of haemolymph MIS with *E. lindoense* miracidia, but a weak level when haemolymph was tested against *E. caproni* miracidia as well as the homologous *E. paraensei* miracidia. Infection with *E. caproni* induced a strong MIT with *E. lindoense* miracidia but only a moderate one when using homologous miracidia. Production of haemolymph MIS to echinostome miracidia was temporary and it began one day post-exposure, reached a maximum at

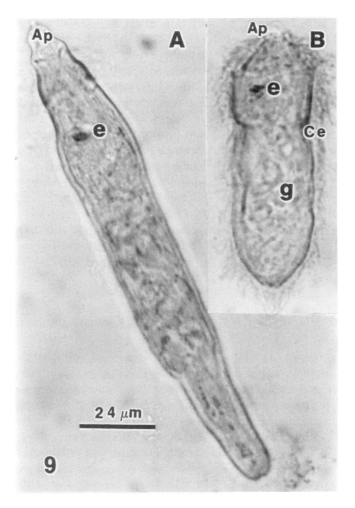

Figure 9 Miracidia of *E. caproni*. In A, the miracidium has been stained intravitally with 0.01% aqueous neutral red. In B, the miracidium was fixed in 10% neutral buffered formalin and stained with alcian blue, pH 2.5. Note the apical papilla (Ap), ciliated epidermis (Ce), eyespots (e) and germinal (g) cells. Scale bar is appropriate for A and B.

10–14 days and declined to the pre-infection level by 28 days PI. Information on the chemical make-up of MIS is not available.

Joky *et al.* (1985) exposed *B. glabrata* to *E. caproni* miracidia to determine the effects of larval exposure on the response of the amoebocyte-producing organ (APD) of the snail. Miracidial stimulation triggered the activity of the APD (this organ is associated with the snail kidney and is the site of amoebocyte formation). Amoebocyte production began soon after miracidial exposure, was maximal at 3 or 4 days post-exposure, and then declined rapidly by day 7 post-exposure.

Behrens and Nollen (1992) studied the responses of *E. caproni* miracidia to gravity, light and chemicals. In a four-tube vertical system, these miracidia showed a strong negative geotaxis which was dominated by a positive phototaxis. A positive phototaxis was also demonstrated using horizontal chambers. Using phi-chambers (for a description of a phi-chamber, see Roberts *et al.*, 1978), these miracidia showed positive chemotaxis to glutamic and aspartic acids, but not leucine. Positive responses were also demonstrated using snail conditioned water (SCW) from *Biomphalaria glabrata* and dilute solutions of sulphuric and acetic acids, but ammonia, Mg^{2+}, and HCl produced no significant chemotaxis. Behrens and Nollen (1992) noted that the responses of *E. caproni* and *S. mansoni* miracidia (both of which develop in *B. glabrata*) were similar, suggesting that miracidia mimic the behavioural patterns of compatible snail species.

5. SPOROCYSTS AND REDIAE

Miracidia penetrate the soft tissues of *Biomphalaria glabrata* and transform into sporocysts, but detailed observations on penetration have not been reported. Sporocysts of this species develop within the heart but studies on the biology of transformation from miracidium to sporocyst are not available. Descriptions of sporocysts are given by Jeyarasasingam *et al.* (1972) who noted that these larvae develop in the ventricle of the heart and can also be found in clusters in the aorta within 1–2 weeks PI of miracidia in juvenile *B. glabrata*. Sporocysts are amorphous sacs containing mother rediae at 2 weeks PI. Details of morphogenesis from the miracidium to mother sporocyst are not available. Six-day-old sporocysts ($n = 20$) averaged 410 μm in length and 153 μm in width; by 13 days PI ($n = 20$), sporocysts averaged 710 μm in length and 278 μm in width. By 2 weeks PI sporocysts contained 7–8 mother rediae and numerous germinal balls; sporocysts lack birth pores and ambulatory buds. SEM or transmission electron microscope (TEM) observations of sporocysts are not available. Temperature used to study early development of *E. caproni* in snails was not stated by Jeyarasasingam *et al.* (1972), but presumably was in the 24–27°C range. Intramolluscan development of *E. caproni* in our laboratory (see Section 6) at room temperature (about 22°C) is slower than that reported by Jeyarasasingam *et al.* (1972).

Descriptions of first generation (mother) and second generation (daughter) rediae are given by Jeyarasasingam *et al.* (1972). First generation rediae developed in sporocysts in the snail ventricle or aorta, and remained in the aorta until about day 10, and then migrated to the digestive gland gonad (DGG) complex of the snail; details of the migration are not available. Length and width of mother rediae (12 days old) averaged (n = 50) 1340 and 212 µm, respectively. Rediae contained a pharynx, saccular gut, ambulatory buds, dorsal birth pore and conspicuous collar. Mother rediae of this age possessed about six daughter rediae and numerous germ balls. Several first-generation rediae were in the snail ovotestis by 15 days PI along with numerous immature second generation rediae. Immature rediae of both generations can not be distinguished, but mature second generation rediae can be determined from mother rediae by the presence of cercariae within the redia. Second generation rediae containing both rediae and cercariae were rare. The size of second generation rediae was variable depending on the age of rediae and factors such as redial crowding in the snail. Daughter rediae (n = 30) at day 30 ranged in length from 1.3 to 3.0 mm and in width from 0.2 to 0.4 mm. The number of mature cercariae in rediae ranged from 10 to 15 per redia plus numerous germ balls. The daughter redia also has a pharynx, saccular gut, collar, birth pore and ambulatory buds (Figures 10 and 11). The redial gut contains orange–red pigment, but the identity of the pigment has not been determined. SEM of the daughter redia (Figure 11) has been studied by Krecji and Fried (1994).

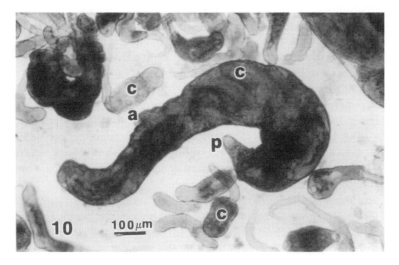

Figure 10 Live daughter redia removed from the digestive gland of an experimentally infected *B. glabrata* snail. Some cercariae (c) in the field as well as in the redia. Redial gut is not evident in this micrograph, but ambulatory bud (a) and pharynx (p) are visible.

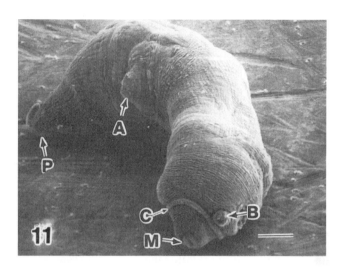

Figure 11 Scanning electron micrograph of a daughter redia of *E. caproni*. Note the presence of an ambulatory bud (A), birth pore (B), collar (C), mouth (M), and papilliform process (P). Scale bar = 91 μm. Micrograph reproduced with the permission of the editors of *Parasitology Research* from Krejci and Fried (1994).

Mohamed (1992) used TEM to study the ultrastructure of developing cercariae of *E. caproni*. He also described the ultrastructure of daughter rediae within the DGG complex of infected *B. alexandrina*. The tegument consisted of a layer with numerous microvilli on the outer surface. Developing cercariae were covered with a thin nucleated primitive temporary epithelium, which was lost when the true tegument was formed beneath it. The mechanism of tegumentary spine formation was not apparent, but there was an indication that the spines were associated with a thickened part of the basal membrane of the tegument.

Beers *et al.* (1995) provided LM and TEM information on daughter rediae of *E. caproni* from *B. glabrata* that had been maintained on a high fat diet of hen's egg yolk. The rediae showed deposits of neutral fat in the tegument and in subtegumentary spaces but the significance of fat accumulation was not determined.

Richard and Voltz (1987) used daughter rediae of *E. caproni* from experimentally infected *B. glabrata* treated in colchicine and stained in Giemsa to determine the karyotype of this echinostome. Eleven pairs of chromosomes (2n=22) were found and the number 4 and 5 pairs were submetacentric; the other pairs were acrocentric. C-banding methods showed a large block of heterochromatin in the number 3 pair and two blocks in pair number 5. Comparisons of studies on karyotypes of other echinostomes are given in Richard and Voltz (1987).

6. CERCARIAE AND METACERCARIAE

B. glabrata infected with miracidia of *E. caproni* and maintained in aerated aquaria at 22°C release cercariae within 5–6 weeks post-infection (Fried *et al.*, 1988). Cercarial longevity is about 24 h at 22°C, but cercariae are maximally active for about 12 h at room temperature and essentially moribund after that. Morphological studies by LM on freshly emerged cercariae have been done by Jeyarasasingam *et al.* (1972) and by Richard and Brygoo (1978). The cercaria is gymnocephalous with a collar and spines and is similar to other cercariae in the 37-collar-spined *E. revolutum* group (see Huffman and Fried, 1990). The tip of the tail is attenuated and bent at an approximate 45 degree angle and the tail has numerous finfolds (see Figure 12). The attenuated tip and finfolds distinguish the 37-collar-spined *Echinostoma* cercariae from *Echinoparyphium* cercariae that lack finfolds and have a straight tail. Mean dimensions of cercariae of *E. caproni* fixed in cold 5% formalin (*n* = 30) from Jeyarasasingam *et al.* (1972) are as follows (in μm): body length, 328; body width, 144; collar length, 80. Cystogenous glands are obvious in live cercariae and occupy most of the body. This cercaria is noted by the absence of para-oesophageal

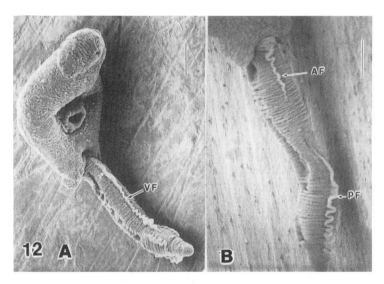

Figure 12 Scanning electron micrographs of cercariae of *E. caproni*. A, ventral aspect of the cercaria. Note the ventral finfold (VF) on the cercarial tail. Scale bar = 70 μm. B, dorsal aspect of the cercarial tail. Note the anterior finfold (AF) and posterior finfold (PF). Scale bar = 51 μm. Micrographs reproduced with the permission of the editors of *Parasitology Research* from Krejci and Fried (1994).

glands (see Fried and Rosa-Brunet, 1991a) that are prominent in other 37-collar-spined *Echinostoma* and the absence of these glands is associated with the inability of *E. caproni* cercariae to penetrate the epidermis of the planarian *Dugesia tigrina*. By comparison, the allopatric species, *E. trivolvis*, has para-oesophageal glands and can penetrate *D. tigrina* (see Fried and Rosa-Brunet, 1991a).

Argentophilic staining of the cercaria of *E. caproni* has shown papillae patterns on the dorsal and ventral surfaces of the body and tail of the cercaria (Jeyarasasingam *et al.*, 1972). Detailed descriptions of argento-philic-staining structures (mainly tegumentary papillae) of *E. caproni* cercariae including details of the collar region have been given by Richard and Brygoo (1978). The argentophilic pattern is probably unique for *E. caproni* cercariae and such studies may be useful to distinguish cercariae of closely related 37-collar-spined *Echinostoma*. Detailed chaetotaxy studies as described for *E. caproni* are not available for the other species in the *E. revolutum* group.

Scanning electron microscopy (SEM) has been used to describe the topography of *E. caproni* cercariae (see Krejci and Fried, 1994). Finfolds of cercariae of the 37-collar-spined *Echinostoma* are delicate and may be lost or improperly preserved during fixation procedures for light micro-scopy. The patterns of the tail finfolds by SEM (see Figure 12) can be seen in the micrographs of Krejci and Fried (1994).

Cercariae of this species encyst in the same *B. glabrata* snails from which they are released or in cohorts in the same aquarium. Studies to determine encystment preferences, i.e. in infected or uninfected snails in the same aquarium, have not been done. Encystment occurs in the kidney–pericardial region within several hours of cercarial entry via the nephri-diopore, but exact details of site localization in the second intermediate host have not been determined. Although the cyst probably forms within 4 h postcercarial entrance, it is not known when the encysted metacercaria is infective to the vertebrate host. Although encystment usually occurs within a snail or amphibian second intermediate host, ectopic encystment on snail mucus has also been reported and such cysts are infective to rodent hosts (Christensen *et al.*, 1980).

Some cercarial preferences to second intermediate snail hosts have been shown. Thus, *Biomphalaria* and *Bulinus* are more compatible as second intermediate hosts than are *Helisoma*, *Physa* and *Lymnaea* (see Christensen *et al.*, 1980). Reasons for snail incompatibility as second intermediate hosts are not known. Considerable information on infectivity of *E. caproni* to second intermediate hosts has been given in Huffman and Fried (1990).

Evans (1985) studied the influence of environmental temperature on the transmission of *E. caproni* cercariae and noted survival and infectivity characteristics of this larval stage for six different water temperatures in the 12–40°C range. As expected, cercarial survival decreased with increasing temperature with the maximum survival time dropping from 75 h at 12°C to about 8 h at 40°C. Changes in environmental temperature affected the infectivity even of newly emerged cercariae, but the most noticeable effect was on the rate at which infectivity declined with increasing cercarial age. Infectivity increased from zero at 12°C to a maximum at 30°C before declining to a low level at 40°C. Maximal infectivity occurred between 10 and 30°C, suggesting that *E. caproni* cercariae are adapted for transmission at water temperatures probably encountered in their native habitats.

To determine information on the dynamics of cercarial transmission of this echinostome, Meyrowitsch *et al.* (1991) examined the effects of

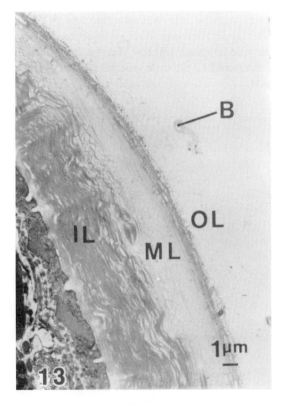

Figure 13 Transmission electron micrograph of a metacercarial cyst of *E. caproni*. Note the presence of an outer layer (OL), middle layer (ML) and inner layer (IL); a bacterium (B) is seen outside the cyst. Micrograph reproduced with the permission of the editor of the *Journal of Helminthological Society of Washington* from Irwin and Fried (1990).

temperature and host density on the snail-finding capacity of *E. caproni* cercariae. They described the effect of temperature (19–36°C) and *Biomphalaria* host density (0.014 to 10 snails l^{-1}) on the snail-finding capacity of this cercaria. As expected, the initial swimming speed of cercariae increased, whereas the length of the infective period decreased with increasing temperature. The *E. caproni* cercarial snail-finding capacity was temperature independent in the 19–36°C range at a snail density of 0.014 snails l^{-1}. Those authors suggested that a relatively low and biologically realistic snail host density should be used in experimental studies if realistic estimates of the dynamics of cercarial transmission were to be obtained.

Encysted metacercariae of *E. caproni* have been studied by LM and SEM (see Irwin and Fried, 1990; Krejci and Fried, 1994) and the cyst is spherical to subspherical with both an outer and inner cyst. The complexity of cyst layers was noted when the wall was examined by TEM (Figure 13) and at least three cyst layers become obvious (Irwin and Fried, 1990). The diameter of the cyst is about 150 μm and the outer cyst wall is relatively smooth when compared to the coarser cyst wall of *E. trivolvis* (see Krejci and Fried, 1994).

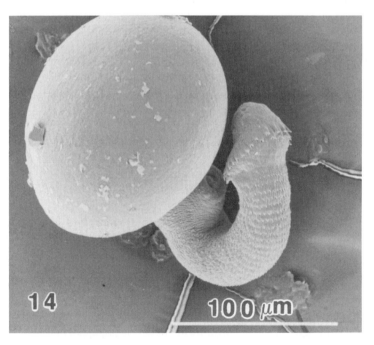

Figure 14 Scanning electron micrograph of the larva of *E. caproni* emerging from its cyst during *in vitro* excystation. Micrograph courtesy of Sam Irwin.

Chemical excystment of this species has been done to obtain basic information on intrinsic and extrinsic excystation (Figure 14), to obtain excysted metacercariae (Figure 15) for *in vitro* and *in ovo* cultivation studies, and for histochemical and immunological studies. The metacercarial cysts can be excysted in an alkaline trypsin–bile medium maintained at 40 ± 1°C and a pH of 8.0 ± 0.2; details of the above mentioned excystation studies on *E. caproni* have been described by Fried (1994).

7. INFECTIVITY OF *E. CAPRONI* IN FIRST AND SECOND INTERMEDIATE SNAIL HOSTS

E. caproni miracidia infect various species of *Biomphalaria* that serve naturally or experimentally as first intermediate hosts of this echinostome (see Section 4). Various gastropods including species of *Biomphalaria* also serve as second intermediate hosts and become infected with the cercarial stage. Additionally, other invertebrates, particularly tadpoles in the genus *Rana*, serve as second intermediate hosts of this echinostome (see Section 6). Numerous studies on snails serving as both first and second intermediate hosts of this echinostome have been cited in Huffman and Fried (1990). Studies not covered in that review or published after the review are presented below.

Experimental infection of the first intermediate host (typically *B. glabrata* snails of the M-line or albino strain) is usually done by exposing snails individually or en masse to miracidia of *E. caproni*. In our laboratory we usually expose *B. glabrata* individually to 10 miracidia in 1–3 ml of water for 8 h at 22–24°C. Snails are then grouped 10 or 20 per 1000 ml of artificial spring water and maintained in aerated aquaria for 4–6 weeks at 22°C. We usually get greater than 50% infectivity of these snails (based on cercarial emergence) by 6 weeks post-infection. Cercariae released from these snails infect other snails in the same aquarium giving rise to encysted metacercariae in the pericardial cavity and kidney.

Little information is available on the biological and histopathological effects of larval *E. caproni* infection on the first intermediate host. Unpublished studies in our laboratory have shown the disruption of the architecture of the digestive gland–gonad complex of *B. glabrata* infected with daughter rediae of *E. caproni* (see Figure 16), but details on gross or histopathological damage caused by the developing echinostomes, or the effects of the infection on snail fecundity, growth or metabolism have not been examined in the *E. caproni–B. glabrata* model. Considerable information on these topics is available for the *S. mansoni–B. glabrata* model and the reader should consult the reviews by Jourdane and Théron (1987)

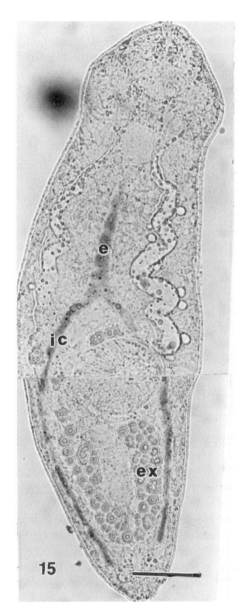

Figure 15 Light micrograph of a freshly excysted metacercaria of *E. caproni* stained intravitally with 0.01% aqueous neutral red. The oesophagus (e) and intestinal caeca (ic) stain orange–red with this procedure. Note the presence of excretory (ex) concretions mainly in the posterior part of the excretory tubules.

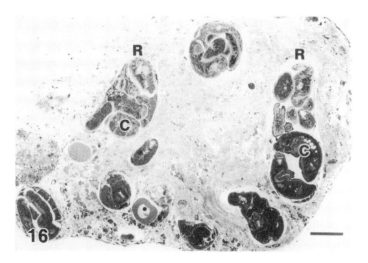

Figure 16 Light micrograph of a section through the digestive gland–gonad complex of *Biomphalaria glabrata* experimentally infected with the intramolluscan stages of *E. caproni*. Note the disruption of the architecture of this complex; R = redia; C = cercaria. Scale bar = 100 μm. Micrograph courtesy of Takahiro Fujino.

and Bayne and Loker (1987) on larval schistosome–snail interactions for further information.

Yousif *et al.* (1989) examined host–parasite relationships between *Biomphalaria alexandrina* and *E. caproni* under laboratory conditions in Egypt. They tested the influence of various biotic and abiotic factors on snail infectivity, i.e. miracidial dosage per snail; individual or en masse infections; age of miracidia; size of snails; maintenance temperature; presence or absence of light; volume and quality of aquarium water. The results of their study (only in abstract) showed a higher infection rate of snails as a result of exposure to larger numbers of miracidia; individual exposure of snails was more effective than mass infection; infectivity of miracidia decreased with miracidial age especially at elevated temperatures; younger snails were more susceptible to infection than older snails; no differences were seen in infectivity rates in snails exposed in light versus dark.

Studies on the experimental infection of *E. caproni* cercariae in *Biomphalaria* have not appeared since the Huffman and Fried (1990) review. Not included in the review was the study by Sullivan (1985) in which juvenile *B. alexandrina* was used as an experimental second intermediate host for *E. caproni*. Recently hatched snails (0.8–1.4 mm in shell diameter) were placed in the same Petri dish with a *B. alexandrina* snail releasing cercariae. Of 20 juveniles used, all were infected with 2–12 cysts per snail at necropsy 48 h PI, showing the suitability of neonatal snails as experimental second intermediate hosts for *E. caproni*. This work is of interest in view of the study by Fried *et al.* (1995) on experimental infection of juvenile *B. glabrata* with cercariae of *Echinostoma trivolvis* and a study on miracidial infection of neonatal snails with *S. mansoni* by Cooper *et al.* (1992).

8. ECHINOSTOMES, SCHISTOSOMES AND SNAILS

Some species of echinostomes and schistosomes share the same planorbid hosts. Thus, *Schistosoma mansoni* and *Echinostoma paraensei* in South America both use *Biomphalaria glabrata* as an intermediate host. In Africa, *S. mansoni* and *E. caproni* both use several species of *Biomphalaria* as intermediate hosts. Interactions between *S. mansoni* and *E. paraensei* in *B. glabrata* have been studied (see review in Loker and Adema, 1995), but less information is available on concurrent *S. mansoni* and *E. caproni* infections in *Biomphalaria*. The available information on *E. caproni* is summarized below.

Jourdane and Mounkassa (1986) observed a change in location of primary sporocysts of *S. mansoni* in *B. pfeifferi* as a result of concurrent infection with *E. caproni*. Interspecific competition was observed between the mother sporocysts of both species of digeneans resulting in a shifting of most *S. mansoni* mother sporocysts out of their usual sites into deeper snail tissues such as the cerebral ganglia, sinuses and kidneys; also noted was a degeneration of some of the mother sporocysts of *S. mansoni* in the normal site (mainly the headfoot region). The authors speculated that if the interspecific competition favoured the echinostome, during early development some of the *S. mansoni* mother sporocysts would be able to protect themselves by migrating to deeper habitats within the snail.

Jourdane *et al.* (1990) studied the influence of intramolluscan larval stages of *E. caproni* on the infectivity of *S. mansoni* cercaraie. During a dual infection of *B. glabrata* with *S. mansoni* and *E. caproni*, the schistosome cercariae released before the resorption of the schistosome sporocysts showed a significant decrease of their infectivity to mice. Under so-called conditions of high interspecific competition, i.e. when the snails were infected with *E. caproni* 8 days after infection with *S. mansoni*, the mean schistosome worm return was 5 times less than that seen in the controls. A significant decrease of the infectivity of schistosome cercariae was also noted when snails, infected exclusively with either sporocysts of *S. mansoni* or rediae of *E. caproni*, were kept in the same tank.

Although a study not involving direct interaction of the echinostome and schistosome in the same snail host, the study by Mounkassa and Jourdane (1990) is of interest as it relates to the dynamics of the haemocytic response of *B. glabrata* during larval development of *S. mansoni* and *E. caproni*. These workers examined haemocyte counts in *B. glabrata* under three conditions: uninfected snails, snails infected with *S. mansoni*, and snails infected with *E. caproni*. The results showed differences in the three experiments as well as in the average amoebocyte density during the experimental period; also noted were changes in the temporal dynamics in the number of circulating haemocytes. The development of *E. caproni* in *B. glabrata* induced a density of circulating haemocytes greater than that in uninfected *B. glabrata* or in *B. glabrata* infected with *S. mansoni*. Mounkassa and Jourdane (1990) suggested that the haemocyte dynamics in the experimental groups could be interpreted by taking into account differences in the immunogenic stimulating capacity of the two digeneans and different physiological functions of the haemocytes during infection.

9. INFECTIVITY, GROWTH AND DEVELOPMENT OF *E. CAPRONI* IN DEFINITIVE HOSTS

Considerable information on infectivity, growth and development of *E. caproni* in definitive (vertebrate) hosts was presented by Huffman and Fried (1990). Information published since that review is summarized below.

Fried *et al.* (1990b) studied single and multiple worm infections of *E. caproni* in golden hamsters. Six of 10 hamsters fed a single metacercarial cyst and 13 of 19 hamsters fed either two or five cysts were infected with adult worms 22 days PI. There were no significant differences in mean length, or wet weight of worms in single-versus multiple worm infections. The mean number of eggs per worm from single worm infections (525) was significantly greater than that from multiple-worm infections (288). The average percentage of fully developed miracidia per worm from single worms (94%) was the same as that from worms in multiple infections. Single worms were capable of self-fertilization and the production of viable eggs. Miracidia derived from single worms were as capable of infecting laboratory-reared *B. glabrata* and producing patent rediae as were those from multiple-worm infections. Interestingly, Nollen (1990) used autoradiographic techniques to confirm the fact that worms of this species grown singly in rodent hosts were capable of self-fertilization (see Section 3).

Christensen *et al.* (1990) examined infectivity, survival and fecundity of *E. caproni* in hamsters and jirds. The *E. caproni*–hamster model had a high level of compatibility, using the criterion of initial worm establishment. This model within the range of a 6–50 metacercarial cyst inoculum in hamsters was characterized by: infectivity that was infection-dose independent; a limited capacity to expel primary infections; the ability to mount regulatory responses to superimposed challenge worm infections. These findings supported previous observations on the *E. caproni*–hamster model (Huffman *et al.*, 1988; Yao *et al.*, 1991; Isaacson *et al.*, 1989. In contrast to the above findings of *E. caproni* in the hamster, Christensen *et al.* (1990) reported that the *E. caproni*–jird model showed a low level of compatibility with reduced and variable primary worm establishment, a limited capacity to expel worms from the primary infection, and a significant capability to mount effective regulatory responses to challenge infections.

Yao *et al.* (1991) studied the effects of crowding on adults of *E. caproni* in experimentally infected golden hamsters. Hamsters were fed either 15 (group A), 50 (group B) or 200 (group C) metacercarial cysts of *E. caproni* and necropsied 7–34 days PI. As expected, worm recovery was greatest in group C and least in group A. Likewise, worms from group C were reduced in size, less developed and more widely distributed in the hamster gut than flukes from either group A or B. Worms in this study showed typical characteristics seen in "worm crowding" studies often described in tapeworm infections, but infrequently documented in digeneans.

Hosier and Fried (1991) examined infectivity, growth and distribution of *E. caproni* in 6–8-week-old female ICR mice. All of 40 mice, each fed 25 cysts of *E. caproni*, were infected 1–20 weeks PI with a mean of 17 worms per host. Some mice had worms at 29 weeks PI indicating that this echinostome could survive in ICR mice for more than 7 months. Growth based on worms fixed in hot alcohol-formalin-acetic acid (AFA) and stained in Gower's carmine showed that the mean body area of worms increased rapidly to 5 mm^2 by week 2, less rapidly to 9 mm^2 by week 12, plateaued until week 24, and then declined. Mean dry weight per worm increased rapidly to 0.5 mg by week 2, less rapidly to 1.4 mg by week 12, and then plateaued until week 24. From 1 to 8 weeks PI most worms localized in the jejunum and ileum; after 8 weeks most worms were in the jejunum and duodenum indicating an anteriad shift as the worms aged. Considerable differences were seen in the growth and distribution of *E. caproni* in ICR mice compared with previous studies on this echinostome in NMRI mice by Odaibo *et al.* (1988, 1989).

Hosier *et al.* (1991) studied the effects of host age on infection of female ICR mice with *E. caproni* cysts. They found no significant difference in the number of adult worms recovered at 3 weeks PI from 1-, 2-, 4-, 5-, 6-, 7-, 12- or 21-month-old ICR mice infected with 25 metacercarial cysts per mouse (a range of 11 to 17 worms was recovered from mice in these age groups). Contrary to a previous study of *E. caproni* in NMRI mice by Odaibo *et al.* (1988, 1989), host age did not affect the establishment of *E.*

caproni in ICR mice. Kaufman and Fried (1994) studied infectivity, growth, distribution and fecundity of a six versus 25 metacercarial inoculum of *E. caproni* in female ICR mice and documented further differences between their study and that of Odaibo *et al.* (1988). It is apparent that certain aspects of the biology of *E. caproni* in ICR mice versus NMRI mice reflect differences in these mouse strains.

Manger and Fried (1993) studied infectivity, growth and distribution of preovigerous *E. caproni* in 6–8-week-old outbred female ICR mice. All 23 mice, each fed 100 metacercarial cysts, were infected on days 2, 4, 6 and 8 PI. To examine worm distribution in the mouse, the small intestine was divided into five equal sections (1–5 beginning with the pylorus). Echinostomes were found mainly in segments 3 and 4 (jejunum and ileum) and never in segment 1 (the duodenum). Distinction of the ovary and ootype was apparent on day 6 PI and uterine coiling was observed on day 8. Only two of 50 worms showed eggs in the uterus on day 8 PI. The large metacercarial cyst inoculum (100 per mouse) and high worm recovery (45 per host) probably inhibited worm development resulting in a delay in maturation. Mice infected with 25 or less cysts of *E. caproni* usually have ovigerous worms by day 7 PI (see Section 3).

10. CONCURRENT STUDIES USING *E. CAPRONI* AND OTHER HELMINTHS

E. caproni has been used in concurrent studies with other helminths and the salient features of these papers are reviewed below.

Andreassen *et al.* (1990) found that superimposing the intestinal tapeworm *Hymenolepis diminuta* on mice with an established *E. caproni* infection resulted in destrobilation and expulsion of the tapeworm. Mechanisms associated with this interaction were not elucidated.

Iorio *et al.* (1991) examined concurrent infections of *E. caproni* and *E. trivolvis* in female ICR mice. In the concurrent infections, 13 (59%) of 22 mice were infected with both species and the percentage of worm recovery was 73% for *E. caproni* and 14% for *E. trivolvis*. There was no difference in worm distribution of either species in single versus concurrent infections. In concurrent infections at 14 days PI, there was a significant decrease in the body area of worms of both species compared to single worm species. In concurrent infections in mice, interspecific competition produced inhibitory effects on worm growth in both species. Huffman *et al.* (1992) studied intra- and interspecific competition between *E. caproni* and *E. trivolvis* in the golden hamster and established five age classes of both species of echinostomes in the hamster. The location of worms in concurrent infections suggested that competition between the two species did not occur and worms of both species were mainly clustered. The host is important in determining the effects of concurrent infections of these two allopatric echinostomes (compare the Huffman *et al.* (1992) study with that of Iorio *et al.* (1991)).

Chemical communication studies in helminths (Haseeb and Fried, 1988) showed that trematodes have a tendency to pair *in vitro* and the pairing is regulated by worm-mediated pheromones. Intraspecific and interspecific pairing of *E. caproni* and *E. trivolvis in vitro* were studied by Fried and Haseeb (1990) at 38°C in a Petri dish bioassay containing an agar substratum overlaid with Locke's solution. Worms of both species at 14-days old, were removed from domestic chicks and used in the bioassays. Each species had an intrinsic pairing pattern with the tendency of *E. trivolvis* to pair being greater than that of *E. caproni*. The interspecific pairing pattern was different from that of either intraspecific pattern. Reasons for differences in interspecific versus intraspecific pairing were not clear in that study and further work is required.

Fujino *et al.* (1996a) did concurrent studies with the intestinal nematode *Nippostrongylus brasiliensis* and *E. caproni* in C3H/HeN mice. Mice infected with *N. brasiliensis* and then secondarily exposed to *E. caproni* metacercarial cysts 8 days later expelled the echinostomes by 16 days after the primary infection. Controls without the primary *N. brasiliensis* infection showed echinostome recovery rates of about 70%. The number of goblet cells in the small intestines of C3H/HeN mice increased rapidly following infection with *N. brasiliensis* larvae. The results of the study suggested that mucins increased by hyperplastic goblet cells associated with the primary *N. brasiliensis* infection caused a rapid expulsion of the secondarily infected *E. caproni* worms from the mouse gut.

11. EFFECTS OF *E. CAPRONI* INFECTION ON PREGNANCY IN MICE

Bindseil *et al.* (1990) noted that infection of female BALB/cABom mice with *E. caproni* reduced mouse fertility as determined on day 18 of pregnancy by counting and weighing infected versus non-infected fetuses. In the absence of histological differences in infected versus non-infected mice, it was suggested that the effect on fertility was not due to parasite-induced lesions, but to some undefined pathophysiological disorders. Of significance was the fact that maternal plasma levels of murine alpha-fetoprotein on day 18 of pregnancy in infected mice were significantly lower than in the non-infected mice.

Bindseil and Hau (1991) showed that infection of BALB/cABom mice with *E. caproni* had a negative influence on pregnancy. The effect of the infection was noted immediately after implantation (day 5 of pregnancy) resulting in fewer fetuses present in infected mice on day 9 than in the controls; ovulation, fertilization and egg implantation were not affected. The infected mice had significantly lower serum progesterone levels on day 5 of pregnancy than did the non-infected controls. The progesterone levels of the infected mice were probably too low to maintain early post-implantation gestation. It is not known how infection with *E. caproni* interferes with progesterone levels in mice, but further investigations along these lines are needed. Bindseil *et al.* (1991) measured maternal plasma levels of pregnancy associated with murine protein 1 (PAMP-1) by the technique of rocket immunolectrophoresis in BALB/cABom mice experimentally infected with *E. caproni*. During the first half of pregnancy the PAMP-1 level was significantly lower in infected versus non-infected hosts. These findings suggested that PAMP-1 was not essential for implantation and early gestation. During the latter half of pregnancy, a rapid increase in PAMP-1 plasma levels was noted in infected mice, possibly suggesting that pituitary gonadal hormone secretion was restored in *E. caproni*-infected mice during the second half of gestation.

The effects of helminths on pregnancy in vertebrate hosts are numerous and complex. Carlier and Truyens (1995) have reviewed the influence of maternal infection on offspring resistance towards parasites and that review should be examined for more insight on this topic.

12. EXCYSTATION, IMPLANTATION, AND *IN VITRO* AND *IN OVO* CULTIVATION OF *E. CAPRONI*

Encysted metacercariae of *E. caproni* can be excysted in the alkaline trypsin-bile salts medium of Fried and Roth (1974; see review in Fried, 1994). Therefore, excysted metacercariae (see Figure 15) of this species are available for experimental purposes and have been used in studies on surgical implantation of metacercariae into the intestines of mice (Chien *et al.*, 1993), cultivation of metacercariae *in vitro* in a defined medium plus serum (Fried and Emili, 1987) and cultivation of metacercariae in chick embryos (*in ovo*) on the chorioallantoic membrane (CAM) and in the allantoic sac (see review in Fried and Stableford, 1991).

Chien *et al.* (1993) surgically implanted encysted or excysted metacercariae into the small intestines of ICR mice. Worm recovery at 10 days post-implantation from mice that had received either 25 encysted or excysted metacercariae was 93%. The fact that encysted metacercariae implanted into the small intestines could produce adult worms suggested that residence in the mouse stomach was not a prerequisite for normal *in vivo* excystation of this species. The ability to surgically implant excysted metacercariae of *E. caproni* into the mouse small intestine can be useful for basic studies on immunology and wound healing of digeneans. In addition to implanting metacercariae, Chien *et al.* (1993) also implanted adults of *E. caproni* into the mouse small intestine.

Chemically excysted metacercariae of *E. caproni* were used by Fried and Manger (1992) to develop a simple acetocarmine procedure to relax, fix, stain and clear these larvae. This procedure clarified the genital anlage and provided morphological details on whole excysted metacercariae comparable to those seen with more elaborate staining protocols. Intravital staining of excysted metacercariae of this echinostome using 0.01% aqueous neutral red was used by Fried (unpublished) to visualize the intestinal caeca in these larvae (Figure 15). The use of cercariae to study flame cells, cystogenous and penetration glands with intravital dyes has been done frequently, but rarely have intravital studies been done on excysted metacercariae.

Fried and Rosa-Brunet (1991b) cultivated excysted metacercariae of *E. caproni* on the chorioallantoic membrane (CAM) of the domestic chick (see review in Fried and Stableford, 1991). Worm development on the CAM lagged about 1-day behind that of development *in vivo* and worms became ovigerous on the CAM on day 8 PI. One worm maintained for 17 days on two successive CAMs reached 6 mm in length, contained about 100 eggs in its uterus and laid an additional 100 eggs on the CAM surface. Findings by Fried and Rosa-Brunet (1991b) suggested that this organism can be useful to study behavioural phenomena of an adult digenean in an ectopic site. Rosa-Brunet and Fried (1992) studied morphogenesis of this organism on the CAM comparatively with organisms grown in domestic chicks and provided confirmatory details in the lag of development *in ovo* first reported by Fried and Rosa-Brunet (1991b). Histopathological studies on worms attached to the CAM showed that some worms attached to the surface of the chorioallantois by the collar spines and acetabulum, whereas others penetrated the chorionic epithelium and encapsulated in the mesenchyme. Pathogenicity to the CAM included hyperplasia of the chorionic epithelium and haemorrhagia, and increased lymphocytes and eosinophils in the mesenchyme. Chien and Fried (1992) inoculated excysted metacercariae of *E. caproni* into the allantois of domestic chick embryos and found that these worms became ovigerous in that site within 9 days PI. Adults of *E. caproni* from the allantois were larger and became ovigerous sooner than worms grown on the chorioallantois. Only worms from the allantois produced eggs with fully developed miracidia. Miracidia were released from these eggs, but an insufficient number was available to attempt infections of *Biomphalaria glabrata*. If eggs derived from *in ovo* worms could produce viable miracidia capable of infecting snails, this would be the first example of a non-progenetic digenean that could complete its life cycle without a vertebrate definitive host. Fried and Emili (1987) reported on the *in vitro* cultivation of excysted metacercariae of *E. caproni* in the defined medium NCTC 135 plus 50% chick serum. Worms survived for up to 10 days in this medium at 37.5°C and increased their body area by about 2.5 times. Although somatic growth was noted, there was no evidence of gonadal growth. Young 7-day-old adults of *E. caproni* removed from domestic chicks and maintained in the NCTC 135–50% chicken serum at 37°C survived for 8 days in these cultures, and showed a marked decrease in body area. The significance of degrowth in the Digenea is unclear and further *in vitro* cultivation studies using excysted metacercariae and adults are needed.

13. PATHOBIOCHEMICAL EFFECTS OF LARVAL AND ADULT *E. CAPRONI* ON THEIR HOSTS

Relatively little information is available on the pathobiochemistry of gastropods infected with larval trematodes and most of the studies are on *Biomphalaria glabrata* infected with larval *Schistosoma mansoni* (see Malek, 1980, for review). Recent studies by one of us (BF) have been concerned with the relationships between *E. caproni* and experimental infections in *B. glabrata*. In the *S. mansoni–B. glabrata* model the major larval stage in the DGG complex is the daughter sporocyst, a passive feeding stage without a gut and wholly reliant on transtegumental feeding for nutrient uptake. In the *E. caproni–B. glabrata* model, the major larval stage in the DGG is the daughter redia, an active feeding stage with a gut and a pharynx. Because of dissimilarities in the biology of the redia and sporocyst, pathobiochemical differences would be expected in *B. glabrata* infected with either *S. mansoni* or *E. caproni*. The following discussion relates mainly to studies on the *E. caproni–B. glabrata* model, but reference to the *S. mansoni–B. glabrata* model is also made for comparisons.

Fried *et al.* (1989) used thin-layer chromatography (TLC) to study the effects of *E. caproni* redial infection on the neutral lipids of the DGG complex of *B. glabrata* and noted a marked reduction in the triacylglycerol fraction of the DGG of experimentally infected snails compared to matched controls. By contrast, Thompson (1987) studied the effects of *S. mansoni* sporocyst infection in the DGG of *B. glabrata* and found that total lipid levels and the triacylglycerols of the DGG were raised after infection with *S. mansoni* larvae. Reasons for differences in the findings between the two studies are not known but probably relate to differences in sporocyst and redial biology. Shetty *et al.* (1992) used gas–liquid chromatography (GLC) to examine differences in the sterol content of the DGG of uninfected *B. glabrata* and of those experimentally infected with *E. caproni* rediae and noted that the major sterol present in both populations was cholesterol, representing 59% in the infected snails and 51% in the control; they also noted a significant reduction in the percentage composition of the phytosterols, campesterol and stigmasterol, in the infected snails compared to controls. Effects of *S. mansoni* infection on the sterol content of *B. glabrata* have not been examined. Fried *et al.* (1993) used GLC to study the effects of larval *E. caproni* infection on the fatty acid composition of experimentally infected *B. glabrata*. The infection reduced the amounts of saturated fatty acids in whole snail bodies, but increased the amounts of these acids in the DGG of the snails. Moreover, infection with larval *E. caproni* increased the proportion of $C_{16:1 \ n-9}$ fatty acids in the DGG of *B. glabrata* leading to a concomitant decrease in $C_{16:1 \ n-7}$ fatty acids as well as various acyl moieties of the $C_{20:1}$ series. Comparable studies on *B. glabrata* infected with *S. mansoni* are not available.

TLC studies on the phospholipid composition of the DGG of *B. glabrata* infected with larval *E. caproni* (Perez *et al.*, 1995) showed that the major phospholipids in the DGG were phosphatidylcholine (PC), phosphatidylethanolamine (PE) and phosphatidylserine (PS). The *E. caproni* infection did not cause significant changes in the levels of PC and PE in the DGG, but there was a significant increase in PS in infected DGGs at 6 weeks PI. This was the first report of significant phospholipid elevation in a snail infected with larval trematodes. Thompson (1987) showed no significant changes in phospholipids in *B. glabrata* infected with *S. mansoni*.

Recently, Beers *et al.* (1995) used TLC to examine the effects of a larval *E. caproni* infection on the neutral lipid composition of the DGG of *B. glabrata* fed a high fat diet (hen's egg yolk). The DGGs of infected snails maintained on the yolk diet showed a significant increase in free sterols and a significant decrease in triacylglycerols compared to uninfected snails maintained on a lettuce–Tetramin diet. The triacylglycerol decrease in the DGGs of infected snails maintained on the yolk diet suggested that neutral fat was utilized by the rediae. Thompson *et al.* (1991) examined the effects of larval *S. mansoni* infection on the neutral lipid composition of the DGG of *B. glabrata* maintained on the hen's egg yolk diet and found that the infected DGG had increased triacylglycerol levels compared with the uninfected controls. They also found that the *B. glabrata* infected with *S. mansoni* and fed the egg yolk diet developed patent infections, i.e. daughter sporocysts with fully developed cercariae, in half the time it took the lettuce-fed controls. This was not the case in the *B. glabrata–E. caproni* system where it took 5 weeks to obtain patent daughter rediae of *E. caproni* regardless of the diet (Beers *et al.*, 1995).

Numerous studies (reviewed in Perez *et al.*, 1994) have shown that *B. glabrata* infected with *S. mansoni* has a lower concentration of carbohydrates than uninfected controls. Perez *et al.* (1994) used high performance thin layer chromatographic (HPTLC) analysis to study sugars in the haemolymph and DGG of *B. glabrata* infected with larval *E. caproni*. The major sugars detected in both the DGG and haemolymph of infected and uninfected snails were glucose and trehalose. There was a significant reduction in glucose concentration in the infected snails at 4, 6 and 8 week PI, but no significant difference in trehalose at 4 weeks PI in infected versus control snails; by 6 weeks PI, trehalose values were significantly reduced in both the haemolymph and DGG of infected snails. The findings of Perez *et al.* (1994) on reduction in glucose in the *B. glabrata–E. caproni* model are in accord with similar findings by Cheng and Lee (1971) on the *B. glabrata–S. mansoni* model. Although trehalose is a major carbohydrate in *B. glabrata* (see Fairbairn, 1958), the effects of *S. mansoni* infection on that sugar have not been studied.

There is little information on the pathobiochemistry of vertebrate hosts infected with adult echinostomes. Horutz and Fried (1995) studied the effects of adult parasitism by *E. caproni* on the neutral lipid content of the intestinal mucosa of experimentally infected mice. A number of studies (see Section 14) on mice infected with *E. caproni* have shown a marked histopathological effect on the host intestine including villous atrophy, fused and eroded villi, elongated crypts of Lieberkuhn, and hypertrophy of the tunica muscularis (Bindseil and Christensen, 1984; Fujino and Fried, 1993a). Horutz and Fried (1995) correlated pathobiochemical events related to neutral lipid content in the gut mucosa with histopathological changes caused by *E. caproni* infection. Using TLC, they found that at 2 weeks PI with *E. caproni,* the mucosa of infected mice showed a marked elevation of the free fatty acid fraction. Horutz and Fried (1995) noted that high levels of free fatty acids in the host intestine may reflect an increase in host membrane breakdown caused by the echinostomes or these flukes may release fatty acid excretory–secretory products that contribute to the increased neutral lipid level. They noted that the importance of elevated free fatty acids in the clinical pathology of the infection remains to be determined, but it is conceivable that an increase in free fatty acids in echinostome infections may be important as a diagnostic tool for echinostomiasis in man.

14. GROSS, HISTOPATHOLOGICAL AND CLINICAL EFFECTS OF ADULT *E. CAPRONI* ON THEIR HOSTS

Some information on the gross and histopathological effects of *E. caproni* in rodent hosts was provided by Huffman and Fried (1990). New studies on this topic in mice, hamsters, domestic chicks and on the chick chorio-allantois appeared since that review and are reported herein.

Simonsen *et al.* (1989) examined the morphology, histology and dynamics of the attachment site (the worm's acetabulum attached to the mucosa of the small intestine) in NMRI mice experimentally infected with *E. caproni*. The site was characteristic microscopically and consisted of a plug of grasped mucosa (the mucosal plug) occupying the cavity of the acetabulum (Figure 17). The area of the intestinal mucosa in which the parasite was located showed villous atrophy and crypt hyperplasia confirming a previous observation by Bindseil and Christensen (1984) on this echinostome in the mouse. Simonsen *et al.* (1989) did not observe differences in the cellular composition of the mucosa at the attachment site compared with other areas of the gut, i.e. no specific cellular host responses at the acetabular attachment site. After mechanical removal of the echinostomes from the intestinal mucosa, the mucosal plug became reduced in size and then disappeared, suggesting that the attachment site

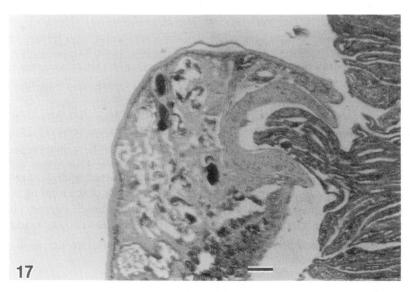

Figure 17 Light micrograph of *E. caproni* in the intestine of a hamster with a plug of mucosa (the mucosal plug) occupying the cavity of the acetabulum. Stained in haematoxylin and eosin. Scale bar = 50 μm.

was a temporary structure formed by the mechanical grasp of the worm's acetabulum as it attached to the intestine.

Weinstein and Fried (1991) reported pathological effects of *E. caproni* in mice incidental to studies on the expulsion of *E. trivolvis* and retention of *E. caproni* in ICR mice. Compared to uninfected mice and mice infected with *E. trivolvis* the small intestines of hosts infected with *E. caproni* showed marked dilation and villous atrophy. The diameter of the mouse gut in hosts infected with *E. caproni* was about three times that of either control mice or mice infected with *E. trivolvis*. The small intestines of *E. caproni*-infected mice showed almost complete goblet cell loss whereas the small intestines of *E. trivolvis*-infected mice showed nearly a twofold increase in goblet cells compared to those of control mice. A marked increase in collagen in the intestinal musculature of the mice infected with *E. trivolvis* was seen compared to uninfected and *E. caproni*-infected mice. It was apparent from this study that marked gross and histopathological differences occurred in the small intestines of ICR mice infected with *E. caproni* compared to those infected with *E. trivolvis*.

Fujino and Fried (1993a) examined pathological, ultrastructural and cytochemical effects of *E. caproni* in the small intestines of C3H mice. The small intestines of infected mice showed villous atrophy with fused or eroded villi. Transmission electron microscopy showed that the microvilli of the enterocytes were sparse and distorted and had reduced alkaline phosphatase activity. The crypts of Lieberkühn were hyperplastic and showed a reduced number of goblet and Paneth cells. As compared with the uninfected controls, there was a reduction in glucose-6-phosphatase activity in the enterocytes of mice infected with *E. caproni*. There was an increase in collagen fibres and in the number of fibroblasts in the subepithelial tissue of the intestines of infected hosts. Considerable ultrastructural differences were seen in the intestines of mice infected with *E. trivolvis* (details not covered herein) compared to what was described for the *E. caproni*–mouse system reflecting the uniqueness of the host–parasite relationship of each species of echinostome in C3H mice. Fujino and Fried (1993b) used six fluorescein-conjugated lectins to examine mucosal glycoconjugates in the small intestines of C3H mice infected with *E. caproni*. The expression of lectin-binding sites and the intensity of the binding of the lectins in the mouse gut were altered by infection with *E. caproni*. In *E. caproni*-infected intestines, binding of the lectins to the villi was markedly reduced compared to uninfected controls. Associated with the reduction in lectin binding was villous atrophy and loss of goblet cells in the small intestines of infected hosts.

Isaacson *et al.* (1989) examined gross and histopathological effects of *E. caproni* infection in golden hamsters (*Mesocricetus auratus*). Gross pathological effects in the hamsters included progressive unthriftiness, watery

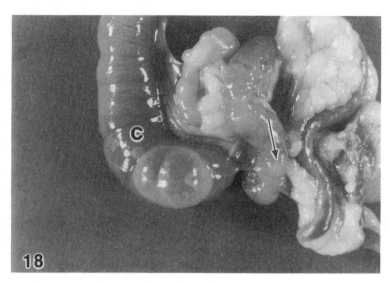

Figure 18 Ballooning of the caecum (C) and small intestine and enlarged lymphatic nodule (arrow) on the surface of the intestine of a hamster infected with *E. caproni*.

diarrhoea, enlarged lymphatic nodules (Figure 18) along the length of the small intestines and oedema of the caecum and the small intestine. Histopathological effects of the infection included intestinal haemorrhagia, erosion of the intestinal villi and lymphocytic infiltration. Fried *et al.* (1990b) noted considerable histopathological changes in the hamster intestine (as described in Isaacson *et al.*, 1989), even in hosts infected with only a single worm (Figures 19 and 20).

Gross and histopathological effects of *E. caproni* have been reported in experimentally infected domestic chicks (Kim and Fried, 1989) and chick embryos (Rosa-Brunet and Fried, 1992). Chicks infected with *E. caproni* for 2 weeks showed dilated ilea, unkempt feathers, watery diarrhoea, and weight loss (Kim and Fried, 1989). Light microscopic examinations of infected guts prepared as stained paraffin sections showed haemorrhagia, atrophic villi and hypertrophied circular musculature with collagen-like fibres. The brush borders of epithelial cells and the goblet cells were absent in the mucosa of infected guts; worms in contact with the host mucosa showed tissue plugs in the acetabulum and oral sucker and the collar spines of worms produced lesions in the host mucosa. Rosa-Brunet and Fried (1992) used paraffin and cryostat sections stained with haematoxylin and eosin to examine the histopathological effects of *E. caproni* on the chick CAM. Some worms were attached to the CAM by their collar spines and the acetabulum, whereas others penetrated the chorionic epithelium and

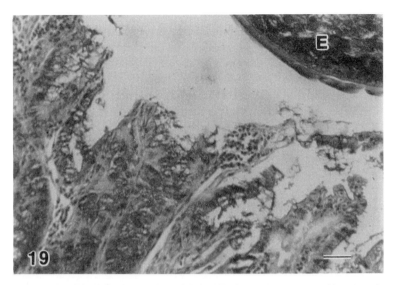

Figure 19 Light micrograph of the histopathological response of hamster intestinal mucosa to *E. caproni* (E). Erosion of intestinal villi and lymphocytic infiltration are the primary response. Stained in haematoxylin and eosin. Scale bar = 50 μm.

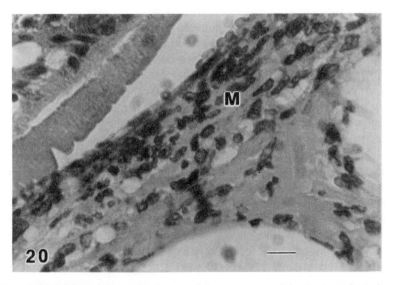

Figure 20 Light micrograph of the cellular response of the hamster intestine to *E. caproni*. Multinucleated leucocyte (M) infiltration at the interface of the parasite and the gut. Stained in haematoxylin and eosin. Scale bar = 20 μm.

encapsulated in the mesenchyme. Histopathological effects on the CAM included hyperplasia and haemorrhagia of the chorionic ectoderm and increased lymphocytes and eosinophils in the mesenchyme; the chorionic endoderm was not affected by echinostome parasitism. The chorioallantoic membrane–*E. caproni* interface should provide an excellent model for histopathological studies at the ultrastructural level.

The clinical signs exhibited by golden hamsters infected with *E. caproni* were dependent on the intensity of the infection (Huffman *et al.*, 1988). Infected hamsters showed decreases in packed cell volume, haemoglobin concentration and body weight. Haemorrhage was seen in association with the damage to the intestinal villi. Relative splenic weights decreased in hamsters infected with *E. caproni* (McMaster *et al.*, 1995).

15. IMMUNOBIOLOGY OF *E. CAPRONI* IN DEFINITIVE HOSTS

Earlier studies on the immunology of *E. caproni* in vertebrate hosts were discussed by Huffman and Fried (1990). Since that review one study on the immunology of *E. caproni* in golden hamsters and four on this echinostome in mice have been reported and are covered herein.

Simonsen *et al.* (1991) used ELISA, SDS-PAGE, Western blot, and IFAT techniques to examine the serum antibody response in golden hamsters (*Mesocricetus auratus*) infected with *E. caproni*. All methods showed that the hamsters responded slowly but developed a positive humoral response to the infection. In most hamsters, an antibody response to worm infection could not be detected before 11–13 weeks PI with 6 or 25 metacercariae/host; responses were weak when compared to previous studies on mice infected with *E. caproni* (see Simonsen and Andersen, 1986; Andresen *et al.*, 1989). IFAT with positive hamster sera on newly excysted metacercariae showed fluorescence at the posterior tip of the worm (in the region of the excretory bladder). This pattern was different from that seen when sera from mice infected with *E. caproni* were used, suggesting a different response to antigens from excysted metacercariae by these two hosts. Simonsen *et al.* (1991) discussed the results of their findings in view of the limited self cure and development of resistance in golden hamsters compared to *E. caproni* infections in mice which develop a strong resistance to this parasite. Simonsen and Andersen (1986) showed that surface antigens of *E. caproni* induced a serum antibody response in experimentally infected mice. Andresen *et al.* (1989) showed that these antigens were released from the surface of juvenile and adult worms maintained *in vitro*. SDS-PAGE and Western blot analyses indicated that four major antigens with molecular weights of approximately 26000, 66000,

75 000 and 88 000 Da, were released from the surfaces of worms. Studies using an ELISA technique showed that the *in vitro* turn-over rate of the surface antigens was high with a half-life of 12 ± 3 min. TEM showed that the tegumentory surface was packed with membrane-bound vesicles, suggesting a high rate of shedding of these antigens. Attempts to immunize mice with detergent-solubilized surface agents did not induce resistance to infection with *E. caproni* metacercariae. Simonsen *et al.* (1990) used TEM to examine the binding of mouse antibodies to the surface antigens of excysted and metacercariae adults of *E. caproni* treated with ferritin-conjugated antibodies in a double sandwich technique. The surface of adult worms was covered with a mouse antibody containing matrix, possibly representing a mouse antibody–worm antigen complex. The complex was lost after maintaining the parasite for 24 h; however, incubation of *in vitro*-maintained antibody-negative adult worms with immune mouse serum led to the formation of a new cover with immune complex. Excysted metacercariae and adults of *E. caproni,* which had not been exposed to mouse antibodies, acquired a layer of immune complex on the worm surface following worm maintenance *in vitro* in immune mouse serum. The antibody–antigen complex was loosely attached to the tegumentary surface of the parasite, where the antigens probably comprised part of the glycocalyx. TEM also indicated that parasite surface antigens occurred in vesicles in the tegument.

Agger *et al.* (1993) examined the antibody response in serum, the intestinal wall and the intestinal lumen of NMRI mice infected with either six (low dose) or 25 (high dose) metacercarial cysts of *E. caproni*. An ELISA test was developed to measure IgM, IgG and IgA antibody responses in mice sera or intestines against a crude adult worm antigen. Significant levels of IgM were measured in the sera of infected mice from day 14 PI, and of IgG and IgA from 28 days PI. Early in the infection, the serum IgM level was higher in the low- rather than high-dose infection groups. Later in the infection, raised levels of IgA were seen in the serum of mice with high- rather than low-dose infections. The onset of the appearance of antibodies in the tissue of the small intestines reflected the situation seen in the serum. For IgM, and to a lesser extent IgG, the highest antibody titres were in the posterior region of the small intestine where the worms were located. The IgA level was uniform throughout the length of the small intestines. High levels of IgA were detected in the lumen of the small intestines at 28 days PI, especially in the anterior region where few worms were found; no specific IgM or IgG was detected in the intestinal lumen at 28 days PI. Agger *et al.* (1993) related their findings to worm intestinal location and to the pattern of *E. caproni* expulsion in NMRI mice.

Graczyk and Fried (1994) used surface glycocalyx antigens from adult worms to develop an ELISA method for detecting anti-*E. caproni* immuno-globulins in experimentally infected ICR mice. Although whole blood, serum and dried blood on filter paper gave similar results, the latter sample was selected for convenience. A concentration of 10 μg ml^{-1} of antigen was optimal in terms of specificity, sensitivity and test speed and it was possible to detect anti-*E. caproni* immunoglobulins at a dilution of 10^{-4}. In a blind trial the ELISA accurately differentiated sera from infected versus uninfected mice. All experimentally infected mice had ELISA-detectable anti-*E. caproni* immunoglobulins reactive on day 8 PI with the surface glycocalyx antigens. Graczyk and Fried (1994) concluded that the *E. caproni* ELISA was fast, easy to perform, reproducible, and required minimal equipment for blood samples; moreover, the assay could be used to detect anti-*E. caproni* immunoglobulins in experimentally infected mice or in laboratory experiments where evidence of infection was required.

How does specific immunological resistance to *E. caproni* operate? The most efficient expression of host resistance is likely to be effected by a multicomponent response. Table 5 lists some of the effector mechanisms which may act against *E. caproni*. The prime candidates for effective immune mechanisms against *E. caproni* are the humoral and cellular components most evident during infection. McMaster *et al.* (1995) demonstrated that increasing doses of the corticosteroid, dexamethasone, increased the percentage infectivity of *E. caproni* in the golden hamster.

Table 5 Possible effector mechanisms which may act against *E. caproni*.

Effector cells	Effector molecules	Likely mechanisms	Hosts	References
Mucosal mast cells	Mucus and antibodies	Trapping and expulsion	C3H mice	Fujino *et al.* (1996b)
Mucosal mast cells	IgG and complement	Membranes attack complex	Mice	Simonsen and Andersen (1986)
Neutrophils	IgG, IgA or IgM	ROI and granule proteins and their expulsion	NMRI mice	Agger *et al.* (1993)
Lymphocytes	Th1 and Th2	Activate macrophages and eosinophils	Golden hamsters	McMaster *et al.* (1995)
Eosinophils	IgG	ECP, MBP and ROI	Golden hamsters	Simonsen *et al.* (1991)

Abbreviations: ECP, eosinophil cationic protein; MBP, major basic protein; ROI, reactive oxygen intermediate.

McMaster *et al.* (1995) also reported that hamsters infected with *E. caproni* showed an increase in the total number of circulatory white blood cells.

16. ELECTROPHORETIC AND POLYMERASE CHAIN REACTION (PCR) STUDIES ON *E. CAPRONI*

Some information on electrophoretic studies on *E. caproni* has been given in Huffman and Fried (1990), mainly on the identity of species of the 37-collar-spined *Echinostoma* in the *E. revolutum* complex. These studies were done by Voltz-Kristensen and collaborators (see Voltz-Kristensen, 1987). Studies not included in Huffman and Fried (1990) or published since that review are given below and are related to attempts to provide unequivocal evidence to distinguish the allopatric species *E. trivolvis*, *E. paraensei* and *E. caproni*. In addition to electrophoretic analyses, the polymerase chain reaction (PCR) has recently been applied to taxonomic problems in the genus *Echinostoma*. Applicable PCR studies are also reviewed herein.

Ross *et al.* (1989) used isoelectric focusing to analyse enzymes and pigments of 15-day-old adults of *E. caproni* and *E. trivolvis* raised in golden hamsters. Clear interspecific differences were detected in worm haemoglobin-like pigments, and in glucose phosphate isomerase, phosphoglucomutase and acid phosphatase; malate dehydrogenase was similar in both species. Those authors suggested that enzyme analysis of adult worms should be helpful in elucidating taxonomic problems within the genus *Echinostoma*. Kristensen and Fried (1991) did an isoenzymatic analysis using thin-layer agarose gel electrofocusing on *E. caproni* and *E. trivolvis* adults. The worms showed characteristic monomorphic phenotypes for phosphoglucomucotase and glucose phosphate isomerase. The fixed allelic variation observed between these two echinostomes was consistent with their recent classification as distinct species. Sloss *et al.* (1995) used starch gel electrophoresis to determine genetic relationships between *E. caproni*, *E. paraensei* and *E. trivolvis*. They examined 10 enzyme systems encoding 12 presumptive structural loci and showed that *E. paraensei* and *E. caproni* were genetically inbred as indicated by the lack of heterozygosity in individual worms. All three echinostomes showed fixed differences indicating that they are distinct species. Fixed differences were seen between *E. paraensei* and *E. caproni* in six enzyme systems and between *E. paraensei* and *E. trivolvis* in five enzyme systems. Phenetic relationships among the three species showed that *E. caproni* was genetically more similar to *E. trivolvis* than to *E. paraensei*.

Morgan (1993) examined ribosomal DNA sequence diversity among

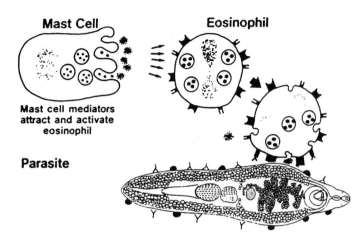

Figure 21 Proposed mechanism for expulsion of *E. caproni* from the rodent small intestine. Mast cell mediators attract and activate eosinophils. Eosinophil binds to IgG via receptors, and contents of granules induce lysis of the parasite.

some well-characterized species of *Echinostoma*, i.e. *E. revolutum*, *E. trivolvis* and *E. caproni*. She was the first to investigate the echinostome genome using internal transcribed spacers (ITS) of the ribosomal DNA cloned and sequenced for adults of the above three echinostome species. DNA sequence divergence in the internal rDNA spacers of *E. revolutum*, *E. trivolvis* and *E. caproni* was very low, 1–3%. Based on sequence variability *E. trivolvis* and *E. revolutum* were more closely related to each other than either was to *E. caproni*. A PCR-based diagnostic assay using the restriction enzyme HINC II successfully distinguished *E. trivolvis* from *E. caproni*. Fujino *et al.* (1995) used the random amplified polymorphic DNA–polymerase chain reaction (RAPD-PCR) technique to demonstrate interspecific polymorphism between 60-day-old adults of *E. trivolvis* and *E. caproni* from experimentally infected golden hamsters. The band patterns generated by the five primers used in the study showed that these two echinostomes were genetically distinct.

Morgan and Blair (1995) obtained DNA sequence data from the nuclear rDNA ITS1, 5.8S and ITS2 of seven nominal species belonging to the 37-collar-spined *E. revolutum* group. The species they examined were *E. revolutum* (Frölich, 1802), *E. caproni* Richard, 1964, *E. liei* Jeyarasasingam *et al.*, 1972, *E. paraensei* Lie and Basch, 1967, two African isolates, *E. sp.* I and *E. sp* II (the latter two obtained from J. Jourdane in France). Additionally, they examined the well-described 28-collar-spined echinostome, *E. hortense* (Asada, 1926). Five of the eight species were distinguishable using ITS data. Sequences from the remaining three taxa, *E. caproni*,

E. sp. II and *E. liei* were identical to one another and the group containing these taxa was distant from the other 37-collar-spined species on a phylogenetic tree. *E. trivolvis* and *E. paraensei* form a second, but less distinct group within the 37-collar-spined group.

17. SUMMARY AND CONCLUDING REMARKS

This review introduces the reader to the biology of the intestinal trematode, *Echinostoma caproni*, a good digenean to work with because it can be easily cycled in the laboratory between the medically important planorbid snail *Biomphalaria glabrata* and laboratory mice or hamsters, the same hosts that are used to maintain *Schistosoma mansoni*. Areas that are in need of additional research are mentioned in each section. All stages in the biology of this echinostome are easily available to the research worker. The adult parasite becomes gravid in 7–10 days in rodent hosts, and by 2 weeks postinfection (PI) produces hundreds of eggs. Eggs can be obtained from the faeces or by teasing the uteri of adult worms. Miracidia can be produced in about 10 days when embryonated eggs are maintained in the dark in artificial spring water at 28°C. Exposure of eggs to incandescent light produces a synchronous hatch of miracidia. Miracidia used to infect laboratory-reared *B. glabrata* become patent in 5–6 weeks PI at 22°C and release cercariae into the water; cercariae infect the same or cohort snails of *B. glabrata* to produce metacercarial cysts in the renopericardial sac of the snails. Cysts are infective to vertebrate hosts from about 1 day to 6 months postencystment in the snails. Metacercarial cysts can be excysted in an alkaline trypsin–bile salts medium at 39°C to obtain excysted metacercariae for histological, histochemical, immunological, *in vitro* cultivation or surgical implantation studies. Details of studies done to date on these topics are presented in the review along with ideas for new research.

The review considers recent work on various other topics including ultrastructural studies on the adults, concurrent studies with other parasites both in the snail and the vertebrate hosts, histopathological and pathobiochemical effects of the infection on both the intermediate and definitive hosts and immunobiology studies in both the intermediate and definitive hosts. Suggestions for new research in these areas are given. Studies using electrophoresis and PCR are covered mainly as they relate to the distinctness of this species within the 37-collar-spined *Echinostoma revolutum* complex. It is concluded that this organism is an exceptionally fine model for various studies in parasitology.

REFERENCES

Agger, M.K., Simonsen, P.E. and Vennervald, B.J. (1993). The antibody response in serum, intestinal wall and intestinal lumen of NMRI mice infected with *Echinostoma caproni*. *Journal of Helminthology* **67**, 169–178.

Andreassen, J., Odaibo, A.B. and Christensen, N. Ø. (1990). Concurrent infections of the trematode *Echinostoma caproni* and the tapeworm *Hymenolepis diminuta* and *Hymenolepis microstoma* in mice. *Journal of Parasitology* **76**, 573–575.

Andresen, K., Simonsen, P.E., Andersen, B.J.A. and Birch-Andersen, A. (1989). *Echinostoma caproni* in mice: shedding of antigens from the surface of an intestinal trematode. *International Journal for Parasitology* **19**, 111–118.

Bakke, T.A. (1988). Abnormalities in adult digeneans, with special reference to *Phyllodistomum umblae* (Fabricius) (Platyhelminthes, Gorgoderidae). *Zoologica Scripta* **17**, 123–134.

Barus, V., Moravec, F. and Rysavy, B. (1974). Antagonistic interaction between *Echinostoma revolutum* and *Echinoparyphium recurvatum* (Trematoda) in the definitive host. *Folia Parasitologica (Praha)* **21**, 155–159.

Bayne, C. J. and Loker, E. S. (1987). Survival within the snail host. In: *The Biology of Schistosomes — from Genes to Latrines* (D. Rollinson and A.J.G. Simpson, eds), pp. 321–346. London: Academic Press.

Beers, K., Fried, B., Fujino, T. and Sherma, J. (1995). Effects of diet on the lipid composition of the digestive gland–gonad complex of *Biomphalaria glabrata* (Gastropoda) infected with larval *Echinostoma caproni* (Trematoda). *Comparative Biochemistry and Physiology* **110B**, 729–737.

Behrens, A.C. and Nollen, P.M. (1992). Responses of *Echinostoma caproni* miracidia to gravity, light and chemicals. *International Journal for Parasitology* **22**, 673–675.

Behrens, A.C. and Nollen, P.M. (1993). Hatching of *Echinostoma caproni* miracidia from eggs derived from adults grown in hamsters and mice. *Parasitology Research* **79**, 28–32.

Berntzen, A.K. and Macy, R.W. (1969). *In vitro* cultivation of the digenetic trematode *Sphaeridiotrema globulus* (Rudolphi) from the metacercarial stage to egg production. *Journal of Parasitology* **55**, 136–139.

Bindseil, E. and Christensen, N.Ø. (1984). Thymus-independent crypt hyperplasia and villous atrophy in the small intestine of mice infected with the trematode *Echinostoma revolutum*. *Parasitology* **88**, 431–438.

Bindseil, E. and Hau, J. (1991). Negative effect on early post-implantation pregnancy and progesterone levels in mice infected with the intestinal trematode *Echinostoma caproni*. *Parasitology* **102**, 387–390.

Bindseil, E., Lotte, L., Andersen, I. and Hau, J. (1989). Reduced fertility in mice double-infected with *Schistosoma mansoni* and *Echinostoma revolutum*. *Acta Tropica* **46**, 269–271.

Bindseil, E., Andersen, L.L.I. and Hau, J. (1990). Negative influence of extragenital *Schistosoma mansoni* and *Echinostoma caproni* infections on fertility and maternal murine alpha fetoprotein levels in the circulation of female mice. *International Journal of Feto-Maternal Medicine* **3**, 236–241.

Bindseil, E., Andersen, L.L.I., Krog, N.L. and Hau, J. (1991). Effect of extragenital *Schistosoma mansoni* and *Echinostoma caproni* infections on serum levels of pregnancy associated murine protein-1 during murine pregnancy. *In Vivo (Athens)* **5**, 175–178.

Carlier, Y. and Truyens, C. (1995). Influence of maternal infection on offspring resistance towards parasites. *Parasitology Today* **11**, 94–99.

Cheng, T.C. and Lee, F.O. (1971). Glucose levels in the mollusc *Biomphalaria glabrata* infected with *Schistosoma mansoni*. *Journal of Invertebrate Pathology* **18**, 395–399.

Chien, W.Y. and Fried, B. (1992). Cultivation of excysted metcercariae of *Echinostoma caproni* to ovigerous adults in the allantois of the chick embryo. *Journal of Parasitology* **78**, 1019–1023.

Chien, W.Y., Hosier, D.W. and Fried, B. (1993). Surgical implantation of *Echinostoma caproni* metacercariae and adults into the small intestine of ICR mice. *Journal of the Helminthological Society of Washington* **60**, 122–123.

Christensen, N.Ø. (1980). *Echinostoma revolutum*: Labeling of miracidia with radioselenium *in vivo* and assay for host finding. *Experimental Parasitology* **50**, 67–73.

Christensen, N.Ø., Frandsen, F. and Roushdy, M.Z. (1980). The influence of environmental conditions and parasite–intermediate host-related factors on the transmission of *Echinostoma liei*. *Zeitschrift für Parasitenkunde* **63**, 47–63.

Christensen, N.Ø., Nydal, R., Frandsen, F. and Nansen, P. (1981a). Homologous immunotolerance and decreased resistance to *Schistosoma mansoni* in *Echinostoma revolutum*-infected mice. *Journal of Parasitology* **67**, 164–166.

Christensen, N.Ø., Nydal, R., Frandsen, F., Sirag, S.B. and Nansen, P. (1981b). Further studies on resistance to *Fasciola hepatica* and *Echinostoma revolutum* in mice infected with *Schistosoma* sp. *Zeitschrift für Parasitenkunde* **65**, 293–298.

Christensen, N.Ø., Fagbemi, B.O. and Nansen, P. (1984). *Trypanosoma brucei*-induced blockage of expulsion of *Echinostoma revolutum* and of homologous *E. revolutum* resistance in mice. *Journal of Parasitology* **70**, 558–561.

Christensen, N.Ø., Knudsen, J., Fagbemi, B. O. and Nansen, P. (1985). Impairment of primary expulsion of *Echinostoma revolutum* in mice concurrently infected with *Schistosoma mansoni*. *Journal of Helminthology* **59**, 333–335.

Christensen, N.Ø., Knudsen, J. and Andreassen, J. (1986). *Echinostoma revolutum*: resistance to secondary and superimposed infections in mice. *Experimental Parasitology* **61**, 311–318.

Christensen, N.Ø., Odaibo, A.B. and Simonsen, P.E. (1988). *Echinostoma* population regulation in experimental rodent definitive hosts. *Parasitology Research* **75**, 83–87.

Christensen, N.Ø., Simonsen, P.E., Odaibo, A.B. and Mahler, H. (1990). Establishment, survival and fecundity in *Echinostoma caproni* (Trematoda) infections in hamsters and jirds. *Journal of the Helminthological Society of Washington* **57**, 104–107.

Cooper, L.A., Ramani, S.K., Martin, A.E., Richards, C.S. and Lewis, F.A. (1992). *Schistosoma mansoni* infections in neonatal *Biomphalaria glabrata* snails. *Journal of Parasitology* **78**, 441–446.

Donovick, R.A. and Fried, B. (1988). Scanning electron microscopy of *Echinostoma revolutum* and *Echinostoma liei* from domestic chicks. *Journal of the Pennsylvania Academy of Science* **62**, 78–82.

Evans, N.A. (1985). The influence of environmental temperature upon transmission of the cercariae of *Echinostoma liei* (Digenea: Echinostomatidae). *Parasitology* **90**, 269–275.

Fairbairn, D. (1958). Trehalose and glucose in helminths and other invertebrates. *Canadian Journal of Zoology* **36**, 787–795.

Fried, B. (1994). Metacercarial excystment of trematodes. *Advances in Parasitology* **33**, 91–144.

Fried, B. and Emili, S. (1987). Comparative studies on infectivity, growth, excystation, and cultivation of *Echinostoma revolutum* and *E. liei* (Trematoda). *Proceedings of the First Autumn School, Parasite–Host–Environment*, Varna, Bulgaria. pp. 193–206.

Fried, B. and Emili, S. (1988). Excystation *in vitro* of *Echinostoma liei* and *E. revolutum* (Trematoda) metacercariae. *Journal of Parasitology* **74**, 98–102.

Fried, B. and Haseeb, M.A. (1990). Intraspecific and interspecific chemoattraction in *Echinostoma caproni* and *Echinostoma trivolvis adults in vitro. Journal of the Helminthological Society of Washington* **57**, 72–73.

Fried, B. and Manger, P.M., Jr (1992). Use of an acetocarmine procedure to examine the excysted metacercariae of *Echinostoma caproni* and *E. trivolvis. Journal of Helminthology* **66**, 238–240.

Fried, B. and Rosa-Brunet, L.C. (1991a). Exposure of *Dugesia tigrina* (Turbellaria) to cercariae of *Echinostoma trivolvis* and *E. caproni* (Trematoda). *Journal of Parasitology* **77**, 113–116.

Fried, B. and Rosa-Brunet, L.C. (1991b). Cultivation of excysted metacercariae of *Echinostoma caproni* (Trematoda) to ovigerous adults on the chick chorioallantois. *Journal of Parasitology* **77**, 568–571.

Fried, B. and Roth, R.M. (1974). *In vitro* excystment of the metacercariae of *Parorchis acanthus. Journal of Parasitology* **60**, 465.

Fried, B. and Stableford, L.T. (1991). Cultivation of helminths in chick embryos. *Advances in Parasitology* **30**, 107–165.

Fried, B. and Weaver, L.J. (1969). Effects of temperature on the development and hatching of eggs of the trematode *Echinostoma revolutum. Transactions of the American Microscopical Society* **88**, 253–257.

Fried, B., Donovick, R.A. and Emili, S. (1988). Infectivity, growth and development of *Echinostoma liei* (Trematoda) in the domestic chick. *International Journal for Parasitology* **18**, 413–414.

Fried, B., Schafer, S. and Kim, S. (1989). Effects of *Echinostoma caproni* infection on the lipid composition of *Biomphalaria glabrata. International Journal for Parasitology* **19**, 353–354.

Fried, B., Irwin, S.W.B. and Lowry, S. F. (1990a). Scanning electron microscopy of *Echinostoma trivolvis* and *Echinostoma caproni* (Trematoda) adults from experimental infections in the golden hamster. *Journal of Natural History* **24**, 433–440.

Fried, B., Huffman, J.E. and Weiss, P.M. (1990b). Single and multiple worm infections of *Echinostoma caproni* (Trematoda) in the golden hamster. *Journal of Helminthology* **64**, 75–78.

Fried, B., Sherma, J., Rao, K.S. and Ackman, R.G. (1993). Fatty acid composition of *Biomphalaria glabrata* (Gastropoda: Planorbidae) experimentally infected with the intramolluscan stages of *Echinostoma caproni* (Trematoda). *Comparative Biochemistry and Physiology* **104B**, 595–598.

Fried, B., Idris, N. and Ohsawa, T. (1995). Experimental infection of juvenile *Biomphalaria glabrata* with cercariae of *Echinostoma trivolvis. Journal of Parasitology* **81**, 308–310.

Fujino, T. and Fried, B. (1993a). Expulsion of *Echinostoma trivolvis* (Cort, 1914). Kanev, 1985 and retention of *Echinostoma caproni* Richard, 1964 (Trematoda: Echinostomatidae) in C₃H mice: pathological ultrastructural and cytochemical effects on the host intestine. *Parasitology Research* **79**, 286–292.

Fujino, T. and Fried, B. (1993b). *Echinostoma caproni* and *E. trivolvis* alter the binding of glycoconjugates in the intestinal mucosa of C3H mice determined by lectin histochemistry. *Journal of Helminthology* **67**, 179–188.

Fujino, T., Takahashi, Y. and Fried, B. (1995). A comparison of *Echinostoma trivolvis* and *E. caproni* (Trematoda: Echinostomatidae) using random amplified polymorphic DNA analysis. *Journal of Helminthology* **69**, 263–264.

Fujino, T., Yamada, M., Ichikawa, H., Fried, B., Arizono, N. and Tada, I. (1996a). Rapid expulsion of the intestinal trematodes *Echinostoma trivolvis* and *E. caproni* from C3H mice after infection with *Nippostrongylus brasiliensis*. *Parasitology Research* **82** (in press).

Fujino, T., Fried, B., Ichikawa, H. and Tada, I. (1996b). Rapid expulsion of the intestinal trematodes *Echinostoma trivolvis* and *E. caproni* from C3H mice by trapping of increased goblet cell mucins. *International Journal for Parasitology* **26**, 319–324.

Graczyk, T.K. and Fried, B. (1994). ELISA method for detecting anti-*Echinostoma caproni* (Trematoda: Echinostomatidae) antibodies in experimentally infected ICR mice. *Journal of Parasitology* **80**, 544–549.

Haseeb, M.A. and Fried, B. (1988). Chemical communication in helminths. *Advances in Parasitology* **27**, 169–207.

Horutz, K. and Fried, B. (1995). Effects of *Echinostoma caproni* infection on the neutral lipid content of the intestinal mucosa of ICR mice. *International Journal for Parasitology* **25**, 653–655.

Hosier, D.W. and Fried, B. (1991). Infectivity, growth and distribution of *Echinostoma caproni* (Trematoda) in the ICR mouse. *Journal of Parasitology* **77**, 640–642.

Hosier, D.W., Fried, B. and Szewczak, J. P. (1988). Homologous and heterologous resistance of *Echinostoma revolutum* and *E. liei* in ICR mice. *Journal of Parasitology* **74**, 89–92.

Hosier, D.W., Ross, D. and Fried, B. (1991). Effects of host age on infection of ICR mice with *Echinostoma caproni* or *Echinostoma trivolvis*. *International Journal for Parasitology* **48**, 137–138.

Huffman, J.E. and Fried, B. (1990). *Echinostoma* and echinostomiasis. *Advances in Parasitology* **29**, 215–269.

Huffman, J. E., Alcaide, A. and Fried, B. (1988). Single and concurrent infections of the golden hamster, *Mesocricetus auratus*, with *Echinostoma revolutum* and *E. liei* (Trematoda: Digenea). *Journal of Parasitology* **74**, 604–608.

Huffman, J.E., Murphy, P.M. and Fried, B. (1992). Superimposed infections in golden hamsters infected with *Echinostoma caproni* and *Echinostoma trivolvis* (Digenea: Echinostomatidae). *Journal of the Helminthological Society of Washington* **59**, 16–21.

Iorio, S.L., Fried, B. and Hosier, D.W. (1991). Concurrent infections of *Echinostoma caproni* and *Echinostoma trivolvis* in ICR mice. *International Journal for Parasitology* **21**, 715–718.

Irwin, S.W.B. and Fried, B. (1990). Scanning and transmission electron microscopic observations on metacercariae of *Echinostoma trivolvis* and *Echinostoma caproni* during *in vitro* excystation. *Journal of the Helminthological Society of Washington* **57**, 79–83.

Isaacson, A.C., Huffman, J.E. and Fried, B. (1989). Infectivity, growth, development and pathology of *Echinostoma caproni* (Trematoda) in the golden hamster. *International Journal for Parasitology* **19**, 943–944.

Jeyarasasingam, U., Heyneman, D., Lim, H.K. and Mansour, N. (1972). Life cycle of a new echinostome from Egypt, *Echinostoma liei* sp nov. (Trematoda: Echinostomatidae). *Parasitology* **65**, 203–222.

Joky, A., Matricon-Gondran, M. and Benex, J. (1985). Response to the amoebocyte-producing organ of sensitized *Biomphalaria glabrata* after exposure to *Echinostoma caproni* miracidia. *Journal of Invertebrate Pathology* **45**, 28–33.

Jourdane, J. and Kulo, S.D. (1981). Etude expérimentale du cycle biologique de *Echinostoma togoensis* n. sp., parasite a l-état larvaire de *Biomphalaria pfeifferi* au Togo. *Annals of Parasitology (Paris)* **56**, 477–488.

Jourdane, J. and Kulo, S.D. (1982). Perspectives d-utilisation d-*Echinostoma togoensis* Jourdane et Kulo, 1981, dans le contrôle biologique de la bilharziose intestinale en Afrique. *Annales Parasitologie Humaine et Comparée* **57**, 429–442.

Jourdane, J. and Mounkassa, J. B. (1986). Topographic shifting of primary sporocysts of *Schistosoma mansoni* in *Biomphalaria pfeifferi* as a result of coinfection with *Echinostoma caproni*. *Journal of Invertebrate Pathology* **48**, 269–274.

Jourdane, J. and Théron, A. (1987). Larval development: eggs to cercariae. In: *The Biology of Schistosomes — from Genes to Latrines* (D. Rollinson and A.J.G. Simpson, eds) pp. 83–113. London: Academic Press.

Jourdane, J., Mounkassa, J.B. and Imbert-Establet, D. (1990). Influence of intramolluscan larval stages of *Echinostoma liei* on the infectivity of *Schistosoma mansoni* cercariae. *Journal of Helminthology* **64**, 71–74.

Kanev, I. (1985). On the morphology, biology, ecology and taxonomy of *E. revolutum* group (Trematoda: Echinostomatidae: *Echinostoma*). PhD Dissertation, University of Sofia, Bulgaria.

Kanev, I. (1994). Life cycle, delimitation and redescription of *Echinostoma revolutum* (Froelich, 1802) (Trematoda: Echinostomatidae). *Systematic Parasitology* **28**, 125–144.

Kanev, I., Fried, B., Dimitrov, V. and Radev, V. (1995). Redescription of *Echinostoma trivolvis* (Cort, 1914) (Trematoda: Echinostomatidae) with a discussion of its identity. *Systematic Parasitology* **32**, 61–70.

Kaufman, A.R. and Fried, B. (1994). Infectivity, growth, distribution and fecundity of a six versus twenty-five metacercarial inoculum of *Echinostoma caproni* in ICR mice. *Journal of Helminthology* **68**, 203–206.

Kim, S. and Fried, B. (1989). Pathological effects of *Echinostoma caproni* (Trematoda) in the domestic chick. *Journal of Helminthology* **63**, 227–230.

Krejci, K.G. and Fried, B. (1994). Light and scanning electron microscopic observations of the eggs, daughter rediae, cercariae and encysted metacercariae of *Echinostoma trivolvis* and *E. caproni*. *Parasitology Research* **80**, 42–47.

Kristensen, A.R. and Fried, B. (1991). A comparison of *Echinostoma caproni* and *Echinostoma trivolvis* (Trematoda: Echinostomatidae) adults using isoelectrofocusing. *Journal of Parasitology* **77**, 496–498.

Kuris, A.M. (1980a). *Echinostoma liei* miracidia and *Biomphalaria glabrata* snails: effect of egg age, habitat heterogeneity, water quality and volume on infectivity. *International Journal for Parasitology* **10**, 21–26.

Kuris, A.M. (1980b). Effect of exposure to *Echinostoma liei* miracidia on growth and survival of young *Biomphalaria glabrata* snails. *International Journal for Parasitology* **10**, 303–308.

Kuris, A. M. and Warren, J. (1980). Echinostome cercarial penetration and metacercarial encystment as mortality factors for a second intermediate host *Biomphalaria glabrata*. *Journal of Parasitology* **66**, 630–635.

Lewis, F.A., Stirewalt, M.A., Souza, C.P. and Gazzinelli, G. (1986). Large-scale laboratory maintenance of *Schistosoma mansoni*, with observations on three schistosome/snail host combinations. *Journal of Parasitology* **72**, 813–829.

Lie, K.J. and Basch, P.F. (1967). The life history of *Echinostoma paraensei* sp. n. (Trematoda: Echinostomatidae). *Journal of Parasitology* **53**, 1192–1199.

Lie, K.J., Jeong, K.H. and Heyneman, D. (1980). Inducement of miracidia immobilizing substance in the hemolymph of *Biomphalaria glabrata*. *International Journal for Parasitology* **10**, 183–188.

Loker, E.S. and Adema, C.M. (1995). Schistosomes, echinostomes and snails: comparative immunobiology. *Parasitology Today* **11**, 120–124.

Malek, E.A. (1980). *Snail Transmitted Parasitic Diseases*, Vol. 1, Boca Raton: CRC Press.

Manger, P.M., Jr and Fried, B. (1993). Infectivity, growth and distribution of preovigerous adults of *Echinostoma caproni* in ICR mice. *Journal of Helminthology* **67**, 158–160.

McMaster, R.P., Huffman, J.E. and Fried, B. (1995). The effect of dexamethasone on the course of *Echinostoma caproni* and *E. trivolvis* infections in the golden hamster *(Mesocricetus auratus)*. *Parasitology Research* **81**, 518–521.

Meyrowitsch, D., Christensen, N.Ø. and Hindsbo, O. (1991). Effects of temperature and host density on the snail-finding capacity of cercariae of *Echinostoma caproni* (Digenea: Echinostomatidae). *Parasitology* **102**, 391–396.

Mohamed, S.H. (1992). Ultrastructure aspects of intramolluscan developing cercariae of *Echinostoma liei*. *Journal of the Egyptian Society of Parasitology* **22**, 479–485.

Moore, S.J., Thorndyke, M.C., Riddell, J.H., Balogun, M.A. and Whitfield, P.J. (1989). Ultrastructural analysis of enterochromaffin cell proliferation in the mouse following infection with the digenean parasite *Echinostoma liei*. *Regulatory Peptides* **26**, 83.

Morgan, J.S. (1993). Ribosomal DNA sequence diversity among well-characterized *Echinostoma* species. MS Thesis, Department of Zoology, James Cook University, North Queensland, Australia, 76 pp.

Morgan, J.A.T. and Blair, D. (1995). Nuclear rDNA ITS sequence variation in the trematode genus *Echinostoma*: an aid to establishing relationships within the 37-collar-spine group. *Parasitology* **111**, 609–615.

Mounkassa, J.B. and Jourdane, J. (1990). Dynamics of the leukocytic response of *Biomphalaria glabrata* during the larval development of *Schistosoma mansoni* and *Echinostoma liei*. *Journal of Invertebrate Pathology* **55**, 306–311.

Nollen, P.M. (1990). *Echinostoma caproni* mating behavior and the timing of development and movement of reproductive cells. *Journal of Parasitology* **76**, 784–789.

Odaibo, A. B., Christensen, N.Ø. and Ukoli, F.M.A. (1988). Establishment, survival and fecundity in *Echinostoma caproni* infections in NMRI mice. *Proceedings of the Helminthological Society of Washington* **55**, 265–269.

Odaibo, A.B., Christensen, N.Ø. and Ukoli, F.M.A. (1989). Further studies on the population regulation in *Echinostoma caproni* infections in NMRI mice. *Proceedings of the Helminthological Society of Washington* **56**, 192–198.

Perez, M.K., Fried, B. and Sherma, J. (1994). High performance thin-layer chromatographic analysis of sugars in *Biomphalaria glabrata* (Gastropoda) infected with *Echinostoma caproni* (Trematoda). *Journal of Parasitology* **80**, 336–338.

Perez, M.K., Fried, B. and Sherma, J. (1995). Comparison of mobile phases and HPTLC qualitative and quantitative analysis, on preadsorbent silica gel plates, of phospholipids in *Biomphalaria glabrata* (Gastropoda) infected with *Echinostoma caproni* (Trematoda). *Journal of Planar Chromatography-Modern TLC* **7**, 340–343.

Pritchard, M.H. and Kruse, S.O.W. (1982). *The Collection and Preservation of Animal Parasites*, p. 141. Lincoln: University of Nebraska Press.

Richard, J. (1964). Trématodes d-oiseax de Madagascar (Notes III). Espèces de la famille des Echinostomatidae. *Annales Parasitologie Humaine et Comparée Paris* **34**, 607–620.

Richard, J. and Brygoo, E.R. (1978). Life cycle of the trematode *Echinostoma caproni* Richard, 1964 (Echinostomatoidea). *Annales Parasitologie Humaine et Comparée (Paris)* **53**, 265–275.

Richard, J. and Voltz, A. (1987). Preliminary data on the chromosomes of *Echinostoma caproni* (Richard 1964) (Trematoda: Echinostomatidae). *Systematic Parasitology* **9**, 169–172.

Richard, J., Klein, M.J. and Stoeckel, M.E. (1989). Neural and glandular localization of substance P in *Echinostoma caproni* (Trematoda: Digenea). *Parasitology Research* **75**, 641–648.

Riddell, J.H., Whitfield, P.J., Balogun, M.A. and Thorndyke, M.C. (1991). FMRF-amide-like peptides in the nervous and endocrine systems of the digenean helminth *Echinostoma liei*. *Acta Zoologica (Stockholm)* **72**, 1–6.

Roberts, T.M., Stibbs, H.H., Chernin, E. and Ward, S. (1978). A simple quantitative technique for testing behavioral responses of *Schistosoma mansoni* miracidia to chemicals. *Journal of Parasitology* **64**, 277–282.

Rosa-Brunet, L.C. and Fried, B. (1992). Growth, development, pathogenicity and transplantation of *Echinostoma caproni* (Trematoda) on the chick chorioallantois. *Journal of Parasitology* **78**, 99–103.

Ross, G.C., Fried, B. and Southgate, V.R. (1989). *Echinostoma revolutum* and *Echinostoma liei*: observations on enzymes and pigments. *Journal of Natural History* **23**, 977–982.

Schaefer, F.W., III, Saz, H.J., Weinstein, P.P. and Dunbar, G.A. (1977). Aerobic and anaerobic fermentation of glucose by *Echinostoma liei*. *Journal of Parasitology* **63**, 687–689.

Shetty, P, Fried, B. and Sherma, J. (1992). Effects of patent *Echinostoma caproni* infection on the sterol composition of the digestive gland–gonad complex of *Biomphalaria glabrata* as determined by gas–liquid chromatography. *Journal of Helminthology* **66**, 68–71.

Shostak, A.W., Dharampaul, S. and Belosevic, M. (1993). Effects of source of metacercariae on experimental infection of *Zygocotyle lunata* (Digenea: Paramphistomidae) in CD-1 mice. *Journal of Parasitology* **79**, 922–929.

Simonsen, P.E. and Andersen, B.J. (1986). *Echinostoma revolutum* in mice: dynamics of the antibody attack to the surface of an intestinal trematode. *International Journal for Parasitology* **16**, 475–482.

Simonsen, P.E., Bindseil, E. and Køie, M. (1989). *Echinostoma caproni* in mice: studies on the attachment site of an intestinal trematode. *International Journal for Parasitology* **19**, 561–566.

Simonsen, P.E., Vennervald, B.J. and Birch-Andersen, A. (1990). *Echinostoma caproni* in mice: ultrastructural studies on the formation of immune complexes on the surface of an intestinal trematode. *International Journal for Parasitology* **20**, 935–942.

Simonsen, P.E., Estambale, B.B. and Agger, M. (1991). Antibodies in the serum of golden hamsters experimentally infected with the intestinal trematode *Echinostoma caproni*. *Journal of Helminthology* **65**, 239–247.

Sirag, K.I., Christensen, N.Ø., Frandsen, F., Monrad, J. and Nansen, P. (1980). Homologous and heterologous resistance in *Echinostoma revolutum* infections in mice. *Parasitology* **80**, 479–486.

Sloss, B., Meece, J., Romano, M. and Nollen, P. (1995). The genetic relationships between *Echinostoma caproni*, *Echinostoma paraensei*, and *Echinostoma trivolvis* as determined by electrophoresis. *Journal of Helminthology* **69**, 243–246.

Sullivan, J.J. (1985). Juvenile snails as hosts for echinostome metacercariae. *Southeast Asian Journal of Tropical Medicine and Public Health* **16**, 343–344.

Thompson, S.N. (1987). Effect of *Schistosoma mansoni* on the gross lipid composition of its vector *Biomphalaria glabrata*. *Comparative Biochemistry and Physiology* **87B**, 357–360.

Thompson, S.N., Mejia-Scales, V. and Borchardt, D.B. (1991). Physiologic studies of snail-schistosome interactions and potential for improvement on *in vitro* culture of schistosomes. *In Vitro Cell and Developmental Biology* **27A**, 497–504.

Thorndyke, M. and Whitfield, P.J. (1987). Vasoactive intestinal polypeptide immunoreactive tegumental cells in the digenean *Echinostoma liei*: possible role in host-parasite interactions. *General and Comparative Endocrinology* **68**, 202–207.

Thorndyke, M.C., Riddell, J.H., Dimaline, R., Balogun, K. and Whitfiel, P.J. (1988). Changes in ileal vasoactive intestinal polypeptide and gastrin releasing peptide-bombesin levels associated with chronic infections of the digenean helminth *Echinostoma liei*. *Regulatory Peptides* **22**, 435.

Ulmer, M.J. (1970). Notes on rearing of snails in the laboratory. In: *Experiments and Techniques in Parasitology* (A.J. MacInnis and M. Voge, eds), pp. 143–144. San Francisco: W.H. Freeman.

Ursone, R.L. and Fried, B. (1995). Light and scanning electron microscopy of *Echinostoma caproni* (Trematoda) during maturation in ICR mice. *Parasitology Research* **81**, 45–51.

Voge, M. (1970). Infection of rats, mice, and hamsters with intestinal helminths. In: *Experiments and Techniques in Parasitology* (A.J. MacInnis and M. Voge, eds), pp. 134–135. San Francisco: W.H. Freeman.

Voltz, A., Richard, J. and Pesson, B. (1985). A genetic comparison between natural and laboratory strains of *Echinostoma* (Trematoda) by isoenzymatic analysis. *Parasitology* **95**, 471–478.

Voltz, A., Richard, J., Pesson, B. and Jourdane, J. (1986). Chemotaxonomic study of the genus *Echinostoma:* comparison between a strain isolated from Cameroon, *Echinostoma* sp, and two African species, *Echinostoma caproni* and *Echinostoma togoensis*. *Annales de Parasitologie Humaine et Comparee* **61**, 617–624.

Voltz, A., Richard, J., Pesson, B. and Jourdane, J. (1988). Isoenzyme analysis of *Echinostoma liei*: comparison and hybridization with other African species. *Experimental Parasitology* **66**, 13–17.

Voltz-Kristensen, A. (1987). Contribution a l'identification par la biometrie et le typage enzymatique de souches experimentales et de populations naturelles de trematodes du genre *Echinostoma*. Doctoral Thesis, University Louis Pasteur de Strasbourg, France, 221 pp.

Weinstein, M.S. and Fried, B. (1991). The expulsion of *Echinostoma trivolvis* and retention of *Echinostoma caproni* in the ICR mouse: pathological effects. *International Journal for Parasitology* **21**, 255–258.

Yao, G., Huffman, J.E. and Fried, B. (1991). The effects of crowding on adults of

Echinostoma caproni in experimentally infected golden hamsters. *Journal of Helminthology* **65**, 248–254.

Yousif, F. and Haroun, N. (1986). Intramolluscan development of *Echinostoma liei* (Trematoda: Echinostomatidae) in its snail host *Biomphalaria alexandrina*. *Journal of the Egyptian Society of Parasitology* **16**, 127–140.

Yousif, F., El-Gindy, H., Roushdy, M. and El-Dafrawy, S. (1989). Host–parasite relations of *Biomphalaria alexandrina* and *Echinostoma liei*. 1. Infection of *E. liei* miracidium to *B. alexandrina*, p. 279, Abstracts of the 10th International Malacological Congress, Tubingen.

Index

Aedes, immunization experiments 64
Africa, East Africa study of cystic
    echinococcosis 204–8
AIDS, *Strongyloides stercoralisi*,
    in humans 273–4
Alaska, St Lawrence Island, study of
    alveolar echinococcosis 216–18
albendazole 292, 294
alveolar echinococcosis *see Echinococcus
    multilocularis*; taeniid cestode zoonoses
Anopheles spp.
    blood meals
        concentration factors 93
        proteolytic enzymes 61, 96–9
        site of *Plasmodium* ookinetes 96
        size and infection intensity 92
    distribution of gametocytes in blood
        59–61
    early vs late infection serum, effect on
        development 78
    parasitaemia and infectivity 68
    Plasmodium development in midgut,
        summary diagram 62
    prevalence of oocyst infection 60–1
anthelmintics 290–2
antigen recognition, T cell immunological
    response 122–7
Aspicularis tetrapterai 143–9
Azadirachta indicai, effect on exflagellation
    of *Plasmodium* 92

Bcg/Ity/Lsh resistance gene,
    Leishmania/Salmonella/Mycobacterium
        infections 127–40
benzimidazoles 290–2, 294
Biomphalaria spp. 312, 316
British Isles, study of cystic echinococcosis
    211–12

canines, taeniid cestode zoonoses
    200–2

China
    echinococcosis
        alveolar 218–21
        cystic 208–10
    coproantigen tests 203
    cystic echinococcosis *see Echinococcus
        granulosus*; taeniid cestode zoonoses
    cysticercosis *see Echinococcus granulosus*;
        Echinococcus multilocularisi; *Taenia
        soliumi*

dog, taeniid cestode zoonoses, detection
    200–2

Echinococcus granulosus
    in endemic communities 203–15
        British Isles 211–12
        East Africa 204–8
        Kathmandu City 214–15
        Northwest China 208–10
        Uruguay 213–14
    global distribution by country 172–3
    summary, host aspects 171
    see also taeniid cestode zoonoses
Echinococcus multilocularis
    in endemic communities 215–22
        Japan 221–2
        Northwest China 218–21
        St Lawrence island, Alaska 216–18
    global distribution by country 173–5
    intermediate hosts, China 220
    summary, host aspects 171
    see also taeniid cestode zoonoses
Echinococcus spp., life cycle and
    transmission 185–9
Echinostoma caproni 311–60
    adults 324–6
    cercariae and metacercariae 333–7
    concurrent studies 343–4
    eggs and miracidia 326–30
    host effects 347–54

immunobiology 354–7
life cycle 315
list of laboratories using 314
list of studies 313
maintenance in laboratory 315–19
PCR and electrophoresis 357–9
snail hosts 316, 337–43
sporocysts and rediae 330–2
in vitro and *in ovo* cultivation 345–7
Echinostoma liei 313
Echinostoma paraensei 357–9
Echinostoma revolutum 313, 358
Echinostoma togoensis 313
Echinostoma trivolvis, golden hamster
    343–4, 357
Eimeria 55
Eimeria vermiformis, T_H1 protective
    response 126, 152
ELISA
    coproantigen tests 203
    serum antibody tests 203
endocytic vesicles 11–12

fitness, and parasite susceptibility
    150–3
flagella, exflagellation 90
fox, taeniid cestode zoonoses 200–1
    detection 202–3, 221

global distributions, taeniid cestode
    zoonoses 172–7
golden hamster
    Echinostoma caproni 316
    Echinostoma trivolvis 343–4, 357

Haemoproteus 86
Heligmosomoides polygyrus 126, 136
HIV infection, *Strongyloides stercoralis*, in
    humans 273–4
HTLV-1 infection, and *Strongyloides
    stercoralis*, in humans 275
hybrid fitness, and parasite susceptibility
    150–3
Hymenolepis, mouse–parasite interactions
    143
Hymenolepis citedelli, and *Peromyscus
    leucopus*, mouse–parasite interactions
    152
Hymenolepis diminuta, with *E. caproni*
    infection 343

immune response in mammals 121–7
immunobiology, *Echinostoma caproni*
    354–7
immunodiagnosis, animal hosts 203
immunodiagnostic screening in humans
    192–6
intracellular survival, protist parasites
    1–34
isoenzyme electrophoresis 260
ivermectin 292–3, 294

Japan, alveolar echinococcosis 221–2

Latin America
    Taenia solium study 224–7
    Uruguay, *Echinococcus granulosus* study
        213–14
Leishmania spp. 1–34
    amastigotes 7, 29–30
    inhibition of nitric oxide synthesis 26
    L. braziliensis 30
    L. donovani 23, 30–2
    mouse–parasite interactions 121, 127–35,
        150
    L. major 30, 126
    L. mexicana 29
    L. pifanoi 30
    lipophosphoglycan (LPG) 5–6
    macrophage protein kinase C 24
    major surface molecules 4–5
    phagocytosis 1–6
    promastigote surface protease (PSP) 4
    promastigotes 7, 22
    resistance to lysosomal toxicity 14–17
leprosy, *Nramp* gene 129–30
Leucocytozoon smithi 59
lipophosphoglycan (LPG), *Leishmania* spp.
    4, 23, 25–7
Listeria monocytogenes 132
Locke's solution 317

macrophages, interference with respiratory
    burst activity 23–6
major histocompatibility complex (MHC)
    mouse–parasite interactions 133–40
    polymorphism, host strategy 138–40
mebendazole 291, 294
medamine 293
Meriones unguiculatus, and *Strongyloides
    venezuelensis* 153
metalloproteases, *Leishmania* spp. 4

Mexico, *Taenia solium*, cysticercosis and taeniasis in communities 224–7
microtubule organizing centre (MTOC) 63
mouse, laboratory, genome polymorphism 147–50
mouse–parasite interactions 119–55
    Echinostoma caproni infection in pregnancy 344–5
    genetic control of pathogens by *Bcg/Ity/Lsh* resistance gene 127–33
    hybrid fitness, and parasite susceptibility 150–3
    Mus m. musculus–M. m. domesticus species complex 138–43
    parasite load 143–7
    Peromyscus leucopus and *Hymenolepis citedelli* 152
Mycobacterium bovis, mouse–parasite interactions 121, 127–33, 150
Mycobacterium lepraemurium, mouse–parasite interactions 135–6, 150

neem (*Azadirachta indica*), effect on exflagellation of *Plasmodium* 92
Nippostrongylus brasiliensis 136
    and *Echinostoma caproni* 344
nitric oxide 26
    and derivatives 12–14
    induced nitric oxide synthase (iNOS) 13
    infectivity of *Plasmodium* spp. 78
Nramp gene, humans 129–30

ovine hydatidosis 201–2
ovothiol A, *L. donovani* 23
oxygen metabolites 12
    neutralization 22–3

parasite susceptibility, and hybrid fitness 150–3
parasitophorous vacuole
    mouse pathogens 128
    Plasmodium spp., secretion of vesicles 64
    Toxoplasma gondii 19–21
    Trypanosoma cruzi 17–19
PCR, and electrophoresis, *Echinostoma caproni* 357–9
Peromyscus leucopus, and *Hymenolepis citedelli*, mouse–parasite interactions 152
phagocytosis
    conventional, *Leishmania* spp. 3–11
    induced, *Trypanosoma cruzi* 7–10
    interference with respiratory burst 21–6
phagolysosomes
    formation 11–14
    toxicity 14–21
pigs, porcine cysticercosis 201–2
Plasmodium spp. 53–99
    fertilization 92–5
    gametocyte biology 58–79
    gametocytogenesis 55–8
    gametogenesis 79–84
    induction mechanisms 84–92
    P. berghei 58, 67–8, 71, 73, 75, 77–8, 90, 93–4, 97
    P. brasilianum 69
    P. chabaudi 79, 132
    P. cynomolgi 59, 66–7, 70
    P. elongatum 87
    P. falciparum 55–77, 90–3
    P. gallinaceum 64, 67, 78, 86, 88, 97–9
    P. knowlesi 68
    P. reichenowi 58, 59
    P. vinckei petteri 71
    P. vivax 66–7, 70, 76
    P. yoelii nigeriensis 69, 83
    post-fertilization development 95–9
porcine cysticercosis 201–2
promastigote surface protease (PSP) 4
protein kinase C, macrophages, *Leishmania* spp. 24
public health, taeniid cestode zoonoses 170–8, 229–30

radiological screening 196–200
reactive oxygen intermediates (ROIs) 12
    neutralization 22–3
reticuloendothelial system (RES) cells 127
rhoptries 21
rodents, taeniid cestode zoonoses 200–2

Salmonella typhimuriumi
    mouse–parasite interactions 121, 127–33, 150
    RES cells 127
Salmonella/Leishmania/Mycobacterium infections, *Bcg/Ity/Lsh* resistance gene 127–40
Schistosoma mansoni, T_H1 protective response 126

screening procedures, taeniid cestode
  zoonoses 191–200
sheep
  cystic echinococcosis 201–2
  hydatidosis 201–2
Strongyloides
  history 252–3
  key for identification in humans 282–3
  list of species and their hosts 254–5
Strongyloides fuelleborni, in humans 296–7
Strongyloides stercoralis, morphology
  255–60
Strongyloides stercoralis, in humans
  autoinfection 265–7
  chronic infection 268–70
  clinical features of infection 276–80
  diagnosis 280–90
  and HIV infection 273–4
  homogonic development 261–3
  host–parasite relationship 267–8
  hyperinfection 270–4
  life cycle 261–7
  parasitic cycle 264
  prevention and control 295–6
  recent trends 280
  treatment 2904
Strongyloides stercoralis, in other mammals
  275–6
Strongyloides venezuelensis, and *Meriones
  unguiculatus* 153
Syphacia obvelata 143–7

T cells
  cytotoxic CD8+ lymphocytes 122
  T helper CD4+ lymphocytes 122
  T_H1 protective response 126
Taenia, rodent–parasite interactions 143–7,
  153
Taenia hydatigena 200, 201
Taenia ovis 200, 201
Taenia solium

cysticercosis and taeniasis in
  communities 223–9
global distribution by country 175–7
life cycle and transmission 186–9
summary, host aspects 171
see also taeniid cestode zoonoses
taeniid cestode zoonoses 169–229
  age prevalence data 178
  in animal hosts 200–1
  community-based studies, problems
    184–5
  economic importance 177
  global distributions 172–7
  history 170
  measurement of human infection 189–200
  pathology, diagnosis and treatment
    179–84
  public health 170–8, 229–30
  screening procedures 191–200
  summary 171
thiabendazole 290–1, 294
Toxoplasma gondii
  attachment to host cells 10–11
  H-2 control 136
  nutritional requirements 29–33
  parasitophorous vacuole 19–21
  tachyzoites 27–8
Trichinella spiralis
  MHC response 133–5
  T cell responses 126
Trichuris muris 126, 136
Trypanosoma brucei 28
Trypanosoma cruzi
  induced phagocytosis 7–10
  nutrition 29–30
  parasitophorous vacuole 17–19

ultrasound based screening 197
United States of America, *Taenia solium*
  study 228–91